Food Taints and Off-Flavours

Food Taints and Off-Flavours

Edited by

M. J. SAXBY
Former Manager
Analytical Chemistry Section
Leatherhead Food Research Association
Surrey

BLACKIE ACADEMIC & PROFESSIONAL
An Imprint of Chapman & Hall
London · Glasgow · New York · Tokyo · Melbourne · Madras

Published by
Blackie Academic & Professional, an imprint of Chapman & Hall,
Wester Cleddens Road, Bishopbriggs, Glasgow G64 2NZ, UK

Chapman & Hall, 2-6 Boundary Row, London SE1 8HN, UK

Blackie Academic & Professional, Wester Cleddens Road, Bishopbriggs, Glasgow G64 2NZ, UK

Chapman & Hall Inc., 29 West 35th Street, New York NY 10001, USA

Chapman & Hall Japan, Thomson Publishing Japan, Hirakawacho Nemoto Building, 6F, 1-7-11 Hirakawa-cho, Chiyoda-ku, Tokyo 102, Japan

DA Book (Aust.) Pty Ltd, 648 Whitehorse Road, Mitcham 3132, Victoria, Australia

Chapman & Hall India, R. Seshadri, 32 Second Main Road, CIT East, Madras 600 035, India

First edition 1993

© 1993 Chapman & Hall

Typeset in 10/12pt Times by Pure Tech Corporation, Pondicherry, India
Printed in Great Britain at the University Press, Cambridge

ISBN   0 7514 0096 3   0 442 30863 9 (USA)

A catalogue record for this book is available from the British Library

Library of Congress Cataloging-in-Publication Data available

Contributors

Mr D. A. Cumming Safeway plc, Beddow Way, Aylesford, Maidstone, Kent ME20 7AT, UK

Mr J. Ho UCLA School of Public Health, Los Angeles, CA 90024–1772, USA

Professor I. J. Jeon Department of Animal Sciences and Industry, Call Hall, Kansas State University, Manhattan, Kansas 66506–1600, USA

Dr D. Kilcast Sensory Analysis and Food Texture Section, Leatherhead Food Research Association, Randalls Road, Leatherhead, Surrey KT22 7RY, UK

Dr S. P. Kochhar SPK Consultancy Services, 48 Chiltern Crescent, Earley, Reading RG6 1AN, UK

Dr H. Maarse TNO Division of Nutrition and Food Research, PO Box 360, 3700 AJ Zeist, The Netherlands

Dr J. Mallevialle Central Laboratory–CIRSEE, Lyonnaise des Eaux et Dumez, 38 Rue du President Wilson, 78230 Le Pecq, France

Mr R. M. Pascal Safeway plc, Beddow Way, Aylesford, Maidstone, Kent ME20 7AT, UK

Dr M. J. Saxby 84 Pixham Lane, Dorking, Surrey RH4 1PH, UK

Dr M. B. Springett Campden Food and Drink Research Association, Chipping Campden, Gloucestershire GL55 6LD, UK

Dr I. H. Suffet UCLA School of Public Health, Los Angeles, CA 90024–1772, USA

Mr K. Swoffer Safeway plc, Beddow Way, Aylesford, Maidstone, Kent ME20 7AT, UK

Dr P. Tice PIRA International, Randalls Road, Leatherhead, Surrey KT22 7RU, UK

Preface

Contamination of food with extremely low levels of certain organic compounds can cause an unpleasant taste. This can result in the destruction of vast stocks of product, and very substantial financial losses to food companies. The concentration of the alien compound in the food can be so low that very sophisticated equipment, operated by experienced staff, is needed to identify the component, and to determine its source. It is vital that every company involved in the production, distribution and sale of foodstuffs is fully aware of the ways in which contamination can occur, how it can be avoided, and what steps need to be taken in the event that a problem does arise.

It is regrettable that many previous instances of serious taints in a food have not been investigated, so that their cause, source and ultimate means of removal have remained unresolved. Sometimes this may have been due to the lack of sufficiently sensitive analytical equipment required for the identification of the compound, but more often it is because the expertise needed to interpret all the clues has not been available. This book provides the background information needed for personnel to recognise how food can become tainted, how to draw up guidelines to prevent this contamination, and how to plan the steps that should be taken in the event of an outbreak. For the reader in academic or research institutions, the book provides a very extensive literature survey as a basis for future research.

The book consists of nine chapters, each written by specialist authors known throughout the world for expertise in their own subject. It sets out to present this unique subject in a way that will be understandable to academic, technical and commercial staff. The importance of detecting the presence of a taint in a foodstuff by organoleptic methods, both as a preventive measure and in the early stages of an investigation, is dealt with in the first chapter. The second chapter surveys what is known about the nature and origins of the chemicals that can cause taints, while the following chapter deals with the critical subject of their analysis. The following three chapters deal with taints in important areas of the food and drink industry. Chapters 7 and 8 are written by experts in the packaging industry and in the retailing sector, respectively. Finally, in chapter 9 the emphasis is on off-flavours and their microbiological formation.

M. J. S.

Contents

6 Oxidative pathways to the formation of off-flavours   150
S. P. KOCHHAR

7 Packaging material as a source of taints   202
P. TICE

1 Sensory evaluation of taints and off-flavours

D. KILCAST

1.1 Taint: definition and perception

In an increasingly competitive commercial environment, the food industry must satisfy consumer demands for high-quality, palatable food and must strive to minimise adverse consumer reaction. Consumer complaints can arise from numerous food defects, but the highly unpleasant nature of food taints can generate severe problems for retailer, producer, ingredient supplier, farmer, equipment supplier and even building contractor.

These problems can include lost production, lost sales, lost consumer confidence, damaged brand image, damaged commercial relationships between supplier, manufacturer and retailer, and expensive litigation proceedings. Food manufacturers readily understand the financial implications of a day's defective production, but not the more widespread implication of tainted production; for example, the tainted production will not be reworkable, and the plant may suffer from extended shut-down whilst defective building materials are replaced. In addition, identifying the source of a taint can be time-consuming and expensive.

Examination of English language dictionaries for definitions of taint produces words such as blemish, contamination, corruption, infection, pollution and defect. Even outside the specific context of food, the undesirable nature of taint is clear. Within the context of food, definitions become more precise. The standard definition of taint (ISO, 1992) is a taste or odour foreign to the product. This definition also distinguishes an off-flavour as an atypical flavour usually associated with deterioration. These definitions are characterised by an important distinction from the dictionary definitions: food taints are perceived by the human senses. This does not diminish the undesirable nature of chemical contamination of foods, but focuses on those contaminants that can be perceived, particularly by their odour or flavour, and which can be perceived at extremely low concentrations, for example parts per million (10^6), ppm, parts per billion (10^9), ppb, or even parts per trillion (10^{12}), ppt.

An alternative means of distinguishing taints and off-flavours is as follows:

Taints: unpleasant odours or flavours imparted to food through external sources

Off-flavours: unpleasant odours or flavours imparted to food through internal deteriorative change

These definitions have the useful practical value of discriminating taint arising from external contamination, from taint arising from internal change. Although the two types of taint can render food equally unpleasant, this distinction is of great assistance in identifying the cause of taint problems. The difficulties in identifying taint problems result from a number of sources.

Consumer descriptions of taint are, with a few exceptions, notoriously unreliable, partly from a lack of any training in analytical descriptive methods but mainly from unfamiliarity with the chemical species responsible for taint. One notable exception is taint resulting from chlorophenol contamination, which is reliably described as antiseptic, TCP or medicinal, this reliability being a consequence of consumer familiarity with products characterised by these sensations. Equally importantly, as already indicated, taint can occur at all stages of the food manufacture and supply chain, and from many different sources at each stage. Consequently, the detective work needed to identify the cause of taint-oriented consumer complaints can be quite different for taints and for off-flavours.

If minimising the risk of taint is so important, then it is natural to seek ways of preventing taint occurrence. In the more specific definition of the term off-flavour, this can in principle be achieved through consideration of ingredients, process, packaging and storage factors, and is often reflected in product shelf-life. These factors can also be of great importance in taint problems occurring from external sources, but of course these external sources of risk also need to be identified. The complexity of the picture that emerges means that the use of chemical analytical methods can be compared to looking for a very small needle in a large field full of haystacks, a task that can be simplified by the use of appropriate sensory methods. Discussion of the uses of sensory test methods must be preceded, however, by an examination of human sensitivity to chemical stimuli and of how tainting species are perceived.

1.2 Thresholds and their measurement

Whether a chemical species can be perceived in a food depends on the chemical structure, its concentration, the type of food and the sensitivity of the human subject. In order to be perceived as a taint, the chemical need not be positively recognised, but must be characterised as a deterioration in flavour quality. It should be noted that chemical contamination does not necessarily produce this deterioration, and can even produce a perceived improvement in flavour.

Characteristic food taints are often detectable at sub-ppm levels, even down to ppt levels. However, the concentrations at which chemical species can be

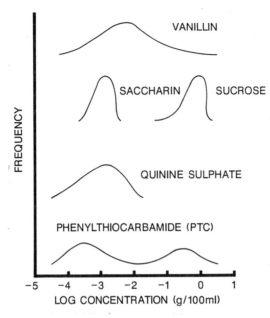

Figure 1.1 Distribution of thresholds for some flavour stimuli. Redrawn and adapted from Blakeslee and Salmon (1935), with permission from Walter de Gruyter, Berlin.

detected varies considerably between individuals. Figure 1.1 shows the frequency distributions of sensitivity to some flavour stimuli (Blakeslee and Salmon, 1935). Not only are the different chemical species perceived at widely different concentrations, but the range of sensitivities to a specific species also depends on its chemical nature. An extreme in this range is exhibited by PTC (phenylthiocarbamide), with a bimodal distribution. It has been reported (Teranishi, 1971) that for pyrazines there is a range of 10^8 in odour thresholds.

The term threshold, commonly defined as the concentration in a specified medium that is detected by 50% of a specified population, is widely used in describing sensory perception of stimuli, but unfortunately is frequently misused and misunderstood. Thresholds indicate the level of stimulus that is sufficient to trigger perception (Moskowitz, 1983) but, contrary to common usage, a number of thresholds can be defined, none of which is invariant. ISO 5492 (ISO, 1992) gives the following definitions for thresholds:

Detection threshold: the minimum value of a sensory stimulus needed to give rise to a sensation

Recognition threshold: minimum value of a sensory stimulus permitting identification of the sensation perceived

Difference threshold: value of the smallest perceptible difference in the physical intensity of a stimulus

Terminal threshold: minimum value of an intense sensory stimulus above which no difference in intensity can be perceived

In dealing with taints we are generally concerned with detection thresholds, but frequently literature data do not define whether detection or recognition thresholds are being quoted.

Compilations of chemical threshold values have been published, but these have been subjected to criticism. Early tables published by Zwaardemaker (1926) were shown to be in error by a factor of 100 (Jones, 1953). Computer-generated tables produced by Fazzalari (1978) have been criticised as being cumbersome and difficult to decipher (Pangborn, 1981). More seriously, these tables do not specify important methodological variables such as number of test subjects, degree of experience of test subjects, nature of instructions to test subjects, test procedure and whether replicated, and details of any statistical analysis. These omissions may serve to explain the wide range of numerical values found by different researchers for the same thresholds (Pangborn, 1981), as shown in Table 1.1.

Table 1.1   Examples of odour and taste thresholds reported for hexanal

Threshold	Medium	Value/range (ppb)
Odour detection	Air	4.5
	Water	0.19–30.0
Odour recognition	Water	4.5–400
Taste detection	Water	0.2–10
	Paraffin	150–300

Sources: Fazzalari (1978) and Pangborn (1981)

Even if threshold measurements utilised the same test methodologies and exercised careful control over experimental variables, variations in measured thresholds must be expected as a result of the enormous range of human sensitivities. Measurements should therefore use as many human subjects as possible, and should also include provision for repeat testing since subject performance improves with practice, as illustrated by data from Pangborn (1959), shown in Table 1.2. Examination of the threshold data in Table 1.1 shows that the medium in which the stimulus is present has a substantial effect on the measured thresholds. Table 1.3 shows flavour detection threshold data, taken from Maarse *et al.* (1988), on some chloroanisoles in different media. Even greater dependency of detection thresholds on the medium was found by Jewell (1976) (Table 1.4). The influence of familiarity would be expected to be less important in measuring detection thresholds than in measuring recognition thresholds.

A further problem associated with the use of threshold data is the incorrect assumption that threshold concentration and potency are synonymous, since

Table 1.2 Influence of practice on individual taste recognition thresholds

Subject Test number	Sucrose		Citric acid		Sodium chloride		Caffeine	
	1	6	1	6	1	6	1	6
1	36	22	1.6	0.10	34	30	1.6	0.20
2	24	8	0.6	0.04	13	3	0.6	0.08
3	36	8	2.1	0.01	34	5	1.6	0.40
4	18	10	1.1	0.10	20	3	1.6	0.40
5	12	6	0.6	0.01	20	7	0.6	0.40
6	18	2	0.1	0.03	20	5	1.1	0.30
7	18	6	2.1	0.04	7	3	3.1	0.60
8	18	4	1.1	0.03	20	9	1.6	0.60
Mean	22	8	1.2	0.05	21	8	1.4	0.40

All thresholds measured as molarity $\times 10^{-13}$
Source: Pangborn (1959)

Table 1.3 Flavour thresholds of some chloroanisoles in different media

Medium	Chloroanisole*		
	2,4,6-Tr CA	2,3,4,6-TeCA	Penta CA
Water	0.02	0.2	3.2
Beer	0.007	–	–
Wine	0.01	–	–
Egg yolk	2.4	2.7	2800
Dried fruit	0.12–0.45	1.0	33
Fruit buns	0.21	1.9	126
Plain buns	1.4	5.8	183

Threshold values given in µg/kg (ppb)
* See Table 1.7 for codes
Souce: Maarse et al. (1988)

Table 1.4 Flavour thresholds of 6–COC, 2,3,6–Tr CA and 2,3,4,6–TeCA in different media

Medium	Material		
	6-COC	2,3,6-TrCA	2,3,4,6-TeCA
Tea	0.03	0.016	15×10^{-6}
Red wine	–	–	25
Whisky	–	100	–
Blancmange	2	500	29

Threshold values given in µg/kg (ppb)
6-COC = 6-chloro-o-cresol
2,3,6-TrCA = 2,3,6-trichloroanisole
2,3,4,6-TeCA = 2,3,4,6-tetrachloroanisole

the perceived intensities of different chemical stimuli do not necessarily increase at the same rate with concentration. Figure 1.2 shows a schematic diagram (from Pangborn, 1981), showing theoretical curves for different stimuli. Stimuli A and B have equal thresholds but dissimilar intensities at higher concentrations, possibly concentrations that may be used in products. Stimuli B and C have different thresholds but equal intensities at higher

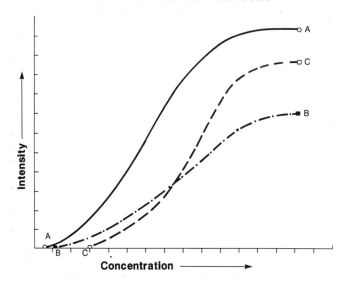

Figure 1.2   Theoretical relationships between threshold and rate of increase of perceived intensity with stimulus concentration. Redrawn and adapted from Pangborn (1981).

concentrations. Pangborn also indicated that these curves could also represent the same stimulus perceived by different subjects.

The wide range of human thresholds to chemical stimuli is a major reason for the difficulties in preventing food taints and in positively identifying the causes of taints. Figure 1.3 shows threshold data from studies of Zoeteman and Piet (1973) on taints in drinking water. The figure shows normalised cumulative distributions of concentrations of 1,3,5-trimethylbenzene, geosmin, dimethyldisulphide and 2-chlorophenol, all differing widely in chemical structure and mean threshold.

Examination of the curve shows that the threshold concentration difference between the 10% most sensitive and 10% least sensitive is over 200-fold, and between the 5% most and 5% least sensitive nearly 2000-fold. The difference at the 1% most and 1% least sensitive rises to approximately 10^6. Since 1 in every 100 consumers complaining to a supermarket that food has, for example, an antiseptic taint, would be regarded as unacceptably high, it can be seen that the food industry must ensure that any tainting species is not only below the mean threshold level but also below the levels that this small but highly important proportion of consumers can detect. Effectively, therefore, the industry must ideally aim for zero levels of tainting species.

In using sensory methods for taint identification and prevention, it is therefore important to utilise human subjects who are known to represent, as far as possible, those highly sensitive consumers, through selection criteria based on threshold measurement. The methodologies used are similar for measurement both of individual sensitivity and of mean thresholds.

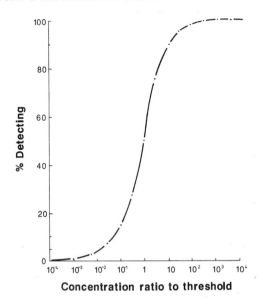

Figure 1.3 Cumulative distribution of sensitivities of 120 people to four different tainting substances (1,3,5-trimethylbenzene; dimethyldisulphide; 2-chlorophenol; geosmin) in water. Data from Zoeteman and Piet (1973), and redrawn and adapted from Williams and Aitken (1983) with permission from Ellis Horwood Limited, Chichester.

Any test procedure must recognise the potentially wide range of individual sensitivities, and ensure that a stimulus range is presented that is sufficiently wide to bracket that range. In addition, the stimulus should be presented in a medium as similar as possible to that which will be encountered in subsequent use. If little information on the likely threshold concentration is available, a wide concentration range can be spanned using a logarithmic presentation sequence, in which the stimulus concentration increases in a constant ratio. An arithmetic presentation sequence, in which the stimulus concentration increases by a constant added increment, covers a smaller concentration range but offers the possibility of more precise threshold data (Table 1.5). In prac-

Table 1.5 Comparison of logarithmic and arithmetic presentation sequences

Logarithmic	Arithmetic
1	1
4	5
16	9
64	13
128	17
512	21
(multiply by 4)	(add 4)

tice, logarithmic sequences are usually more appropriate when a wide range of sensitivities is anticipated. Detailed descriptions and critiques of threshold measurement methods can be found in Brown *et al.* (1978), Pangborn (1981) and Moskowitz (1983).

The coded stimuli are presented to the subjects, who note the samples for which they can detect a stimulus or recognise a stimulus, depending on the purpose of the test. In tests in which the stimuli are presented in increasing order of concentration, there exists the risk of 'error of anticipation', i.e. subjects incorrectly anticipate the presence of a stimulus. Conversely, if a decreasing order of concentration is presented, subjects may exhibit the 'error of habituation', i.e. a tendency to continue to report false positives. These errors can, in principle, be averaged out using a randomised or balanced order of presentation. Many taint stimuli, however, are characterised by extremely persistent flavour carry-over, and a high level of such a stimulus can desensitise the palate and render subsequent judgements highly unreliable. Ascending presentation orders are therefore commonly used, but with modifications to minimise the 'error of anticipation'. One of the simplest ways of achieving this is to include random blank samples in the presentation sequence. A more sophisticated method is to use the 'choose 1 of n' procedure, where n is usually 2 or 3 samples at each level. For two stimuli at each level, one sample contains the stimulus and the other is a blank. The sequences are shown in Table 1.6.

Table 1.6 Sample presentation sequences

Straight sequential	0, 1, 2, 3, 4, 5, 6, 7, . . .
Modified sequential	0, 1, 0, 2, 3, 0, 0, 4, 0, 5, . . .
Choose 1 of n : $n = 2$	0; 0, 1; 2, 0; 0, 3; 0, 4; . . .
$n = 3$	0; 0, 0, 1; 2, 0, 0; 0, 3, 0; . . .

0 = blank (no stimulus)
1, 2, 3, . . . = increasing stimulus concentration

It is important to note that, regardless of the method used, care should be taken to minimise the risk of subject fatigue and carry-over effects by limiting the number of stimuli presented in any given session and by allowing sufficient time between sample tastings.

The importance of these factors will depend on the nature of both the stimulus and subject response, but no more than 4–8 stimuli should be presented in a single session, and about 1–2 minutes should be allowed between stimuli. If the stimuli are presented in foods, palate cleansers should also be specified and used. As indicated previously, the detection or recognition threshold is commonly defined as the concentration that is detected or recognised by 50% of a specified population, or more precisely as the concentration detected or recognised on 50% of occasions by each of a specified group of individuals (Land, 1989).

There are other methods for threshold assessment that are used for chemical stimuli, but these are rarely used for food taints. *Ad libitum* mixing, for example, in which the subject adjusts the concentration of a comparison stimulus to apparent equality with a standard (Pangborn, 1981) is only possible with mixable materials and would be seriously compromised by carry-over effects. Methods based on signal detection procedures, in which responses to a blank or a stimulus are classified as hit, miss, false alarm or correct rejection, require a large number (*ca.* 200) of measurements per subject (Green and Swets, 1966).

A short-cut signal detection method, R-index, is less demanding but nevertheless requires about 20 judgements per subject (O' Mahoney, 1979, 1986).

1.3 Sensory descriptions of taints and off-flavours

The problems associated with the verbalisation of perceived sensory phenomena are often underestimated. A common problem encountered when carrying out consumer acceptance testing is that whilst it is relatively easy to elicit reliable information on level of liking, it is difficult to elicit reliable information on the sensory properties that contribute to these levels. A major concern in quantitative descriptive testing is the selection of individuals who are capable of verbalising sensory perceptions, especially in the presence of many other sensory stimuli.

The difficulties in obtaining reliable and consistent descriptions for chemical tainting species were well illustrated by Thomson (1984). Boar taint is an unpleasant odour sometimes detected when fat from uncastrated male pigs (boars) is heated. The taint was reported (Hansson *et al.*, 1980) to be caused by 5-androst-16-en-3-one (androstenone), 3-methylindole (skatole) and indole, constituting 36%, 33% and 7% of the boar taint, respectively. The sensory descriptions most commonly associated with androstenone are urinous, animal or sweaty (Griffiths and Patterson, 1970; Amoore, 1977); with skatole, mothballs, faecal or musty; and with indole, mothballs, musty/earthy or paint (Harper *et al.*, 1968). In a survey of untrained subjects, Thomson (1984) found that only 25% of the terms described androstenone as sweaty, animal or urinous, and about the same proportion were descriptions such as solvent-like. In total, 32 subjects used 22 different descriptive terms. Similarly, only 28% and 44% of the terms used to describe skatole and indole, respectively, were mothballs or camphor. In total, 40 different terms were used to describe skatole and 44 terms to describe indole. It is also interesting to note that faecal was only used 4 times to describe skatole and not at all to describe indole, despite the ready association of these terms and chemical species in the literature (Whitfield *et al.*, 1982).

It is generally recognised that description of sensory terms is more readily achieved when subjects are familiar with those stimuli as a result of past

experience. This is most easily seen in the case of chlorophenols, since exposure to oral mouthwashes (e.g. TCP) results in a ready production of terms
such as TCP, medicinal and antiseptic. However, even within a given class of
chemical compounds, it should be noted that relatively minor changes in
chemical structure can give rise to substantially different perceptions. Table
1.7 shows data on the odour description of chloroanisoles, together with the
detection thresholds measured in water for a selection of these materials
(Griffiths, 1974). Relatively small structural differences can result in major
changes, not only in odour thresholds but also in the sensory descriptions.
Although chloroanisoles as a class are often described as musty, other descriptive terms are often equally appropriate and, in the case of compounds such
as 2,3,5-TrCA, 3,4,5-TrCA and 3,4-DCA, medicinal and sweet/fruity descriptors are more appropriate. Mustiness was more apparent in chloroanisoles
with higher degrees of chlorination and with chlorination in the 2,6 posi-
tions. Later work (Griffiths and Fenwick, 1977) reported the sensory
changes produced by replacing the chlorine substituents by methyl groups.
The 2,6 compounds retained the musty odour character, but thresholds were
higher.

Table 1.7 Odour description (% of terms used) and odour detection thresholds for chloroanisoles

Stimulus	Musty and related	Medicinal and related	Solvent/ alcoholic	Sweet/ fruity	Detection threshold in water (ppm)
Penta CA	86	3	–	3	4×10^{-3}
2,3,4,5-TeCA	39	9	30	9	
2,3,4,6-TeCA	83	7	–	3	4×10^{-6}
2,3,5,6-TeCA	44	12	32	4	
2,3,4-TrCA	56	–	24	20	
2,3,5-TrCA	19	37	–	41	
2,3,6-TrCA	80	–	–	12	3×10^{-10}
2,4,5-TrCA	41	28	–	21	
2,4,6-TrCA	96	–	–	4	3×10^{-8}
3,4,5-TrCA	27	12	12	48	
2,3-DCA	30	20	15	20	
2,4-DCA	56	–	–	44	4×10^{-4}
2,5-DCA	25	20	20	30	
2,6-DCA	62	23	15	–	4×10^{-5}
3,4-DCA	16	–	3	72	
3,5-DCA	25	16	–	50	
2-CA	–	35	43	22	
3-CA	–	22	30	48	
4-CA	4	37	41	18	
Anisole	–	30	27	47	5×10^{-2}

Penta CA = 2,3,4,5,6 pentachloroanisole
- TeCA = tetrachloroanisoles
- TrCA = trichloroanisoles
- DCA = dichloroanisoles
- CA = monochloroanisoles
Source: Griffiths (1974)

In general, searching the literature for reliable descriptor sets for chemical stimuli is a tedious task, but one that should not be confined to the food literature. For example, chemical companies with an interest in pollution control have attempted to characterise odours associated with the petrochemical industry (Hellman and Small, 1973, 1974). In a study of 101 petrochemicals, odour thresholds, odour qualities and hedonic characteristics were recorded (Hellman and Small, 1974). Although the reported descriptors were limited to one or two per compound, the hedonic rating (pleasant/neutral/unpleasant) is potentially useful. However, with only a few exceptions, the compounds studied were not those commonly implicated in food taints. A consequence of the non-unique nature of sensory descriptors is that inspection of the indexes of taint data sets can produce a relatively large number of compounds characterised by a specific descriptor.

In Table 1.8 data on the possible compounds that may be associated with three common sensory descriptors are shown. (These data are taken from the Index of Chemical Taints, available to members of the Contaminants Working Group at the Leatherhead Food Research Association.) Each descriptor may be associated with distinctly different classes of chemical structures, adding to the difficulties experienced by the chemical analyst in attempting a positive

Table 1.8 Possible chemical compounds related to specific sensory descriptors

Descriptor	Chemical compound
Musty	2,6-Dimethyl-3-methoxypyrazine
	2-Methoxy-3-isopropylpyrazine
	2,4-Dichloroanisole
	2,6-Dichloroanisole
	2,3,6-Trichloroanisole
	2,4,6-Trichloroanisole
	2,3,4,6-Tetrachloroanisole
	Pentachloroanisole
	2,4,6-Tribromoanisole
	Geosmin
	2-Methylisoborneol
	1-Octen-3-ol
	Octa-1,3-diene
	α-Terpineol
	4,4,6-Trimethyl-1,3-dioxan
Painty	Heptan-2-one
	Trans,trans-hepta-2,4-dienal
	Trans-1,3-pentadiene
	2-(2-Pentenyl) furan
Plastic	Benzothiazole
	Methyl acrylate
	Methyl methacrylate
	Trans-2-nonenal
	Oct-1-en-3-one
	Trans-1,3-pentadiene
	Styrene

identification. In addition, the descriptor for a particular compound may change with concentration and the medium. As the concentration of *trans*-2-nonenal in water increases from 0.2 ppb to 1000 ppb, for example, the descriptors change from plastic to woody to fatty to cucumber (Parliment *et al.*, 1973). Even the relatively easily identified chlorophenols are frequently described as plastic when present in orange juice at threshold concentrations (Kilcast, Crawford and Marchant, unpublished results). An additional difficulty in using such information is that a specific description and chemical compound may be a result of either external contamination (taint) or internal degradation (off-flavour). Diagnosis could clearly be hindered if an incorrect assumption is made. For example, oct-1-en-3-one can be present from autoxidation of fats (Hammond and Seals, 1972; Swoboda and Peers, 1977) or from plastics containing the plasticiser diisooctylphthalate.

Descriptive information is clearly essential in diagnosing the source of taints and off-flavours, but descriptive information commonly available is limited and, in general, incomplete. More comprehensive sets of descriptions from larger numbers of human subjects would ease diagnosis, if available in suitable form.

1.4 Principles of sensory evaluation of food

Many methods can be used for the sensory analysis of foods, but some care is needed in selecting appropriate methods for taint assessment, and, of equal importance, it should be recognised that many of these methods require modification both in use and in interpretation.

Sensory testing methods can be grouped into two broad classes: (i) analytical tests; and (ii) affective (or hedonic) tests (IFT, 1981). Although each class sometimes contains the same test procedure, the purpose of the test is quite different. In analytical tests, which can be subdivided into discriminative and descriptive categories, the senses of human subjects are used to provide information on characteristics of the food. The panel of human subjects is consequently used as the equivalent of an analytical instrument, and steps are taken to minimise the various forms of bias that can influence their performance, and to reduce the effect of natural biological variation. Affective tests (hereafter referred to as hedonic tests), in contrast, are used to assess the effect of the food on human response, normally in terms of preference or acceptability. Natural human reactions are therefore required, which reflect biases normally encountered and also biological variation that can result from both physiological and psychological sources. A major practical distinction between the tests that fall into each category is that analytical tests utilise small numbers of carefully selected and trained assessors, whereas hedonic tests use relatively large numbers of untrained assessors, the members being chosen to try to reflect the likely response of a larger population. It is normally important not to seek hedonic information from trained assessors or, conversely, analytical information from untrained consumers. The unique problems of taint, however, sometimes require a relaxation of this otherwise stringent requirement, as discussed in section 1.5.

The following sections describe the available sensory test methods only briefly; further details can be found in various standards (e.g. ISO, 1985) and reference texts (e.g. Stone and Sidel, 1985; Piggott, 1988).

1.4.1 Analytical tests

1.4.1.1 Discriminative, or difference, tests.   These tests are used to determine whether or not a sensory difference exists between two samples, and also as a means of measuring thresholds and sensitivity, as described in section 1.2. Three types of difference tests are used most frequently: (i) paired comparison; (ii) triangular; and (iii) duo-trio. Two other tests, two out of five and 'A' or 'not A' are used less commonly, and will not be described here. A short-cut signal detection test (R-index) has been reported and may have future applications.

Paired comparison test.   Two coded samples are presented either sequentially or simultaneously in a balanced presentation order (i.e. AB and BA). There are two variations on the test. In the simple difference variant the panellists are asked if there is a difference between the two samples, having been previously informed that there may or may not be a difference. In the directional difference variant, the panellists are asked to choose the sample with the greater or lesser amount of a specified characteristic. The panellists may be allowed to record a 'no-difference' response, or they may be asked to make a choice (forced-choice procedure). The forced-choice procedure is more correct statistically, but no-difference responses can provide a useful source of information. The probability of a selection being made purely by chance is 0.5, and responses are analysed in terms of statistical significance levels calculated using the binomial distribution. This has traditionally been carried out using statistical tables calculated at fixed levels of significance (e.g. 1% and 5%), but calculation of exact significance levels (Lewins and Wilson, 1985) should be used where possible (see section 1.5.1). Numbers of assessors recommended by ISO 6658 (ISO, 1985) are shown in Table 1.9.

Triangular test.   Three coded samples are presented to the panellists, two of which are identical, using all possible sample permutations, i.e.

ABB   AAB
BAB   ABA
BBA   BAA

Table 1.9 Minimum recommended numbers of assessors (ISO, 1985)

Test	Number of assessors		
	Experts*	Selected*	Assessors*
Analytical tests			
Paired comparison	7	20	30
Triangular	5	15	25
Duo-trio	–	–	20
Ranking	2	5	10
Classification	1	1	–
Rating	1	5	20
Scoring	–	5	20
Simple descriptive	5	5	–
Profile	5	5	–
Hedonic tests			
Paired comparison	–	–	100
Ranking	–	–	100
Rating	–	–	50 (2 samples)
			100 (> 2 samples)
Scoring	–	–	50 (2 samples)
			100 (> 2 samples)

* as defined in ISO 6658

The panellists are asked to select the odd sample in either no-difference or fixed-choice procedures. As the possibility of choosing the odd sample purely by chance is 1/3, the test is more powerful than the paired comparison, and therefore fewer panellists are required (Table 1.9). The increased number of samples can, however, result in problems with flavour carry-over when using strongly flavoured samples, making identification of the odd sample more difficult. Difficulties can also be encountered in ensuring presentation of identical samples of some foods. Statistical significance values can again be calculated from the binomial distribution. No-difference responses can either be ignored and the number of judgements reduced accordingly or, alternatively, one-third of the no-difference responses can be added to the correct responses.

Duo-trio test. The panellists are presented with a sample that is identified as a standard, followed by two coded samples, one of which is the same as the standard and the other different. They are asked to identify the sample that is the same as the standard. The sample presented as the standard may be the same for each panellist (fixed reference standard) or may be balanced between the two types. The test has the same power as the paired comparison test and is analysed using the same procedures. Presentation of an identified reference reduces the need for expert or selected assessors, and use of a fixed reference standard can reduce problems when flavour carry-over associated with the test sample is anticipated—of potential importance in taint testing. An additional advantage of the duo-trio test is that if there are difficulties in preparing identical portions of a food sample (e.g. pizza), the panellists can be asked to identify the sample that is most similar to the reference.

Difference from control.   This is sometimes used when a control is available. The panellists are presented with an identified control and a range of test samples. They are asked to rate the samples on suitable scales anchored by the points 'not different from control' to 'very different from control'. This type of test is more commonly classified as similarity/dissimilarity scaling and is analysed using multidimensional scaling methods (Schiffman and Beeker, 1986).

R-index test.   This method is a relatively recent development (O'Mahoney, 1979, 1986), and is a short-cut signal detection method. The test samples are compared against a previously presented standard, and rated in one of four categories. For difference testing these categories are: (i) standard; (ii) perhaps standard; (iii) perhaps not standard; and (iv) not standard. The test can also be carried out as a recognition test, in which case the categories are: (i) standard recognised; (ii) perhaps standard recognised; (iii) perhaps standard not recognised; and (iv) standard not recognised. The results are expressed in terms of R-indices, which represent probability values of correct discrimination or correct identification. The method is claimed to give some quantification of magnitude of difference, but its use has not been widely reported in the literature. In a recent paper, however, an application to taint testing has been described (Linssen *et al.*, 1991).

1.4.1.2 Descriptive tests.

Classification.   This method is used to sort items into a pre-defined number of categories, using a small number of expert or selected assessors. The categories used in classification are nominal only.

Ranking.   Ranking tests are used to sort several samples into order of intensity of a specific characteristic. Rank totals are calculated over all panellists, and differences are interpreted by statistical tests, e.g. Friedman test. The tests are often carried out as a screening test prior to more precise assessments. No magnitude information is produced and different sets of results cannot be compared.

Rating.   These tests involve classification into categories in the form of an ordered scale. The categories need to be clearly defined and understood by panellists. Unlike ranking tests, estimates of magnitude of attributes are produced. The scales can take various forms, for example graphic or descriptive, and uni-polar or bi-polar. It should not be assumed that the magnitude of differences between adjacent categories is the same throughout the scale.

Scoring.   Scoring is a form of rating using a numerical scale, in which the scores bear meaningful mathematical relationships. Coded samples are

not   very

Figure 1.4 Example of an unstructured line scale.

presented simultaneously or sequentially to panellists in a balanced presentation order. The scales can be category scales, unstructured line scales (or graphic scales) with verbal anchors (Figure 1.4), or in a form of scaling known as magnitude estimation, ratio scales.

Analysis of scored data from two samples can be carried out on two samples using simple t-tests or, when more than two samples are being assessed, analysis of variance and multiple comparison tests. Checks should be made to ensure that the data are normally distributed; if not, non-parametric tests should be used.

Simple descriptive tests.   Purely qualitative in nature, these tests require each panellist to assess one or more samples independently to identify and describe the sensory attributes. Following the independent assessments, the results are collated by the panel leader and a list of descriptive terms is produced based on frequency of use. The order of appearance of individual attributes is also frequently recorded. An additional step that is frequently carried out is for the panel leader and the panel to discuss the descriptive terms and to generate a list of terms that is agreeable to all panellists. In order to reach agreement, several discussions may be needed, and it is important that the panellists can also agree on a definition of each descriptive term.

Sensory profile tests.   Profile tests are a means of quantifying the sensory attributes of foods that have been established using descriptive tests. Two main classes of tests are used. Consensus profiling, based on a procedure termed quantitative descriptive analysis (QDA), uses the agreed list of attributes generated by the simple descriptive tests already described. Following agreement of attributes and their definitions, suitable scoring scales (usually unstructured line scales) are constructed with appropriate anchors, and the panellists are trained in scoring reproducibly the intensity of the chosen attributes using a training set of samples. Training can be lengthy for foods characterised by large attribute numbers. Once satisfactory reproducibility has been achieved, the test samples are scored in replicated tests using appropriate statistical designs. The data are analysed statistically using appropriate methods, for example analysis of variance and multiple difference testing, but multivariate analysis methods such as principal components analysis are increasingly common. Graphical methods are often used to display results.

The second profile method is free choice profiling, in which the assessors use the individual attribute list generated as the first stage of a simple qualitative test. Each assessor constructs an individual profile based on his own attributes, and a consensus profile is constructed mathematically using a technique known as generalised Procrustes analysis. This procedure reduces panel training times considerably, but the consensus data are difficult to interpret reliably.

1.4.2 Hedonic tests

Hedonic tests are used to measure liking, usually in terms of acceptability or preference. Although employing some of the mechanics of analytical tests, hedonic tests demand the use of sufficient numbers of respondents who are, ideally, typical of a larger consumer population. In addition, the tests may need to be carried out in an appropriate consumption situation, e.g. at the respondent's home or in a restaurant. The most common tests used are: (i) paired comparison tests (usually paired preference); (ii) ranking tests; or (iii) scoring tests using hedonic category scales.

A consequence of the major conceptual difference between analytical and hedonic tests is that the two types of test should not be carried out with the same assessors or respondents. A notable exception to this general rule, however, lies in the adaptation and application of sensory methods in testing for taint, as will be described subsequently.

1.4.3 Operation of sensory testing procedures

The use of formal sensory testing procedures in the food industry depends on practical operating constraints. Few organisations can apply sensory testing procedures rigorously as laid down in standards, but there are some key guidelines that should be followed where possible:

(i) The testing must be carried out in a suitable environment. In particular, analytical testing requires an environment free from distraction, noise and odours, in which the panellists can give their tasks sufficient concentration. Individual booths, lighting control and temperature control may be necessary.

(ii) Sources of external bias must be minimised. This will involve, among other possible steps, minimising panellist interaction (e.g. by using individual booths), coding samples with different random number codes for each test, balancing presentation order, careful questionnaire design, and using statistical tests that are appropriate to the data and the purposes of the test.

(iii) Panellists possessing the appropriate level of training must be used for the tests. For example, there is no value in operating profile tests using panellists who are untrained or unfamiliar with the products to be profiled.

(iv) Panel managers must be prepared to allow the conclusions from the panel test to override their own personal judgements and expectations. This is particularly important if the confidence in panel operations is to be maintained through a reporting chain to senior management. The only exception to this is when the panel manager possesses special information or expertise that can complement panel findings.

1.5 Sensory testing for taint

1.5.1 Test selection and modification

Any attempt to select sensory testing procedures for taint testing encounters a fundamental problem. Since a taint that could spell commercial disaster may only be detectable by a few per cent of consumers, can sensory tests that, for practical reasons, use only small numbers of panellists, be designed to guard against this occurrence? Although no procedure of practical value can guarantee that a taint will be detected, steps can be taken to minimise the risk of not identifying a taint stimulus. The most important of these are the following.

(i) Where the identity of the tainting species to be tested for is known, use panellists who are known to be sensitive to that species. As explained in section 1.2, threshold measurements can be used as a guide to sensitivity, but it is also useful to measure sensitivity at the higher concentrations that may be present in practice. Unfortunately, it cannot be assumed that a panellist sensitive to one specific tainting species will also be sensitive to other tainting species. In principle, therefore, a taint panel would need to be selected on the basis of high sensitivity to several different species. In general this is not feasible, particularly when panels are to be used to guard against unknown taints, and a compromise must be found by using panellists who are known to be generally sensitive and reliable in their judgements.

(ii) If a high-sensitivity panel is not attainable, use as many panellists as possible in the hope of having someone present who is sensitive to the taint. In section 1.4, recommended minimum numbers of panellists of various levels of expertise have been given (Table 1.9). In testing for unknown taints, it is important that at least these minimum numbers are used, and increased if possible. For example, a triangular test should not use less than 15 panellists, and for key tests at least 30 should be used.

At this stage, a note of caution is necessary. Practical operating constraints often limit the availability of personnel for forming sensory panels. Frequently, therefore, a small number of panellists is used to make replicate judgements; for example a triangular test can use a panel of five to carry out three tests. This is allowable in conventional difference testing, since the statistical analysis is, strictly, based on the number of independent judgements and not on the number of panellists. This procedure is, however, of no value for taint testing; if the five panellists used are insensitive to a taint, then no amount of replication will increase the chances of detecting that taint.

(iii) Use of a high-sensitivity test procedure. Difference tests are generally more suitable than profile-type tests, being more rapid whilst not requiring intensive training. In addition, a difference test against an appropriate, untainted control is a relatively easy task for the panellist. The publication of taint tests using the R-index method (Linssen *et al.*, 1991) demonstrates some potential value but this is relatively unproven.

In selecting high-sensitivity difference tests, potential problems of flavour carry-over must be addressed. The triangular test is statistically powerful, with a 1/3 chance of identifying the odd sample by chance, but a balanced presentation order will involve presenting two tainted samples to all the panellists. Examination of triangular taint test data has shown that in presentations containing two tainted samples (TTC, TCT, CTT) identification of the odd sample is less easily achieved than in permutations containing two control samples (CCT, CTC, TCC) (Wilson and Kilcast, 1984). Limiting the permutations to those containing two control samples may introduce unacceptable bias into the test, and should only be used in selection or threshold measurement procedures.

Paired comparison tests and duo-trio tests, although less powerful, have the advantage that only one taint sample is presented to each panellist. The former test is not appropriate for taint testing, but the constant reference duo-trio test (with the control sample as reference) is becoming used with increasing frequency. The main disadvantage of the duo-trio test, however, is that the lower statistical power requires the use of more panellists than the triangular test, with consequent practical problems. Other sensory tests have shown few applications in taint testing.

(iv) Maximise the information content of the test. Some purists in the field of sensory analysis (e.g. Stone and Sidel, 1985) maintain that identification of a difference is the only information that should be elicited from panellists, the reason being that any attempt to elicit other information will require different mental processes that may invalidate the test. As a minimum requirement, descriptive information on the nature of any identified difference must be recorded.

At the Leatherhead Food Research Association, two other types of information are elicited. Firstly, since taints are by definition disliked, preference information is recorded. As indicated previously, this is a unique exception to the general rule that hedonic and analytical tests must not be mixed. The preference information is not interpreted as a likely measure of consumer response, but is used purely as a directional indicator in conjunction with descriptive information. Secondly, panellists are asked to rate how confident they were in their choice of the odd sample on a 3-point category scale. (An example of a typical triangular test questionnaire is shown in Figure 1.5.)

Name: ....

Date: ....

TRIANGULAR TEST

Of the three samples in front of you, two are the same and one may be different. Please taste them in the order given and try to select the odd one.

Sample codes:

Can you detect any difference?	
If so, which is the odd sample?	
Which sample(s) do you prefer?	
Please describe any difference(s):	

How confident are you in your choice of sample?

Absolutely sure	
Fairly sure	
Not very sure	

Figure 1.5 Example of triangular test questionnaire.

Confidence levels weighted toward one end of the scale or the other can help resolve indeterminate results by indicating to what extent panellists may be guessing. Such a scale may be formalised by assigning scores to the scale points.

An important point to note when using such ancillary data, however, is that these data are only valid from panellists who have correctly identified the odd sample. Data from panellists who have made incorrect identifications are invalid and must not be used.

(v) Use statistical tests that are appropriate to taint testing. A fundamental problem is apparent here, as discussed by O'Mahoney (1982, 1986). Conventional hypothesis testing involves testing the experimental data against a null hypothesis (H_0) that no trend, or difference, exists in the data. A probability value that represents a difference occurring by chance is calculated. If this value is low, it is unlikely that the null hypothesis is true, and the alternative hypothesis (H_1) is accepted, which states that a difference is present. On the other hand, a high value indicates that the result could have occurred by chance, and the null hypothesis is not rejected. A probability value of 0.05 (5% significance) in a difference test can then be interpreted as saying that a difference does appear to exist, but with a 5% (1 in 20) probability that the result could have been due to chance. If we require more assurance that we really have found a difference, a lower significance level of 1% could be used, giving a 1 in 100 probability of a chance result.

Unfortunately, the greater the assurance of a real difference that is sought, the greater the risk of not identifying a real difference that is present (Type II error). By increasing the significance level to 10%, 15% or even 20%, the risk of not identifying a real difference diminishes, but the risk of incorrectly identifying a difference (Type I error) increases. The choice of an appropriate cut-off point depends on the required level of risk; even 1% would be too high a risk in medical experiments, and values of 0.1% or 0.01% may be more appropriate. In sensory testing, however, and in particular in taint testing, the consequences of incorrectly saying a difference exists are relatively minor, compared with the consequences of not identifying a difference and allowing tainted product to reach consumers. Consequently, levels of up to 20% should be used to minimise this risk, but accepting that by using a 20% cut-off there will be an expectation that overall 1 in 5 will be incorrect.

It should be noted that, in interpreting probability levels, there is little practical difference between probabilities of 4.9% and 5.1%, but that if a rigid cut-off of 5% were used, different interpretations would result. Consequently, it is preferable to calculate exact significance values and use common sense in their interpretation.

O'Mahoney (1982) has made the observation that, for the purposes of sensory evaluation, there is a need for statistical tests designed to reject the alternative hypothesis (H_1) rather than the null hypothesis (H_0).

(vi) Regardless of the results of statistical tests, take careful note of minority judgements, particularly from panellists of established reliability, and re-test for added assurance. Such procedures have been built into guidelines produced for taint testing with pesticides (MAFF, undated).

1.5.2 Diagnostic taint testing

Unfortunately for the food industry, taint problems continue to be so widespread that considerable effort and expense continue to be necessary for diagnosing taint problems. These problems frequently involve insurance claims or litigation and, in such cases, correct sensory (and also chemical analysis) procedures must be rigorously adhered to.

The first indication of a taint problem is usually through consumer complaints of sensory quality. One consequence of the commonly low level of taint detection is that the complaints may come in at a low rate over an extended period, and recognition of a taint problem may not be immediate. In addition, investigating a sensory quality complaint arising from a unit item is undesirable as the safety of an item removed from its packaging may be questionable. In such cases, examination should be restricted to odour and, if feasible, chemical composition.

Examination of batches of suspect product should be carried out as a means of investigation, but again care must be taken to guard against possible safety problems. The suspect product to be tested should be from the same batch coding as the complaint material, and as far as possible should have gone through the same distribution channels. In addition, suitable control material of similar age should be available. In circumstances in which the complaint pattern suggests non-uniform distribution within a production batch, testing can be carried out to a suitable statistical sampling plan, but such testing can often prove prohibitively time-consuming and expensive.

Consumer descriptions of most taints cannot be relied on as a means of focusing chemical analysis investigations, and sensory testing of suspect batches should be carried out to generate reliable descriptive information. However, as has been discussed in section 1.3, care is needed in relating descriptions to possible chemical species.

If the presence of a taint in complaint batches can be established, efforts must be made as quickly as possible to isolate affected product and to identify the source of the taint. Sensory testing can be used to investigate whether the problem is associated with a single transport container, production run, ingredients batch or packaging material batch. If the problem appears to be continuing over a period of time, however, possible sources such as new building materials, process line components or water-borne contamination must be examined. If ingredients (including water supply) are suspected as continuing sources of taint, small test batches of product can be prepared and compared against appropriate controls. Materials suspected as sources of taint can be tested using taint transfer tests, as described in the next section.

Particular care must be taken in gathering evidence and setting up test procedures if, as must frequently be assumed, insurance claims or, even more importantly, litigation are likely. Companies supplying tainted materials may face litigation by their customers, and in turn may enter into litigation against their own suppliers. It is frequently advantageous to contract testing work out to a third party organisation in order to establish impartiality in generating data to be used as evidence. Care should, however, be taken to establish the scientific credentials and expertise of such organisations. A number of suggestions can be made in initiating such investigations if the time scales and costs of litigation are to be minimised.

(i) Act quickly to identify the nature and source of the taint.
(ii) Use both sensory and chemical analysis to establish both the occurrence and identity of the taint—do not rely on one type of information only.
(iii) Store both suspect and control samples (under deep freeze if necessary) for future testing.
(iv) Carry out sensory testing following the general guidelines given in ISO 6658 and use as many assessors (preferably sensitive) as possible.
(v) Extract as much information from the tests as possible, but do not compromise the test quality.
(vi) Have the tests carried out and interpreted on a double-blind basis, especially if the tests are to be sub-contracted.
(vii) Ensure that the names and addresses of panellists are held, as presentation of sensory data in a court of law may require the presence of the individual panellists as witnesses.

1.5.3 Preventive (taint transfer) testing

Preventive testing is a powerful, but frequently misapplied, means of limiting problems arising from the introduction of new materials and changes in environmental conditions. The tests seek to expose food or food simulants to potential taint sources in an exposure situation that is severe but not unrealistic. Severity factors of up to ten times are usually used, but higher factors can be used for critical applications. An outline protocol for such tests is shown in Figure 1.6.

The design of the exposure system varies considerably depending on the nature of the test. For example, taint testing of pesticide residues requires a full-scale field trial with rigidly defined crop growing, pesticide application and crop sampling procedures (MAFF, undated). In testing packaging systems, the model system may need to simulate either direct contract or remote exposure, while when testing process line components, factors such as product residence time and product temperature must be considered.

Factors to be considered when designing model systems for testing materials such as flooring, paints and packaging materials include the following:

- type of food/food simulant
- ratio of the volume or surface area of the material to the volume of the vessel
- ratio of the volume or surface area of the material to the volume or surface area of food/food simulant
- stage of exposure (e.g. stage during curing of a flooring material at which exposure is to start)
- length of exposure
- temperature and humidity at exposure
- exposure method (e.g. direct contact or vapour phase transfer)

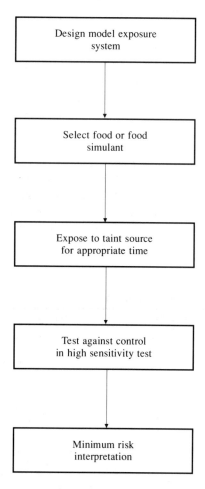

Figure 1.6 Experimental protocol for preventive (taint transfer) testing.

- exposure lighting conditions (especially when rancidity development may occur)
- ventilated or unventilated exposure system
- temperature and length of storage of food/food simulant between exposure and testing
- sensory test procedure and interpretation

Choice of appropriate foods/food simulants is an important consideration, with two alternative approaches. In situations in which a specific ingredient or product is known to be at risk, the test can be designed around it. Where the purpose of the test is more general, however, simple foods or food simulants are often used. Solvent or adsorptive properties are probably the most important considerations in selecting appropriate general simulants. Oils and fats will tend to absorb water-insoluble tainting species, and materials such as butter are known to be sensitive to taint transfer. High surface area powders with hydrophilic characteristics have also been found to be sensitive to taint transfer, and tend to absorb water-soluble taints. Use of such materials will simulate a large proportion of the solvent and adsorptive characteristics of real foods. An additional requirement for suitable simulants, however, is that they should be relatively bland to enable easy detection, and also of reasonably high palatability. This latter consideration unfortunately renders some simulants recommended for packaging migration tests (EC, 1985), for example 3% acetic acid, unsuitable for taint transfer testing. Still mineral water can be used to simulate aqueous liquids, and 8% ethanol in water to simulate alcoholic drinks. In the author's laboratory, however, the characteristic ethanol flavour has been found to be rather unpleasant, and a bland vodka is now used, diluted down to 8% ethanol. Some suitable materials for general purpose use are given in Table 1.10.

Table 1.10 Foods/food simulants for taint transfer testing

Type	Food/simulant	Comments
Fat	Chocolate	Bland variety (e.g. milk) preferred
	Unsalted butter	Mixed prior to sensory testing *or* outer surfaces only tested for severe test
Hydrophilic powder	Sugar	High surface area preferred (e.g. icing sugar)
		Test as 5% solution
	Cornflour	Test as blancmange formulation (can get textural variation)
Combined	Biscuits	High-fat (e.g. shortbread)
	Milk	Full cream. For short-term exposure tests only, or rancidity problems can interfere

Figure 1.7 shows a schematic diagram of a simple model experiment used at the Leatherhead Food Research Association to carry out taint transfer testing on flooring materials and paints. The use of a jam jar lid and a 3-litre beaker gives a surface:volume ratio of 1:100; if required this can be changed by using different lid areas and beaker volumes. The beaker can be raised on supports to give unforced ventilation if required, or a pump can be used to simulate the effect of forced ventilation.

In practice, sophisticated procedures such as those described here are often impractical for quality control procedures, and are adapted as short-cut screening tests.

A number of standard procedures for taint transfer testing have been published, mainly aimed at food packaging materials (BSI, 1964; OICC, 1964;

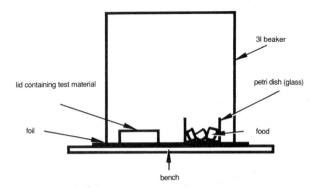

Figure 1.7 System employed at the Leatherhead Food Research Association for testing the tainting potential of building materials.

DIN, 1983; ASTM, 1988). The British Standard and the American Standard deal with taint transfer from packaging films in general, and the OICC standard ('Robinson test') deals specifically with taint transfer to cocoa and chocolate products, although it is frequently used for other products. The German DIN standard also refers to food packaging, but contains much useful information for setting up tests on other materials. All the published methods are, however, deficient in their use of sensory testing methods, and the more correct and more sensitive procedures described here and based on ISO 6658 should be used to replace these where possible. In seeking to maintain high food quality and to minimise the risk of taint problems, Marks & Spencer has developed Codes of Practice referring to the use of packaging films, plastics and paints (Goldenberg and Matheson, 1975). These guidelines stress the importance of testing by the packaging supplier before dispatch, and by the food manufacturer before use. This important principle is, unfortunately, rarely recognised by the food industry in general. Food manufacturers frequently rely on suppliers to provide some general form of certification or test evidence that a material is free from taint, but the material is seldom tested under the conditions in which it will be used. Information provided by suppliers can be regarded as useful screening information, but users must protect themselves by re-testing under more realistic and rigorous conditions.

1.5.4 Sensory quality assurance taint testing

The different test procedures described here need to be used in order to ensure high sensitivity. In a quality assurance (QA) environment, however, the use of relatively large numbers of panellists is usually impractical, and the required throughput often mitigates against tests that can only compare one sample against a control. In such circumstances, it is often necessary to use short-cut screening procedures as part of a more general QA system, and to have in place an alerting system and a back-up system for examining suspect foods or materials.

The importance of having effective systems in place for guarding against taint is gradually being reflected both in recommendations for good quality management in the food industry and in legislation. Many companies are now seeking accreditation under ISO 9000/EN 29000, and guidelines prepared for the food industry (Leatherhead Food Research Association, 1989) have highlighted the need for systems to prevent the occurrence of taint. Perhaps more importantly for the future, companies may need to demonstrate that they are taking measures to prevent taint as a defence under 'due diligence' clauses. In practice, it is likely that satisfactory measures will depend on the resources available to a company. Examples of steps that can be taken are as follows:

(i) When panel numbers are limited, carry out screening for general sensitivity and for specific sensitivity to critical taints, especially chlorophenols.

(ii) Use rating procedures or category scales for large numbers of samples.

(iii) Test incoming ingredients and materials to limit the testing that will be necessary during subsequent processing. If necessary, carry out small-scale processing on incoming ingredients if they are unsuitable for direct testing. For example, some dairies use small-scale pasteurisers to treat incoming unpasteurised milk before testing.

(iv) Be prepared to take action on the basis of a warning by only one sensitive panellist, if necessary, and do not ask panellists to confirm their judgements by re-testing. There have been reports of slight fatiguing desensitisation by chloroanisoles (Griffiths, 1974), but unpublished work carried out at the Leatherhead Food Research Association has shown severe chlorophenol-associated desensitisation that can remain for several hours.

1.5.5 Storage and shelf-life testing

The sensory characteristics of all foods change on storage. The changes can be relatively rapid, for example in fresh produce, or very slow, for example in canned foods. In the case of foods such as cheeses, fermented meats and alcoholic beverages, these changes are essential in improving the sensory quality of the product. In general, however, changes on storage are undesirable, producing unpleasant changes in appearance, odour, flavour and texture. These changes are vitally important in determining the effective shelf-life of the product. Shelf-life has been defined (IFT, 1974) as 'the period between manufacture and retail purchase of a food product during which the product is of satisfactory quality'. It should be noted that although the sensory characteristics will be of great importance in defining 'satisfactory quality', factors such as product safety, company policy on quality, and marketing constraints also play important roles.

Monitoring sensory quality over the intended storage time of the product can be achieved using the general sensory methods described in section 1.4; a general discussion of the use of these methods is outside the scope of this chapter. (Fuller descriptions can be found in Labuza (1982) and Labuza and Schmidl (1988).) One aspect of direct relevance, however, is that the change in sensory quality with time is often manifested by the appearance of off-flavours, and these can often be confused with taint from external sources. These off-flavours commonly arise from microbial action, from enzymic spoilage or from oxidative changes (Goldenberg and Matheson, 1975). The design of sensory test methods for establishing the nature of such changes must address several important issues:

(i) Shelf-life represents an aspect of quality as perceived by consumers, but extensive use of hedonic tests using consumers is frequently impractical. Discriminative and descriptive tests are more easily used, but the relevance of differences or ratings generated by such analytical methods to hedonic characteristics is often difficult to assess without correlation with hedonic information. One means of achieving this is to follow changes on storage using analytical methods, and to submit the products for hedonic testing at those periods when changes that have been shown by experience to be important are identified. In practice, however, the analytical data are frequently related to feedback from consumers via complaint procedures.

(ii) The sensory methods should be sufficiently robust to be used with confidence over the long storage periods that will be required for many foods, and, in particular, should be independent of personnel changes on panels, which will inevitably occur. Profile testing suffers from the need to train

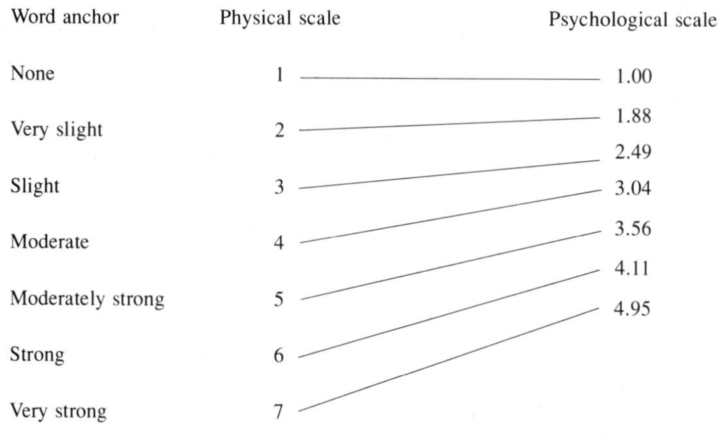

Word anchor	Physical scale	Psychological scale
None	1	1.00
Very slight	2	1.88
		2.49
Slight	3	3.04
Moderate	4	3.56
		4.11
Moderately strong	5	4.95
Strong	6	
Very strong	7	

Figure 1.8 Off-flavour scale with transformation to a psychological scale (after Gacula and Washam, 1986).

replacement panellists to the standard of the rest of the panel, and maintaining panel expertise can be difficult if a specific product is only tested at infrequent intervals, say greater than 4-week periods. For this reason, difference testing is often a preferred option, in spite of its lower information content. A common compromise is to use a rating system to score an undefined 'off-flavour' attribute. Such a system is less demanding on panel training, particularly if descriptive anchor points are used on the scale. In a study of the use of words to describe the intensity of off-flavours, Gacula and Washam (1986) developed a 7-point category scale, and related the scale points to psychological scale values (Figure 1.8). The authors stressed, however, that more studies were needed to establish the generality of such transformations.

(iii) Storage changes are often characterised by the appearance of off-flavours that were not present at the start of the storage trial. This can produce considerable difficulties when using profile methods to evaluate the storage characteristics of unfamiliar products, since the panellists will have had no training in quantifying these new off-flavour attributes. This again results in difference tests being used for new product types. An alternative procedure would be to use free-choice profiling, since panel training requirements are minimal and new attributes can be quantified relatively quickly on an individual basis. Analysis of data by generalised Procrustes analysis may not be appropriate, however, and a simpler means of interpreting the off-flavour data may be needed. A potential defect in using the scale shown in Figure 1.8 is that confusion may occur if several distinct off-flavours appear on storage.

(iv) An important requirement for all sensory storage tests, and particularly those that do not employ highly trained panels, is the availability of suitable standard material representing fresh product (i.e. zero storage time). This is normally achieved in one of two ways, both of which can carry substantial practical difficulties. Firstly, fresh product can be supplied for each testing period. This option is only feasible if production is being maintained on a routine basis, and if consistent production quality can be assumed. These factors often give rise to problems at early stages of product development cycles. Secondly, product from the same batch as that under test can be stored in conditions under which changes are insignificant compared with those in the product under test. The methods will depend on the product under test, but generally will rely on a lowering of temperature or changes in product environment using modified packaging, humidity or gas atmosphere.

(v) Testing products over long storage periods often conflicts with the need to introduce new lines into retail outlets with minimal delays. This problem is exacerbated by the increasing need to have 'sell by' or 'use by' labelling information on foods. This poses particularly severe problems when developing novel foods for which there is little information available on storage characteristics through in-house experience or from the published literature. A commonly used, and abused, procedure is to use accelerated shelf-life testing (ASLT) techniques (Labuza, 1982; Labuza and Schmidl, 1985), in which the product/packaging combination is stored under some specified abuse condition, usually raised temperature and also raised humidity. Such tests are a perfectly valid means of investigating the effect of such abuse conditions on product quality, but problems arise when attempts are made to extrapolate from changes under abuse conditions back to storage under normal (ambient) conditions. This can be carried out using the Q_{10} approach (Labuza, 1982; Labuza and Schmidl, 1985), where Q_{10} is defined as the ratio of the reaction rate constants at temperatures differing by 10°C or, equivalently, the time for an unwanted change to reach unacceptable levels when the food is stored at a temperature higher by 10°C. More commonly, however, many companies have developed a rule-of-thumb approach based on extensive experience with a product range. For example, the snack food industry often relates storage for one week at 37°C, to storage for four to six weeks at 'ambient', loosely taken to be 20–22°C. The basis of this approach lies in the development of oxidative rancidity in the fat component.

The dangers in ASLT lie in the adoption of this rule-of-thumb approach for food systems that are quite dissimilar to those for which the rule was developed, and in ignoring other factors, not normally related to storage changes, that may change, for example on raising storage temperature. In fat-containing foods, for example, the ratio of solid/liquid fat ratio will change, with consequent changes in solvent and migration characteristics. Practical experience shows that this can induce shelf-life-limiting changes that do not normally occur under ambient storage conditions. Particular care must be taken when increasing the structural complexity of foods, as interactions can occur that introduce new shelf-life-limiting factors.

1.6 Ethical considerations

Any sensory evaluation operation using human subjects as a means of acquiring information on the sensory characteristics of foods must have in place ethical procedures designed to protect panellists from hazards associated with consuming unsafe food; these procedures must form part of general safety practices operated by the company management. Consuming or testing food that may be contaminated with tainting species carries a specific toxic risk, and additional measures may be needed to protect panellists against such risks, and also company staff against subsequent litigation. Suitable procedures will depend heavily on company safety policies, but a number of general requirements can be defined.

(i) Establish a risk classification system for the types of foods or ingredients to be presented to the panellists. A simple three-way system may classify the materials as: (a) standard foods/ingredients/processes; (b) non-standard foods/ingredients/processes for which evidence of safety is available; and (c) non-standard foods/ingredients/processes for which there is little or no information on safety. In many cases, including when testing for the presence of unknown taint, no safety information is available, and any tests should strictly be placed at the highest risk level. However, provided there is no evidence that substantial levels of contamination have occurred, toxicity concerns at the very low contamination levels characteristic of taint problems mean that a lower-risk classification can often be given.

(ii) Establish an ethical procedure through a well-defined management line for authorisation of sensory testing. This may typically involve the sensory analysis manager and company safety officer having the authority to approve tests classified in the lower risk region. In order to approve tests classified as higher risk, senior management approval must be sought, and it may be necessary to have in place a committee of external advisers with expertise in specific areas of interest and also with medical knowledge.

(iii) Include in the procedure provision for advice on microbiological hazards. Although this requirement may form part of a risk classification system, particular care must be taken even with (low-risk) commercial foods stored past normal shelf-life, at elevated temperatures. If in doubt, microbiological tests must be carried out in advance of any sensory testing.

(iv) Develop an information and consent procedure designed to give the panellists as much information on the risks associated with the tests as possible (whilst not introducing unacceptable bias), and to request their written consent. It is important to ensure that participation in taint testing is purely voluntary. Information given to the panellists may include:

- risk classification
- chemical name of any contaminants, if known
- natural occurrence in foods, and their use (if appropriate)
- any available toxicity data, either in numeric terms (e.g. LD_{50}) or relative to appropriate materials more familiar to the panellist
- approximate quantity that may be ingested during a test, and the anticipated number of tests over the course of the projects; if appropriate, panellists may be instructed to expectorate the samples and not to swallow, but some involuntary swallowing must be expected.

A typical consent form used at the Leatherhead Food Research Association is shown in Figure 1.9. It should be stressed that such a procedure is not a form of insurance but is primarily a means of demonstrating to panellists that ethical aspects are being given the consideration that is appropriate for such tests. Should any panellists suffer ill-health following the tests, it is unlikely that a signed consent form would offer much legal protection, and it remains the duty of the company to take all necessary steps to minimise risks to panellists, whether they are drawn from company staff, from part-time panellists or from the consumer population.

FLAVOUR OF CHLOROPHENOLS AND CHLORO-CRESOLS IN MILK

Tasting Panel, Class B

This part of Project Y1116 is an experiment to establish the concentration at which
6-chloro-o-cresol, 2 chlorophenol or 4, 6-dichloro-o-cresol can be detected in milk by tasters or
consumers of the milk. Samples of milk containing different concentrations of any one test
substance will be given to you and you are asked to taste them and to say whether you can
detect differences in flavour.

The maximum quantity of test substance which you would consume in the most extreme
tasting test, if you drank all of the milk provided, would be 0.03 µg (a). For comparison, if you
were to gargle with greatly diluted TCP you would probably swallow at least 400 µg (b). The
toxicities of the test substances are lower than that of TCP.

Taste panels will probably be held on four to eight occasions over a period of 1 month.

. . . .
Manager, Sensory Evaluation Section
4th December 1990

(a)   30 ml each of samples containing 1 ppb, 0.1 ppb and 0.01 ppb of test substance.

(b)   5% solution of a medicine containing 0.8% TCP: 1 ml swallowed involuntarily during and
      after gargling.

I, ... understand the programme which has been explained to me in
detail. I declare that I do not know of any reasons of health or diet which preclude me from
this programme, and I wish to participate. I also understand that participation is voluntary and
that I may withdraw at any time, without prejudice.

Signature . . . .   Date . . . .

Figure 1.9 Example of consent form used at the Leatherhead Food Research Association.

References

Amoore, J. E. (1977). Specific anosmia and the concept of primary odours. *Chemical Senses and
    Flavour* **2**, 267–281.
ASTM (1988). *ASTM Standards on Sensory Evaluation of Materials and Products.* American
    Society for Testing and Materials, Philadelphia.
Blakeslee, A. F. and Salmon, T. N. (1935). Genetics of sensory thresholds: individual taste reactions
    for different substances. *Proc. Natl. Acad. Sci. (US)* **21**, 84–90.
British Standards Institution (BSI) (1964). *Methods of Test for the Assessment of Odour from
    Packaging Materials used for Foodstuffs.* British Standard 3755:1964.
Brown, D. G. W., Clapperton, J. F., Meilgaard, M. C. and Moll, M. (1978). Flavour thresholds of
    added substances. *J. Amer. Soc. Brew. Chem.* **36**, 73–80.
Deutsches Institut für Normung (DIN) (1983). *Testing of Container Materials and Containers for
    Food Products.* Deutsches Institut für Normung DIN 10955, Berlin.

European Communities (1985). Council Directive of 19 December 1985 laying down the list of simulants to be used for testing migration of constituents of plastic materials and articles intended to come into contact with foodstuffs. *Official Journal of the European Communities* No. L372/14.

Fazzalari, F. A. (ed.) (1978). *Compilation of Odour and Taste Threshold Values Data.* American Society for Testing and Materials DS 48A, Philadelphia.

Gacula, M. C. Jnr. and Washam, R. W. II (1986). Scaling Word Anchors for measuring off flavour. *J. Food Quality* **9**, 57–65.

Goldenberg, N. and Matheson, H. R. (1975). 'Off-flavours' in foods, a summary of experience: 1948–74. *Chemistry and Industry*, 551–557.

Green, D. M. and Swets, J. A. (1966). *Signal Detection Theory and Psychophysics.* John Wiley and Sons, Inc., New York.

Griffiths, N. M. (1974). Sensory properties of the chloroanisoles. *Chemical Senses and Flavour* **1**, 187–195.

Griffiths, N. M. and Fenwick, R. (1977). Odour properties of chloroanisoles—effects of replacing chloro- by methyl groups. *Chemical Senses and Flavour* **2**, 487–491.

Griffiths, N. M. and Patterson, R. L. S. (1970). Human olfactory responses to 5α-androst-16-en-3-one—principal components of boar taint. *J. Sci. Food Agric.* **21**, 4–6.

Hammond, E. G. and Seals, R. G. (1972). Oxidised flavour in milk and its simulation. *J. Dairy Sci.* **55** (11), 1567–1569.

Hansson, K. E., Lundstrom, K., Fjelkner-Modig, S. and Persson, J. (1980). The importance of androstenone and skatole for boar taint. *Swedish J. Agric. Res.* **10**, 167–173.

Harper, R., Bate-Smith, E. C., Land, D. G. and Griffiths, N. M. (1968). A glossary of odour stimuli and their qualities. *Perfumery and Essential Oil Records* **59**, 22–37.

Hellman, T. M. and Small, F. H. (1973). Characterisation of odours from the petrochemical industry. *Chem. Eng. Prog.* **69**, 75.

Hellman, T. M. and Small, F. H. (1974). Characterisation of odour properties of 101 petrochemicals using sensory methods. *J. Air Pollution Control Association* **24** (10), 979–982.

HMSO (1990). *Food Safety Act 1990.* Her Majesty's Stationery Office, London.

Institute of Food Technologists (1974). Shelf life of foods. A scientific status summary by the Institute of Food Technologists' Expert Panel on Food Safety and Nutrition. *J. Food Sci.* **39**, 1–4.

Institute of Food Technologists (1981). Sensory Evaluation Guide for Testing Food and Beverage Products. *Food Technology*, November, 50–59.

ISO (1985). *Methods for Sensory Analysis of Food. Part 1. General Guide to Methodology.* ISO Standard 6658–1985.

ISO (1992). *Glossary of Terms Relating to Sensory Analysis.* ISO Standard 5492–1992.

Jewell, G.G. (1976). The relationships between disinfectant and musty taints in foods and the presence of chlorophenol derivatives. *Leatherhead Food RA Technical Circular No. 616.*

Jones, F. N. (1953). Olfactory thresholds in the International Critical Tables. *Science* **118**, 333.

Labuza, T. P. (1982). *Open Shelf Life Dating of Foods.* Food and Nutrition Press, Westport, CT.

Labuza, T. P. and Schmidl, M. K. (1985). Accelerated shelf-life testing of foods. *Food Technology* **39** (9), 57–64 and 134.

Labuza, R. P. and Schmidl, M. K. (1988). Use of sensory data in the shelf-life testing of foods. Principles and graphical methods for evaluation. *Cereal Foods World* **33** (2), 193–206.

Land, D. G. (1989). Taints—cause and prevention. In *Distilled Beverage Flavour, Recent Developments.* Eds J. R. Piggott and A. Paterson. Ellis Horwood, Chichester.

Leatherhead Food Research Association (1989). *Quality Systems for the Food and Drink Industries.* Guidelines for the use of BS 5750 Part 2 1987 in the manufacture of food and drink (ISO 9002 : 1987; EN 2999002 : 1987).

Lewins, S. C. and Wilson, L. G. (1985). Exact significance tables for sensory analysts—triangle, paired comparison, duo-trio and two-from-five difference tests. *Leatherhead Food RA Technical Notes No. 20.*

Linssen, J. P. H., Janssens, J. L. G. M., Reitsma, J. C. E. and Roozen, J. P. (1991). Sensory analysis of polystyrene packaging material taint in cocoa powder for drinks and chocolate flakes. *Food Additives and Contaminants* **8** (1), 1–7.

Maarse, H., Nijssen, L. M. and Angelino, S. A. G. F. (1988). Halogenated phenols and chloroanisoles: occurrence, detection and prevention. Characterisation, production and application of food flavours. *Proceedings of the 2nd Wartburg Aroma Symposium.* Akademie-Verlag, Berlin.

MAFF (undated). *Taint tests with Pesticides.* Working Document 10/5. Ministry of Agriculture, Fisheries and Food, Rothamsted.

Moskowitz, H. R. (1983). *Product Testing and Sensory Evaluation of Foods: Marketing and R&D Approaches.* Food and Nutrition Press, Westport.

Office International du Cacao et du Chocolat (OICC) (1964). *Transfer of Packaging Odours to Cocoa and Chocolate Products.* Analytical Methods of the Office International du Cacao et du Chocolat. Verlag Max Glättli, Zurich.

O'Mahoney, M. A. P. D. (1979). Short-cut signal detection measures for sensory analysis. *J. Food Sci.* **44**, 302–303.

O'Mahoney, M. A. P. D. (1982). Some assumptions and difficulties with common statistics for sensory analysis. *Food Technology* **36** (11), 76–82.

O'Mahoney, M. A. P. D. (1986). *Sensory Evaluation of Food: Statistical Methods and Procedures.* Marcel Dekker Inc., New York.

Pangborn, R. M. (1959). Influence of hunger on sweetness preferences and taste thresholds. *Amer. J. Clin. Nutr.* **7** 280–287.

Pangborn, R. M. (1981). A critical review of threshold, intensity and descriptive analyses in flavour research. In *Flavour '81.* Ed. P. Schreier. Walter de Gruyter, Berlin.

Parliment, T. H., Clinton, W. and Scarpellino, R. (1973). *Trans*-2-nonenal: coffee compound with novel organoleptic properties. *J. Agric. Food Chem.* **21**, 485–487.

Piggott, J. R. (1988). *Sensory Analysis of Foods.* 2nd edn. Elsevier Applied Science, London.

Schiffman, S. S. and Beeker, T. F. (1986). Multidimensional scaling and its interpretation. In *Statistical Procedures in Food Research*, Ed. J. R. Piggott. Elsevier Applied Science, London, pp. 255–292.

Stone, H. and Sidel, J. L. (1985). *Sensory Evaluation Practices.* Academic Press Inc., Florida.

Swoboda, P. T. and Peers, K. E. (1977). Volatile odorous compounds responsible for metallic, fishy taint formed in butterfat by selective oxidation. *J. Sci. Food Agric.* **28**, 1010–1018.

Teranishi, R. (1971). Odour and molecular structure. In *Gestation and Olfaction*, Eds G. Ohloff and A. F. Thomas. Academic Press, London, pp. 165–177.

Thomson, D. M. H. (1984). The sensory characteristics of three compounds which may contribute to boar taint. In *Progress in Flavour Research 1984.* Ed. J. Adda. Elsevier, Amsterdam.

Whitfield, F. B., Last, J. H. and Tindale, C. R. (1982). Skatole, indole and *p*-cresol: components in off-flavoured frozen French fries. *Chemistry and Industry*, 662–663.

Williams, A. A. and Aitken, R. K. (eds) (1983). *Sensory Quality in Foods and Beverages.* Ellis Horwood, Chichester, p. 23.

Wilson, L. G. and Kilcast, D. (1984). Investigations into triangular taint testing procedures. Part I Effect of flavour carryover. *Leatherhead Food RA Technical Note No. 16.*

Zoeteman, B. C. J. and Piet, G. J. (1973). Drinkwater is nog geen water drinken. H_2O **6**, 174–189.

Zwaardemaker, H. (1926). In *International Critical Tables.* Ed. E. W. Washburn. Volume 1, pp. 358–361, McGraw-Hill, New York.

2 A survey of chemicals causing taints and off-flavours in foods

M. J. SAXBY

2.1 Introduction

A taint or an off-flavour is caused by the presence of a chemical which imparts a flavour that is unacceptable in the food. This chapter surveys these taints, and includes a number of specific examples. It is possible to differentiate between a taint and an off-flavour by defining the former as the presence of a substance that is totally alien to all foods, and the latter as the chemical reaction of a naturally occurring component in the food, giving a compound with an undesirable taste. For instance, dimethyl trisulphide is a desirable flavour component of cooked cabbage, but this same compound is most objectionable when found in red prawns (Whitfield *et al.*, 1981). However, a disinfectant taste due to a chlorophenol is unacceptable in any foodstuff.

The degree of perception of a flavour varies enormously between individuals, sometimes extending over several orders of magnitude. At a level of 3 parts per million (ppm) about 50% of the population can detect limonin in orange juice, but at a level of about 0.3 ppm only around 5% of the same group of people can detect it; at 30 ppm almost everyone can taste it.

In order to be able to study the effect of chemical structure on the perception of taste and flavour, it is convenient to define this for a given group of people. According to ISO 5492 (ISO, 1992), the odour or taste threshold is defined as the lowest concentration of a compound detectable by a certain proportion of a given group of people; the proportion of the group is usually 50%. This definition is more easily explained with the help of Figure 2.1. The graph is a plot of the percentage of people within a given group who can detect the concentration in arbitrary logarithmic units of a given compound. In the hypothetical example shown, 50% of the group can detect the compound when its concentration reaches one unit. This relationship has further consequences; it shows that when the concentration of the compound is ten times greater than the mean threshold, about 10% of the group are still unable to detect it. The other end of the graph is of greater significance to the food manufacturer, by showing that even when the concentration of the compound is ten-fold lower than the mean threshold, about 5% of the group can still detect it. This means that food manufacturers must continually submit their products to a trained taste panel, to prevent contaminated foodstuff containing very low levels of a

Table 2.1   Comparison of odour thresholds in water

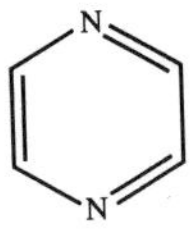

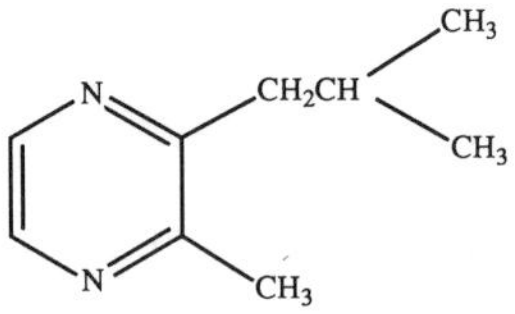

Pyrazine
300 mg/l

2-Isobutyl-3-methylpyrazine
2×10^{-8} mg/l

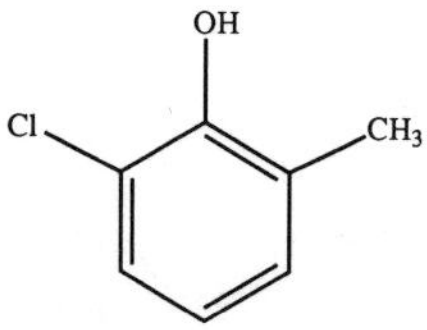

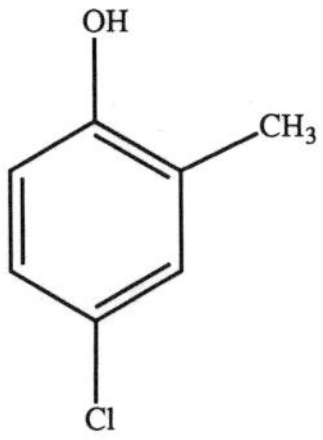

6-Chloro-*o*-cresol
0.08 µg/l

4-Chloro-*o*-cresol
120 µg/l

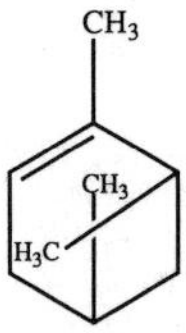

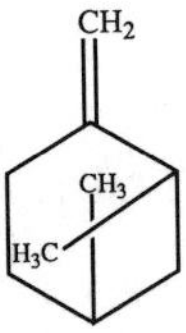

Alpha-pinene
6 µg/l

Beta-pinene
140 µg/l

tainting compound from reaching the general public. Although the graph is of a hypothetical compound, the range of detectable levels is representative of many compounds.

The effect of chemical structure on odour thresholds is important, and some examples of this are shown in Table 2.1. The odour and taste thresholds of an organic compound in different media often vary by several orders of magnitude, with the lowest thresholds being found in water as a matrix. Corresponding values in edible oils may be 10- to 100-fold higher. Very few values are available for most compounds in solid foodstuffs, probably because of the

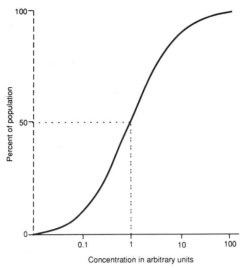

Figure 2.1 Variation of taste threshold within a given population.

difficulty in obtaining an even distribution of the compound throughout the product. However, Whitfield *et al.* (1988) have published some interesting figures for bromophenols in water and prawn meat. These are shown in Table 2.2. These results show that the panellists could tolerate about 100-fold higher levels of the bromophenols in the prawn meat compared with water, without detecting any taint.

Table 2.2 Flavour thresholds (μg/l) of bromophenols in water and prawn meat

Compound	Water	Prawn meat
2-Bromophenol	3×10^{-2}	2
2,6-Dibromophenol	5×10^{-4}	6×10^{-2}

Reproduced by permission of Elsevier Applied Science Publishers Ltd.

The level of a tainting compound in a foodstuff may also have a considerable effect on the taste. The compound *trans*-2-nonenal is formed through the autoxidation of some fats, and incidentally is also one of the natural flavour compounds in cucumber. The taste descriptions at different levels are shown in Table 2.3 (Parliment *et al.*, 1973).

The position of a double bond also alters the flavour perception, as shown by comparison of the results in Table 2.3 with *trans*-6-nonenal, which has a taste threshold of 0.07 μg/l in milk and a description of stale (Parks *et al.*, 1969). The effect of the position, and the geometry of the double bond on flavour is dramatically shown by the work of Meijboom and Jongenotter

Table 2.3 Variation of taste description of *trans*-2-nonenal with concentration in water.

Concentration (μg/l)	Taste
0.2	Plastic
0.4–2.0	Woody
8–40	Fatty
1000	Cucumber

Reprinted with permission from *J. Agr. Food Chem.* **21**, 485. Copyright (1973) American Chemical Society.

(1981), who studied the flavour of a number of unsaturated aldehydes in paraffin oil (Table 2.4).

Table 2.4 Flavour description of unsaturated aldehydes dissolved in paraffin oil

Aldehyde	Flavour description
Trans-3-hexenal	Green, pine needles
Cis-3-hexenal	Green beans, green tomato
Trans-2-heptenal	Bitter almonds
Cis-6-heptenal	Green, melon
Trans-2-octenal	Nutty
Trans-5-octenal	Cucumber
Cis-5-octenal	Cucumber
Trans-2-nonenal	Starch, glue
Trans-7-nonenal	Melon

Adapted from Meijboom and Jongenotter (1981)

2.2 Taints derived from known chemicals

2.2.1 Phenols

Although simple phenols substituted only by alkyl groups have much higher taste thresholds than their corresponding halogenated derivatives, nevertheless they have been identified as the source of carbolic-type taints in foods, probably for three reasons: (i) they are very volatile; (ii) they are present in high concentration in tar distillates; and (iii) they appear to have a fugitive effect on high protein foods, so that once absorbed into the product they are not lost to the atmosphere.

Some excellent hard-wearing flooring materials are manufactured from resins containing a complex mixture of alkyl phenols, but their success depends on the use of a precise ratio with the other monomers forming the polymer. If the phenols are in excess, not only will the texture of the floor be poor, but the phenols will evaporate over a period of time and may be absorbed on to nearby food materials. Staff who are contemplating re-laying the floor in their factory, or even repairing a part of it, are advised to have a sample portion laid by the contractors. Taste transfer tests can then be carried out, before accepting the contract for the whole job.

Other industrial materials which may contain free phenols include a wide range of water-proofing and acid-resistant materials, which may be applied to large storage tanks. Damp-proofing fluids applied under a concrete floor may contain volatile phenols, which, if present, will gradually volatilise through the cement and lead to a taint in foodstuffs manufactured in the vicinity.

One compound that has been fairly well documented is 4-methyl-phenol (p-cresol). It has a taste threshold of 2 μg/l and an odour threshold of 200 μg/l, both measured in water (Whitfield et al., 1982). This compound has been detected in fried potatoes with an off-flavour, described as 'pigsty' and 'drain-like' (Whitfield et al., 1982). Examination of the potato tubers in storage sheds revealed that tubers in contact with the cold wall of the building contained p-cresol. Those parts of the storage sheds that are particularly cold are subject to moisture condensation and provide a suitable environment for the growth of soft-rot microorganisms. Several species of Clostridium were detected in the rotten potatoes.

p-Cresol has also been detected as a metabolite of C. difficile and C. scato-logenes (Elsden et al., 1976).

Several decades ago the Leatherhead Food Research Association was asked to investigate a smoky flavour in some chocolate ice cream. At that time the firm operated a procedure in which ice cream left over from a previous production was mixed with similar remnants containing other flavours. Subsequently the mixture was treated with a strong flavouring material such as chocolate, and a suitable dark colour. At that time, both gas chromatography and mass spectrometry were relatively unsophisticated, so that examination of the extracts was difficult. Patient study of all the data produced eventually yielded evidence for the presence of guaiacol (2-methoxyphenol) and the corresponding 2-ethoxyphenol. Both compounds adequately accounted for the smoky flavour, but their origin was less obvious until it was recognised that they were microbiologically-derived degradation products of vanillin (Figure 2.2) and the so-called synthetic flavouring compound ethyl vanillin.

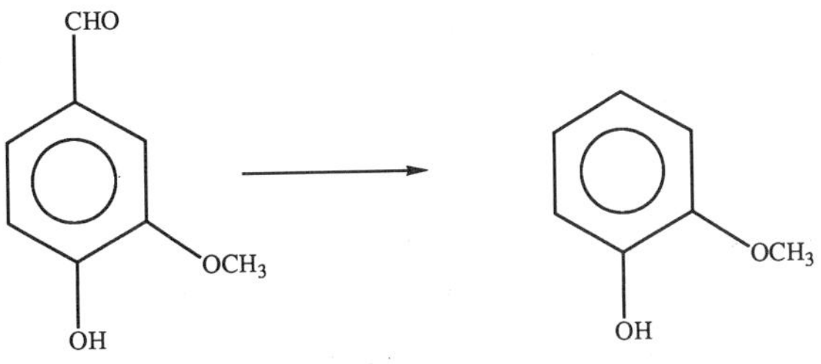

Figure 2.2 Degradation of vanillin to guaiacol.

Subsequent samples of a number of confectionery products, which suffered from the same complaint and which were flavoured with the same compounds, were also shown by mass spectrometry to contain the same two phenols.

This degradation was subsequently confirmed and reported by Lefebvre *et al.* (1983). Braus and Miller (1958) reported the presence of guaiacol as a degradation product of lignin in bottle corks. In Australia, research workers at the Wine Research Institute were confronted with a taint in wine (Simpson *et al.*, 1986). When they suspected the corks as a possible source of the offending compound, they identified guaiacol (by mass spectrometry) as the cause. They then developed a technique for its quantitative determination in methylene chloride extracts of the wine, and the associated corks. They also measured the odour threshold in dry white wine as 20 µg/l. Guaiacol has a taste threshold in water of 50 µg/l (Dietz and Traud, 1978).

2.2.2 Chlorophenols

Many simple chlorophenols impart a disinfectant taste to foodstuffs at levels below 1 ppb. This taste is described by some people as soapy, medicinal or 'TCP'. A food manufacturer who has the misfortune to suffer such a taint problem in his/her product, and who receives complaints from the public concerning the above named descriptions can be fairly sure that the offending chemical will be chlorophenol. One of the most potent chlorophenols causing problems in foods is a chemical commonly known as 6-chloro-*o*-cresol, often abbreviated to 6-COC. One source of this chemical was traced many years ago to the effluent from a factory producing herbicides based on chlorophenoxy-alkanoic acids. Strong evidence was presented for the transference of the 6-COC by aerial transport to a biscuit factory some 8 km distant (Goldenberg and Matheson, 1975). As a result of this incident a report was published, which showed that the taste threshold of 6-COC in biscuit is 0.05 µg/kg (Griffiths and Land, 1973).

Around the same time a report appeared (Patterson, 1972) regarding the occurrence of a disinfectant taint in poultry, which was also identified as 6-COC. The source of this compound was traced to certain batches of cresylic acid used as a disinfectant in broiler houses and a factory. The potential of 6-COC to ruin large quantities of foodstuffs was not recognised by some disinfectant manufacturers, and this chemical appeared by error in a number of products, including a retail dustbin cleaner selling on supermarket shelves in the vicinity of a wide range of foods, leading to disastrous results (Newton, 1973). Drain cleaners widely used in the meat industry were also found to be the source of 6-COC (CSIRO, 1982). One meat manufacturer regularly used drain cleaners to emulsify trace levels of fat in waste water until the producer of the cleaner inadvertently replaced dichlorobenzene with chlorophenols, leading to the tainting of several tonnes of product. Users of tray cleaners in the bakery industry were also troubled with this problem some years ago.

Chlorophenols are of course generated by the chlorination of free phenols and, where the phenol is *o*-cresol, one of the products will inevitably be 6-COC. This reaction was studied at the Leatherhead Food Research Association by Saxby *et al.* (1983), who showed that significant quantities of 6-COC are formed when ppm quantities of *o*-cresol and hypochlorite are mixed. This reaction had disastrous results for one firm who inadvertently used a cleaning agent based on mixed phenols and a hypochlorite sterilising agent. Another firm whose main product was canned meat was in the habit of washing the carcasses with water from a private source; for safety this water was also chlorinated. Unknown to the owners of the firm, the water sporadically contained trace levels of cresols, which were converted to chlorocresols and which then contaminated the canned product.

Chlorination of mixed phenols containing the parent phenol yields 2-chlorophenol, which has a mean taste threshold of 2 ppb in milk. It has also been detected in tainted beer (Wenn *et al.*, 1987), where the authors proposed a unique mechanism for its formation. Nylon tubing is widely used in the brewing industry and may be cleaned with chlorine-based sanitisers, but Wenn *et al.* (1987) discovered that it forms an *N*-chlorinated derivative, which acts as chlorinating agent in the presence of simple phenols. A review paper by Harper and Hall (1976) refers briefly to disinfectant taints in milk, which were attributed to the reaction of chlorine with phenolic-based udder washes. The presence of chlorophenols in water has been attributed to the combined chlorination and decarboxylation of hydroxyacids commonly found in natural waters (Larson and Rockwell, 1979).

Although less well recorded in the scientific press, both 2,4-dichlorophenol and 2,6-dichlorophenol are a common cause of disinfectant taints in foods. They have taste thresholds in water of 0.3 and 0.2 ppb, respectively (Dietz and Traud, 1978). Unpublished evidence has shown conclusively that wooden floors in transport containers may from time to time be contaminated with mixed dichlorophenols; when these containers are used for the delivery of food raw materials and finished products, contamination of the product is almost inevitable. Spillage of crude mixtures of dichlorophenols in containers is certainly one source of this taint, but Tindale and Whitfield (1989) discovered a second mechanism. Apparently it is the practice to store pallets loaded with cartons near processing areas where disinfectants containing high-pressure-steam and high levels of available chlorine are used. The combination of temperature, water, chlorine and phenol derived from the decomposition of lignin in the fibreboard or pallets provides the necessary conditions for the formation of chlorophenols.

Manufacturers of vending machines have sometimes experienced disinfectant taints in coffee, and this has been related to the presence of dichlorophenols, which may have been formed by the same mechanism as proposed by Wenn *et al.* (1987).

2.2.3 Bromophenols

Although chlorophenols have been known to be the major source of disinfectant taints for many years, it is only more recently that bromophenols have been implicated. A major contribution to the understanding of these compounds was made by Whitfield and coworkers at CSIRO in Australia (Whitfield *et al.*, 1988). Not only did they identify 2,6-dibromophenol as the cause of an iodoform-like taint in prawn meat, but they also found 2- and 4-bromophenol, 2,4-dibromophenol and 2,4,6-tribromophenol. All the compounds were fully characterised by mass spectrometry and the synthesis of the title compound was described. It was concluded that the bromophenols are natural compounds found on the sea-bed, associated with algae, which form a major portion of the diet of prawns. Bemelmans and den Braber (1983) found an offensive odour and taste in marinated herring, described as iodine-like and phenolic. After isolation of the volatiles from the fat, mass spectrometry identified the compound as 2-bromophenol at levels much above the odour threshold of 0.1µg/kg in water. However, naturally occurring compounds cannot always be used as a source of bromophenols. A serious incident involving the presence of bromophenols in a food, which was investigated at the Leatherhead Food Research Association, was traced to a series of chemical reactions, in which fish was cured in salt, bleached with hydrogen peroxide (a practice allowed in certain countries) and then packed in oak barrels. When bromide was found as an impurity in the salt, it was suspected that the hydrogen peroxide oxidised the bromide to bromine, which then brominated phenols present in the oak barrel.

2.2.4 Haloanisoles

Musty taints in foodstuffs were a problem for many years and a wide range of compounds were postulated as the cause of the taint. Absolute proof of the identity of the compounds responsible often remained unknown until modern analytical techniques like mass spectrometry became available. Following an outbreak of a musty taint in chickens in the United Kingdom, the problem was presented to the Food Research Institute, Norwich. Pioneering work by Curtis and coworkers (1974) led to the isolation of the volatile compounds from the affected broiler carcasses using a Likens–Nickerson continuous steam distillation extraction apparatus (Gee *et al.*, 1974), and identification of a series of chloroanisoles as the compounds responsible for this taint. These workers then went on to determine the source of the compounds and discovered that the chickens were housed on litter consisting of wood shavings which contained a range of chlorophenols, particularly tri- and tetra-chlorophenols. They demonstrated that these chlorophenols can be microbially methylated by numerous organisms to the corresponding chloroanisoles (Parr *et al.*, 1974). Chloroanisoles possess odour thresholds in water that were probably lower than any previously recorded (Table 2.5) (Curtis *et al.*, 1972).

Table 2.5 Odour thresholds of some chloroanisoles in water

Compound	Structure	Odour threshol in water (pg/l)
2,3,6-Trichloroanisole		0.3
2,4,6-Trichloroanisole		30
2,3,4,6-Tetrachloroanisole		4000

Adapted and reprinted by permission from *Nature*, Vol. 235, p. 223. Copyright © 1972 Macmillan Magazines Ltd.

At the time of this incident most soft wood was treated with technical pentachlorophenol in order to prevent the formation of mould, which imparts a purple stain (sap-stain) to the wood, severely hindering its use in the manufacture of furniture. In 1968 and 1969, over 11 000 tons of chlorophenols were used in the United States for timber preservation and, since impregnation of these compounds into the wood is low, it is likely that shavings used as chicken litter would contain enhanced levels of these compounds. Furthermore, technical pentachlorophenol (PCP) contains significant quantities of tetra-chlorophenol and trichlorophenol, and it was these compounds which were the precursors to the highly odorous chloroanisoles. Since their discovery, chloroanisoles have continued to be the cause of numerous outbreaks of musty taints in foods throughout the world.

Most pallets are made from softwood and therefore became the source of chloroanisoles for many years. Both raw materials and finished products stored on affected pallets for any extended period of time were subject to this contamination. Products with a significant fat content are more likely to become tainted, as the chloroanisoles are soluble in non-polar solvents; consequently, there are several reports of musty taints in cocoa raw materials and finished chocolate products. However, there have also been reports of non-fatty basic food products, which were held on pallets for long periods of time, becoming affected. Fortunately, as the use of PCP was discontinued in most of the timber industry, so the threat of musty taints originating from pallets receded. Even when pallets became eliminated from the possible causes of musty taints in foods, reports continued to be published, indicating that the problem had not been eliminated. In particular there were sporadic reports of a musty taint in dried fruit packed in fibreboard cartons and exported from Australia to Europe (Whitfield et al., 1985). The taint was again traced to the presence of chloroanisoles, principally 2,4,6-trichloroanisole and 2,3,4,6-tetrachloroanisole. Cartons were manufactured from fibreboard with a high content of recycled wastepaper, which contained relatively high levels of the corresponding chlorophenols. Following recommendations, the dried fruit industry used only virgin fibreboard for the cartons, which resulted in a drastic reduction in the incidence of this problem, but did not completely eliminate it. Analysis showed that the cartons became contaminated after manufacture, and it transpired that the packaging was stored in an area which was cleaned with disinfectants containing a high level of available chlorine. It was concluded (Tindale and Whitfield, 1989) that the chlorine reacted with phenols in the board, derived from the decomposition of wood lignin. Since the use of cleaning agents is essential in the processing industry, the authors studied the formation of chlorophenols by reaction of fibreboard and timber with a number of Australian cleaning agents and were able to recommend one commercial product. Whitfield et al. (1991a) studied the effect of relative humidity and chlorophenol content on the conversion of three chlorophenols to the corresponding chloroanisoles in fibreboard cartons. Positive correlations were found between storage humidity and fungal population, and the production of chloroanisoles in the fibreboard.

Whitfield et al. (1987) have reported the odour thresholds of three chloroanisoles in dried fruit and buns, while Frijters and Bemelmans (1977) have determined the flavour thresholds of six different chloroanisoles in coagulated egg yolk. Buser et al. (1982) reported the presence of a musty taint in wine and traced the origin of 2,4,6-trichloroanisole to the corks. They also reported that a proportion of the tasters could detect the taint down to levels of 10 ng/l of 2,4,6-trichloroanisole in wine. Recently a group of workers from Nestlé in Lausanne (Spadone et al., 1990) has reported the presence of 2,4,6-trichloro-

Table 2.6 Odour descriptions and thresholds of dichloroanisoles in water

Compound	Odour description	Odour threshold in water (ng/l)
2,4-Dichloroanisole	Musty, sweet, fruity, scented	400
2,6-Dichloroanisole	Musty, medicinal, phenolic	40

Adapted from Griffiths (1974). Reprinted by permission of Kluwer Academic Publishers

anisole in Brazilian coffee, which had the so-called 'Rio off-flavour'. Organoleptic trials showed that the taste threshold of this compound in freshly brewed coffee was 1–2 ng/l (ppt), but they were unable to trace its source.

The presence of dichloroanisoles as a cause of a taint in a food is less rarely encountered, but their odour thresholds have been reported by Griffiths (1974) and are shown in Table 2.6.

Since the use of chlorophenols in the treatment of soft wood has been largely discontinued, one major source of musty taints has been reduced. However, the problem still troubles the food industry and often affects raw materials and finished products that are carried in large international containers. It has been the repeated experience of the author, that whenever a large consignment of a food product with a musty taint was found to contain chloroanisoles, and the corresponding container could be located, then without fail chloroanisoles were found in wood shavings taken from the floor of that container. In one instance, raw material was carried in two containers, both of which were contaminated with chloroanisoles. However, the ratio of each chloroanisole isomer in the wood taken from the two containers was markedly different and this was reflected in the composition of the chloroanisoles found in the food material, so much so that it was possible to identify the container from which each sample was taken. Examination of the wood taken from the floor of these containers generally reveals that the extracts contain not only chloroanisoles, but also the corresponding chlorophenols, from which it can be concluded that microbial methylation is also the mechanism for the formation of the chloroanisoles.

Loos *et al.* (1967) have reported that 2,4-dichloroanisole can be formed from 2,4-dichlorophenoxyacetic acid.

Research workers at CSIRO, North Ryde, Australia developed a general method for the determination of three important chloroanisoles both in foodstuffs, and in packaging materials (Whitfield *et al.*, 1986). According to the procedure, the chloroanisoles are isolated from the contaminated material in a combined distillation–extraction apparatus. The resulting extracts are analysed by capillary column chromatography–mass spectrometry. Quantitation is achieved by comparison of peak areas of the analyte with the internal standard 3,5-dimethyl-2,4,6-trichloroanisole. The extraction efficiency and reproducibility are good.

Bromophenols are also subject to microbial methylation to give bromoanisoles, which are as potent as chloroanisoles in their ability to cause musty taints in foods. Apart from their previously described occurrence, it is likely that 2,4,6-tribromophenol can be formed by reaction of a biocide like bromo-chlorodimethylhydantoin with phenol, which is then methylated to give the corresponding tribromoanisole. The odour threshold of 2,4,6-tribromoanisole is claimed to be 8 pg/l in water (unpublished data, 1985). Cant (1979) reported the presence of an iodine-tasting taint in casein, which he apparently identified as a dibromoanisole but, as the compound was not synthesised for final ident-ification, its exact structure must be considered as still unproven.

2.2.5 Compounds containing sulphur

The reaction between the unsaturated ketone, mesityl oxide (4-methylpent-3-en-2-one) and a sulphur compound, particularly hydrogen sulphide, leads to the formation of a compound with an odour of the urine from a tom cat (Pearce *et al.*, 1967). This derivative was positively identified by mass spectrometry as 4-mercapto-methylpentan-2-one (Patterson, 1969), formed according to the following scheme.

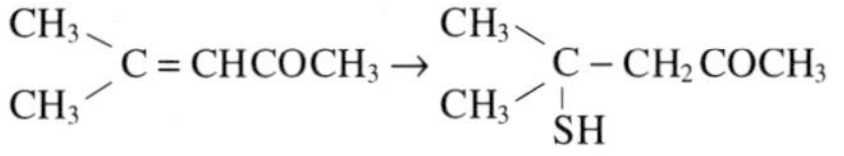

$$\underset{CH_3}{\overset{CH_3}{\diagdown}} C = CHCOCH_3 \rightarrow \underset{CH_3}{\overset{CH_3}{\diagdown}} \underset{SH}{\overset{|}{C}} - CH_2COCH_3$$

Some decades ago, this compound caused considerable problems in the food industry, particularly in the meat and vegetable trade. In one incident carcasses hung in a chill room, which had recently been painted with a polyurethane paint, were found to possess this catty taint (Patterson, 1968). The thinners used in this paint consisted of a mixture of xylene and an industrial compound known as pentoxone (4-methoxy-4-methylpentan-2-one), which, on further examination, was found to contain nearly 0.5% mesityl oxide. The contami-nated meat was found to contain both xylene and pentoxone, together with the compound responsible for the distinct catty odour. It is assumed that the hydrogen sulphide was generated by decomposition of an amino acid such as methionine or cystine. This same taint also appeared in a very wide variety of canned goods including ox-tongue, soup, luncheon meat, sardines, carrots and processed peas (Anderton and Underwood, 1968). Although no mesityl oxide could be detected, can lacquers were suspected as one source of this unsatu-rated ketone (Spencer, 1969), although it may have been that the equip-ment then available was insufficiently sensitive to detect the very low levels that may have been present. The odour threshold of 4-mercapto-4-methylpen-tan-2-one in bland foodstuffs is thought to be about 0.01 μg/kg (Reineccius, 1991).

Some years later a number of manufacturers of canned creamed rice had the same problem of a catty taint in their products. At the country of origin, rice was packed in paper bags, which were designated with lettering using ink that contained mesityl oxide. This solvent migrated into the rice during transport and then reacted with trace levels of hydrogen sulphide in the milk added during manufacture. Many oil-seed meals and flours are extracted with acetone in order to remove aflatoxins or to extract gossypol from some cotton-seed products. Some oil-seed meals that were extracted with acetone subsequently developed objectionable odours and flavours, and this was attributed to the presence of mesityl oxide derived from the use of technical acetone, followed by reaction with hydrogen sulphide. Maarse and de Brauw (1974) reported the synthesis of 2-methyl-2-mercaptopentane and showed that the presence of a carbonyl group in the molecule is not essential for the catty aroma. This compound was detected in the air around a Dutch factory producing dodecyl mercaptan, and is another potential compound causing a catty taint in foodstuffs.

The reaction between hydrogen sulphide and mesityl oxide is an example of the more general interaction of a sulphur compound with an unsaturated ketone. This reaction was studied in more detail by Badings et al. (1975). Since many foodstuffs contain both unsaturated carbonyls and volatile sulphur compounds, the authors investigated the addition of hydrogen sulphide and methanethiol to four unsaturated aldehydes and one unsaturated ketone. The products had a range of flavours variously described as onion, cabbage, rhubarb and floral, with mean taste thresholds at or below 1 µg/l in water. Although none of these compounds has been reported as the cause of a specific food taint, it seems plausible that they have been implicated in cases where the compound has remained unidentified.

Certain foodstuffs, if excessively or repeatedly fumigated with methyl bromide, can develop a sulphury note. This is due to the methylation of methionine residues in which the sulphur atom is expelled in the form of dimethyl sulphide. Dimethyl sulphide has been identified as the cause of taint in a variety of foodstuffs, including nuts (Bills et al., 1969), chicken (Cooper et al., 1978), bread (Burns-Brown et al., 1961) and beer (Harrison, 1970). The odour threshold has been measured as 0.33 µg/l in water (Guadagni et al., 1963) and its taste threshold in beer is 60 µg/l (Harrison, 1970). Dimethyl trisulphide has an odour threshold in water of 0.01 µg/l (Buttery et al., 1976a) and has been detected in red prawns with an onion or metallic taint (Whitfield et al., 1981). 2,4-Dithiapentane has a garlic taste and has been found to be the cause of taints in prawns (Whitfield and Freeman, 1983) and cheese (Sloot and Harkes, 1975). The odour threshold in oil is 3 µg/l.

$$CH_3SCH_2SCH_3$$

2,4-dithiapentane

In Australia and New Zealand, dairy cattle are liable to ingest the cruciferous weed *Coronopus didymus* and this transmits a sharp odour to the corresponding milk and a burnt flavour to the butter. A number of tainting compounds in the milk were isolated by vacuum distillation, and the principal compound responsible for the flavour defect was identified by mass spectrometry as methyl benzyl sulphide (Park *et al.*, 1969). The flavour threshold of methyl benzyl sulphide in milk is $10\,\mu g/kg$.

2.2.6 Alcohols

In general, alcohols are not normally the cause of taints in foods. However, certain cyclic alcohols do have threshold detection limits which are sufficiently low to cause very considerable problems to the water industry. Musty earthy odours have for a long time caused a problem in potable waters, and have been associated with the presence of blue-green algae. Early attempts were made to extract and identify the compound responsible for this unpleasant odour (Dougherty *et al.*, 1966) but, despite nuclear magnetic resonance (NMR) and mass spectral evidence, they incorrectly assigned the molecular formula. Later this compound was correctly identified (Gerber, 1968) as *trans*-1,10-dimethyl-*trans*-9-decalol, otherwise known as geosmin from the Greek 'geo'—earth—and 'osme'—smell. Geosmin has been identified in contaminated water on many occasions (Gerber, 1983).

Geosmin

Extensive studies on the production of geosmin and many other compounds from selected moulds and actinomycetes on agar and bread were made by Harris *et al.* (1986). Recently geosmin was found to be the compound responsible for an earthy off-flavour in wheat flour that had been allowed to become excessively damp (Whitfield *et al.*, 1991b). The compound has also been detected in fish (Dupuy *et al.*, 1986), Navy beans (Buttery *et al.*, 1976b) and clam (Hsieh *et al.*, 1988). These references illustrate the fact that there are numerous reports of the contamination of both water and foodstuffs with geosmin. There are many unpublished incidents where products, for which the production involves the use of large amounts of water or steam, have been similarly contaminated. The reported taste thresholds of geosmin in water and fish are $0.05\,\mu g/l$ (Medsker *et al.*, 1968) and $6\,\mu g/kg$ (Yurkowski and Tabachek, 1974), respectively.

A compound often associated with geosmin is 2-methylisoborneol.

2-Methylisoborneol

This compound is also reported to be the cause of earthy-musty odours in the public water supply (Wood and Snoeyink, 1977) and a similar off-flavour in canned mushrooms (Whitfield *et al.*, 1983). The compound was identified by continuous extraction of the tainted product with trichlorofluoromethane, followed by analysis of the extract by coupled gas chromatography–mass spectrometry. It has an odour threshold in water of 29 ng/l (Persson, 1979).

2.2.7 Hydrocarbons

Sorbic acid is widely used in the food industry as a preservative to prevent the formation of moulds and yeasts. In some countries levels of more than 1000 ppm are allowed in a number of products, including cheese. For many years it has been known that sorbic acid degrades in cheese (Melnick *et al.*, 1954). Although sorbic acid is added with the purpose of preventing mould growth, nonetheless it does appear that mould can still appear on the surface of the product, presumably where the concentration of inhibitory agent is at its lowest. When this occurs, extensive degradation of the sorbic acid tends to occur. This reaction was investigated in some depth by Marth *et al.* (1966), who discovered that the *Penicillium* species in particular was responsible. These authors investigated the mechanism of this reaction by fortifying skim milk with potassium sorbate and inoculating with mould spores of *Penicillium roqueforti*. The mixture was incubated for 36 h, and the volatile compounds isolated and separated by gas chromatography. One major component was identified by infrared spectrophotometry as 1,3-pentadiene by the decarboxylation mechanism shown in Figure 2.3. Although it was noted by the authors that 1,3-pentadiene has a hydrocarbon-like odour, it was probably not realised the extent to which this compound could give rise to taints in food.

Horwood *et al.* (1981) reported a distinctly unpleasant flavour of Feta cheese that had been treated with sorbic acid. The off-flavour was described

$$CH_3 CH = CH CH = CH COOH \longrightarrow CH_3 CH = CH CH = CH_2$$

Figure 2.3 Degradation of sorbic acid in foods.

as plastic, paint and kerosene. The gas chromatogram of the headspace vapour showed the presence of a number of volatile compounds but, in particular, they identified one very large peak as *trans*-1,3-pentadiene by mass spectrometry. In a Feta cheese slurry they found that the odour threshold was 4 ppm, which, although relatively high, adequately accounted for the taint, since they found levels in the cheese which were far in excess of this figure. Daley *et al.* (1986) reported a further occurrence of a kerosene taint in a cheese-spread; this was similarly identified as *trans*-1,3-pentadiene together with a minor amount of the *cis* isomer.

Reports in the scientific literature on the occurrence of pentadiene in foodstuffs treated with sorbic acid seem to be confined to cheese. However, it has been the experience of the Leatherhead Food Research Association that any food product which has been treated with sorbic acid and which has a high moisture content is susceptible to the decarboxylation reaction. On several occasions, cakes containing sorbic acid, and with a high moisture content were submitted to the Leatherhead Laboratory with the request that the source of kerosene taint should be identified; on every occasion 1,3-pentadiene was identified by coupled gas chromatography–mass spectrometry.

Some years ago the concentration of residual vinyl chloride in poly-vinylchloride (PVC) polymer was limited by law when it was realised that migration might lead to toxic levels in food. This led to a general reduction in the amount of monomer in a wide range of plastics. At one time residual amounts of styrene in polystyrene cartons were sufficiently high to allow migration into foodstuffs, sometimes causing a taint in the product. Butter in particular, when packed in polystyrene containers, was susceptible to this problem. However, during 1982 some spiced buns were submitted to the Leatherhead Food Research Association, with the request that the cause of the plastic taint should be investigated. Analysis revealed the presence of styrene, from which it was concluded that the baked product had been stored in polystyrene boxes; this, however, proved to be incorrect. The source of the styrene monomer was subsequently investigated in detail by the Flour Milling and Baking Research Association, who discovered that two essential elements were necessary for its formation: (i) a particular species of yeast called *Hypo-pichia burtonii*; and (ii) a cinnamon flavouring. It was concluded that the styrene was formed by the action of the yeast on cinnamaldehyde by the reaction shown in Figure 2.4. Thurston (1986) developed a simple test for

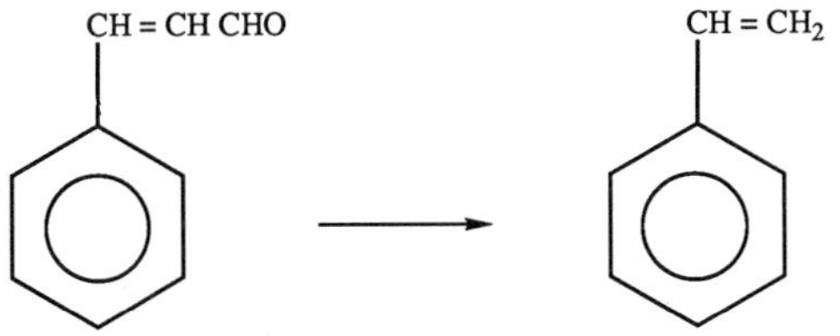

CH = CH CHO   CH = CH₂

Figure 2.4 Formation of styrene from cinnamaldehyde.

confirming the presence of this specific wild yeast, by incubating the suspect yeast with ferulic acid and detecting the formation of 4-vinylguaiacol.

Potato chips fried in cotton-seed oil and subsequently exposed to light develop a distinct off-flavour, which is given the name 'light-struck' (Fan *et al.*, 1983). These authors set out to determine the nature of the compound responsible for the off-flavour, its precursor and the reaction mechanism. When the headspace volatiles from the degraded oil were examined by gas chromatography–mass spectrometry, 1-decyne was identified as the major photodegradative off-flavour component. Cotton-seed oil contains the cyclic fatty acid, sterculic acid, which, through the formation of a hydroperoxide and cleavage of the molecule, leads to the formation of the decyne. In model experiments, chlorophyll was found to be a photosensitiser of the reaction.

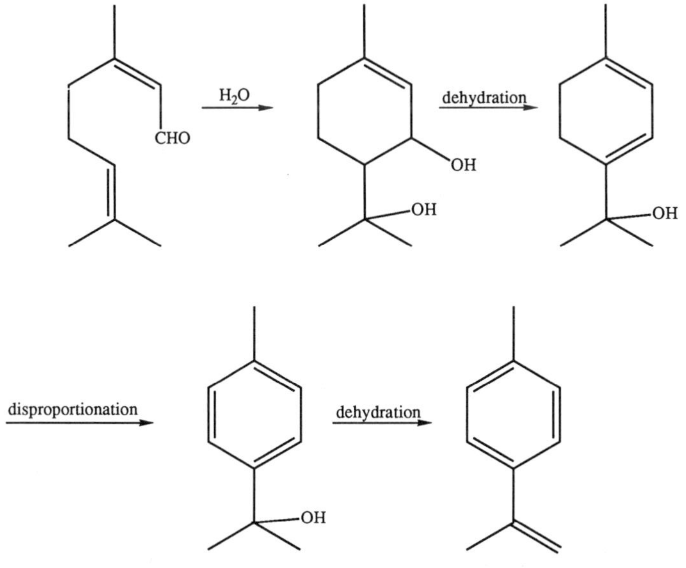

Figure 2.5 Formation of alpha-*p*-methylstyrene from citral. Adapted and reprinted from Peacock and Kuneman, *J. Agr. Food Chem.* **33**, 330. Copyright (1985) American Chemical Society.

Citral, a major component of lemon oil, decomposes to give two isomers of p-menthadien-8-ol; in acidic conditions these isomers are converted to p-cymen-8-ol. This cyclic alcohol decomposes by disproportionation and redox reactions to p-cymene and alpha-p-dimethylstyrene (Figure 2.5), which are responsible for the off-odour in deteriorated lemon (Kimura et al., 1983). The mechanism for this reaction was elucidated by Peacock and Kuneman (1985), who showed that the addition of isoascorbic acid to a carbonated beverage containing 3 ppm citral strongly inhibits the formation of the dimethylstyrene.

Spillage of crude oil at sea is all too common and a number of compounds present in petroleum can produce oily taints in fish. These compounds may consist of saturated and unsaturated paraffins and aromatic hydrocarbons. Fish and molluscs that become contaminated with the oil spill are often tainted to such a degree that they can no longer be sold for human consumption, causing severe financial losses to local fishermen. It has been reported that the tainting of fish is largely due to the petroleum and chemical industries (Ogata and Miyake, 1970). An oily taste in fish can be caused by direct ingestion of the hydrocarbons dissolved or suspended in the water, or by feeding on invertebrates living on the sea-bottom. After a spillage the heavy fraction of the oil sinks to the sea-bottom, and contamination of the fish will continue long after the visual signs of the spillage have disappeared.

A series of straight- and branched-chain hydrocarbons from undecane to dodecane were the cause of a taint in brown trout contaminated with diesel oil (Mackie et al., 1972). The relationship between contamination of fish by aliphatic hydrocarbons, and crude oil spillage was investigated by Motohiro and Inoue (1973). They isolated the n-paraffins from the tainted flesh of salmon, mullet and sea bream caught in the waters polluted by oil from a tanker wreck, and examined the extracts by gas chromatography. The chromatograms were remarkably similar to the corresponding one from the contaminating oil. Eels cultured in waste water from an oil refinery were tainted and were found to contain a number of aromatic hydrocarbons, including toluene (Ogata and Miyake, 1973).

The work of Motohiro and Iseya (1976) showed that the aromatic hydrocarbon fraction is far more important than the aliphatic hydrocarbons in causing taints in fish. The crude oil caused a taint in scallop muscle at a level of 100 mg/kg, whereas a level of 400 mg/kg of tetradecane was necessary before it could be tasted. Xylene, however, was detectable at 100 mg/kg.

2.2.8 Esters and ethers

Sorbic acid is not only used in foodstuffs as a mould inhibitor, but is also allowed in some wines together with sulphur dioxide. Wines which contain 30 ppm of sulphur dioxide will typically contain levels of around 80 ppm of sorbate. At this level there have been reports of an unpleasant taste in the wine, which has been described as 'like geranium flowers' (Würdig et al., 1974).

$$CH_3 \, CH = CH \, CH = CH \, COOH \xrightarrow[\text{bacteria}]{\text{lactic}} CH_3 \, CH = CH \, CH = CH \, CH_2 \, OH$$

$$\xrightarrow{\text{acid}} CH_3 \, CHOH \, CH = CH \, CH = CH_2 \xrightarrow{\text{ethanol}}$$

$$CH_3 \, CH \, (\, O \, C_2 \, H_5 \,) \, CH = CH \, CH = CH_2$$

Figure 2.6 Formation of 2-ethoxyhexa-3,5-diene in wine.

Crowell and Guymon (1975) investigated the reactions leading to the formation of this off-flavour and, following extraction of the wine with methylene chloride and separation by gas chromatography, they identified the compound responsible by mass spectrometry. They proposed the structure as 2-ethoxyhexa-3,5-diene and, through a series of known rearrangement reactions, the reaction mechanism illustrated in Figure 2.6 was suggested. The authors point out that although sorbic acid is often used to replace part of the sulphur dioxide in wine, bottlers must take care since, although both additives inhibit yeast growth, only sulphur dioxide is an effective inhibitor of bacterial growth, which is an essential part of the mechanism for the formation of this ethoxy hexadiene.

A fruity off-flavour in milk, which may present a problem to the dairy industry, was investigated by Witter (1961), who discovered that it can be caused by growth of the organism *Pseudomonas fragi*. By the action of a lipase, cleavage of butyric and hexanoic acids from the milk fat occurs and this is followed by esterification with ethanol to form the corresponding ethyl esters. These two esters have been quantified by gas chromatography and the levels correlated with sensory analysis of the tainted milk (Wellnitz-Ruen *et al.*, 1982). Samples of milk judged to be fruity contained levels around 0.02 ppm of ethyl butyrate and 0.2 ppm ethyl hexanoate.

2.2.9 Amines

Off-flavours that develop in stored dried milk products have been the subject of a vast number of research publications over a period of many decades. One particular flavour is described as 'grape-like' and may occur in stale non-fat dried milk. Parks *et al.* (1964) extracted the tainted product with hexane and submitted the extract to gas chromatography. Using the technique of sniffing the effluent from the chromatograph, they isolated the compound responsible for the taint; this compound was subsequently identified by infrared spectroscopy as *o*-aminoacetophenone. At a level of 5 µg/kg, all the members of a taste panel could detect this compound in dried milk powder and, by interpolation, the mean taste threshold can be calculated at about 0.4 µg/kg. The same compound has been identified as the major cause of a stale flavour in beer (Palamand, 1974).

One of the weeds prevalent in Australia and New Zealand is *Lepidium hyssopifolium*, which has the common name of pepperwort or peppercress. If this weed, or one from a related species, is eaten by dairy cattle it is likely to cause a faecal taint in the product. High-vacuum distillation and infrared spectroscopic analysis demonstrated the presence of skatole as the principal compound responsible for the flavour defect (Park, 1969).

Skatole and indole were found in tainted frozen French fries manufactured from potatoes stored in sheds where soft-rot microorganisms had attacked the potatoes (Whitfield *et al.*, 1982). From the odour and taste values of these compounds, only skatole (with an odour threshold of 10 µg/l in water) was considered to be responsible.

It has been widely reported that in the early stage of spoilage of chilled fish muscle, a musty, potato-like off-odour develops. Miller and coworkers inoculated sterile fish with *Pseudomonas perolens* and collected the volatile compounds produced on a Porapak Q trap (Miller *et al.*, 1973). The volatiles were eluted and analysed by gas chromatography–mass spectrometry. Amongst a variety of compounds that were identified, they concluded that 2-methoxy-3-isopropylpyrazine was primarily responsible for the observed disagreeable smell, with the very potent odour threshold in water of 2 ng/l. Daise *et al.* (1986) describe a similar malodour in retail cuts of beef inoculated with *Pseudomonas*, but no effort was made to identify the compound responsible for this taint nor to relate it to the work of Miller and coworkers.

A mousy taint in wine has been identified as being due to the presence of 2-acetyl-1,4,5,6-tetrahydropyridine and the corresponding 3,4,5,6-isomer (Strauss and Heresztyn, 1984). The odour threshold of the former compound is around 2 µg/l in water.

Over a period of years, odour problems have arisen in certain cutting fluids used in engineering workshops. The smell has been described as musty or 'like a dirty dishcloth'. A bacterium that produced the same odour when grown in a culture medium was isolated from the fluids. Mottram *et al.* (1984) isolated the compound responsible for the malodour and identified it as 2,6-dimethyl-3-methoxypyrazine, which they also chemically synthesised for comparison purposes.

It is well known that the spoilage of fish leads to the degradation of trimethylamine oxide to the corresponding amine. Butler and Fenwick (1984) reviewed the relationship between a fishy taint in eggs and the presence of trimethylamine, and concluded that the use of fish meal and rapeseed meal as feedstuffs accounts for this problem.

2.2.10 Chlorinated hydrocarbons

Organochlorine pesticides based on chlorinated hydrocarbons are now much less widely used than in the past, when their persistence and potential damage to the environment was less well understood. When these compounds were widely used there were several reports that certain of them could cause un-platibility when present in foodstuffs. Because analytical techniques were less sophisticated several decades ago, there is some doubt about the precise identity of the compounds responsible for the taint. De Laveur and Carpentier (1968) reported a musty taint in chickens, and associated this with the presence of hexachlorocyclohexane (HCH) in the meat. Whitewash was thought to be the source of the HCH. Milne (1953) fed birds with a diet containing 5 mg/kg of HCH and confirmed that the flesh of the resulting animals was severely tainted.

In order to unravel the likely causative agent of the musty taint due to the presence of HCH, it should be remembered that the technical product contains a number of isomers, of which only the gamma-isomer is an effective pesti-cide. It was the experience of the Leatherhead Food Research Association that it was the beta-isomer that was particularly responsible for the musty taint.

Gamma-HCH   Beta-HCH

Since modern preparations of HCH (Lindane) contain very low levels of the non-active isomers, particularly the beta-isomer, this problem is very much less likely to occur than in the past. Watson and Patterson (1982) reported an incident in which a woody, almond-like taint was found in a batch of pork. They undertook the usual type of laboratory examination, in which the meat was extracted in a Likens–Nickerson apparatus against diethyl ether and the organic solution was examined by gas chromatography–mass spectrometry. This revealed the presence of 1,4-dichlorobenzene in the meat at levels from 5–20 mg/kg. Although no taint thresholds were reported, there can be little doubt that this compound was the cause of the disagreeable flavour in the meat. The authors were unable to establish the source of the dichlorobenzene, although they did find the compound in the feedstuff, and in the wood shavings used as litter. From time to time, similar samples were submitted to the Leatherhead Food Research Association for examination and it is the opinion of the author that the most likely source of dichlorobenzene in meat is the use of this compound in commercial degreasing agents in some drain-cleaners. It should be emphasised that when these drain-cleaners are used strictly according to the manufacturer's instructions, no taint occurs in the meat; it is only when misused that any problem arises.

In the publication by Watson and Patterson (1982), mention is made of the use of 1,4-dichlorobenzene in slow-release toilet blocks. Companies should pay particular attention to the use of these products in toilets near the manufacturing area, since it is the experience of the author that their use caused a very serious problem with one major food manufacturer.

2.2.11 Carbonyl compounds

In the USA it was found that foam spray-dried milk prepared during hot weather develops an off-flavour, which was examined at the Dairy Products Laboratory in Washington (Parks et al., 1969). The volatiles were recovered from the affected milk powder and the unsaturated aldehyde, trans-6-nonenal, was identified. Its flavour threshold was found to be less than 0.07 µg/l in fresh whole milk. The decomposition of unsaturated oils is a frequent cause of off-flavours in fatty foods. A fishy off-flavour that develops in autoxidised oils containing linolenic acid has been investigated by Meijboom and Stroink (1972), who collected the carbonyls from heated peroxidised soybean oil as their dinitrophenylhydrazones. They positively identified 2-trans-4-cis-7-cis-decatrienal by mass spectrometry and comparison with the synthetic compound, although no accurate taste thresholds were reported. The factors that affect the formation of a fishy flavour in bacon were reported by Coxon et al. (1986), who conclude that a high content of fish meal in the feedstuff was the primary cause of the problem, although no specific compound was identified as the chemical source of the taint. A report from the Food Research Institute, Norwich, revealed that the main requirements for the formation of fishy flavours in butterfat were the presence of copper salts and tocopherol (Swoboda and Peers, 1977a). Following vacuum distillation of tainted butterfat, they identified eleven unsaturated aldehydes and ketones, from which they drew the conclusion that the 2,4-unsaturated aldehydes were primarily responsible for the fishy flavour, whereas vinyl ketones gave rise to a metallic taint. In a further publication (Swoboda and Peers, 1977b), octa-1-en-3-one and the more potent octa-1-cis-5-dien-3-one were identified as the major cause of the metallic taint in butterfat. The respective taste thresholds of the two compounds in oil are around 80 and 0.5 µg/kg. Both compounds are formed from polyenoic acids.

In the production of skim milk, both butterfat and fat-soluble vitamins like vitamin A are removed. Fortification of the resulting non-fat dried milk with vitamin A palmitate may lead to the development of a hay-like flavour. This reaction was studied by Suyama et al. (1983), who isolated the steam volatile compounds from oxidised vitamin A palmitate and identified 19 compounds by mass spectrometry, of which beta-ionone and dihydroactinidiolide were considered to be responsible for the hay-like off-flavour.

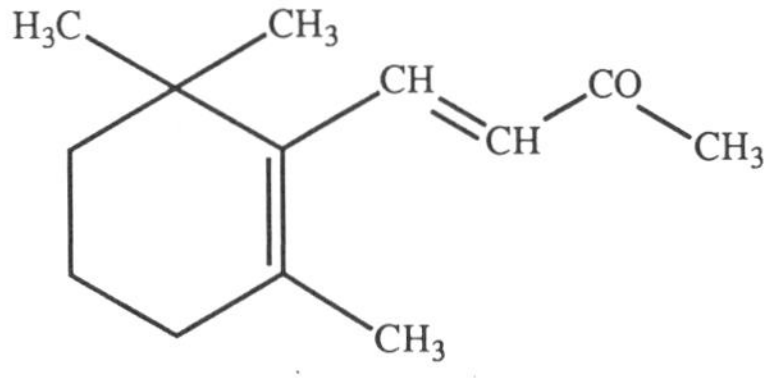

Beta-ionone

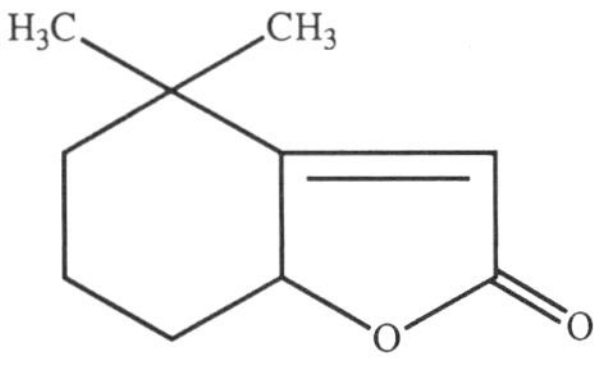

Dihydroactinidiolide

A floral off-flavour in stored carrots has been known for many decades and an investigation carried out at the end of World War II (Tomkins *et al*, 1944) showed that the presence of oxygen was necessary for the defect to develop; a clue to its origin was discovered when it was noted that the colour from the carotene disappeared at the same time. Many years later, Ayers and coworkers (1964) correctly identified alpha- and beta-ionone, and beta-ionone-5,6-epoxide, all of which have a violet-like odour. The odour thresholds of many aldehydes and ketones are recorded in two major publications (Buttery *et al*., 1973; Hall and Anderson, 1983).

Heptan-2-one has been identified as one of the compounds responsible for the musty taste that develops in rancid desiccated coconut and coconut oil (Kellard *et al*., 1985).

2.2.12 Furans and oxygen-ring compounds

Flavour reversion is a well-known problem in the edible oil industry; in soyabean oil it leads to a distinctive beany/grassy flavour prior to the onset of rancidity. Chang *et al*. (1966) isolated and identified 2-pentylfuran as the major component responsible for the reversion flavour in soyabean. The mechanism for the formation of this compound from linoleate is shown to take place through a nine-stage autoxidation process (Ho *et al*., 1978). The flavour threshold of 2-pentylfuran in oil is 1 mg/l (ppm). In a further publication Chang *et al*. (1983) describe the identification of both the *cis* and *trans* isomers of 2-(1-pentenylfuran) and 2-(1-pentenylfuran) in reverted soyabean oil. The synthesis of all the four isomers are described in the above publications.

Plastic film is widely used for packaging foodstuffs and occasionally leads to tainting of the product. McGorrin *et al*. (1987) reported a particularly intense musty odour that appeared in one batch of a packaging film, which was sufficiently volatile for the compound to be detected in the headspace vapours. This particular compound was conclusively identified as 4,4,6-trimethyl-1,3-dioxane, which was apparently formed by reaction of 2-methyl-2,4-

pentanediol (a coating used to help the ink to adhere) with formaldehyde from an unknown source.

Some years ago an off-flavour problem described as cinnamon-like was noticed in bread produced in the western United States. Buttery *et al.* (1978) used the usual sophisticated analytical techniques to identify coumarin as the cause of the taint and to trace the source of contamination to the presence of clover seeds in the flour.

2.2.13 Fatty acids

There are many reports of the formation of an intense bitter taste in stored soy beans and soy flour. A major component of soy bean is linoleic acid, which, when submitted to enzymic oxidation with lipoxygenase and peroxidase, generated fatty acids that had a strong bitter taste (Baur *et al.*, 1977). The main components of the bitter-tasting fraction were identified by mass spectrometry of the trimethylsilyl ethers as a mixture of 9,12,13-trihydroxyoctadec-10-enoic acid and 9,10,13-trihydroxyoctadec-11-enoic acid. Moll *et al.* (1979) developed a quantitative method for the determination of these acids in legumes, based on isolation by thin-layer chromatography and analysis of the silyl derivatives of the corresponding methyl esters. Biermann *et al.* (1980) report that two monohydroxy octadienoic acids are much more intensely bitter than the trihydroxy acids.

References

Anderton, J. I. and Underwood, J. B. (1968). Catty odours in food. *BFMIRA Technical Circular No 407.*

Ayers, J. E., Fishwick, M. J., Land, D. G. and Swain, T. (1964). Off-flavour in dehydrated carrot stored in oxygen. *Nature* **203**, 81.

Badings, H. T., Maarse, H., Kleipool, R. J. C., Tas, A. C., Neeter, R. and ten Noever de Brauw, M. C. (1975). Formation of odorous compounds from hydrogen sulphide and methanethiol and unsaturated carbonyls. *Proc. Int. Aroma Res. Zeist*, 63.

Baur, C., Grosch, W., Wieser, H. and Jugal, H. (1977). Enzymatic oxidation of linoleic acid: Formation of bitter tasting fatty acids. *Z. Lebensm. Unters.-Forsch.* **164**, 171.

Bemelmans, J. M. H. and den Braber, H. J. A. (1983). Investigation of an iodine-like taste in herring from the Baltic Sea. *Wat. Sci. Tech.* **15**, 105.

Biermann, U., Wittmann, A. and Grosch, W. (1980). Vorkommen bitterer Hydroxyfettsauren in Hafer und Weizen. *Fette Seifen Anstrichsmittel* **82**, 236.

Bills, D. D., Reddy, M. C. and Lindsay, R. C. (1969). Fumigated nuts can cause an off-flavour in candy. *Mfg Confect.* **49,** 39.

Braus, H. and Miller, F. D. (1958). Composition of whisky. Steam volatile compounds in fusel oil. *J. Ass. Off. Agric. Chem.* **41**, 141.

Burns-Brown, W., Heseltine, H. K., Devlin, J. J. and Greer, E. N. (1961). *Milling*, 401.

Buser, H.-R., Zanier, C. and Tanner, H. (1982). Identification of 2,4,6-trichloroanisole as a potent compound causing cork taint in wine. *J. Agr. Food Chem.* **30**, 359.

Butler, E. G. and Fenwick, G. R. (1984). Trimethylamine and fishy taint in eggs. *World Poultry Sci. J.*, 39.

Buttery, R. G., Guadagni, D. G. and Ling, L. C. (1973). Flavor compounds: Volatilities in vegetable oil and oil–water mixtures. Estimation of odor thresholds. *J. Agr. Food Chem.* **21**, 198.

Buttery, R. G., Guadagni, D. G., Ling, L. C., Seifert, R. M. and Lipton, W. (1976a). Additional volatile components of cabbage, broccoli and cauliflower. *J. Agr. Food Chem.* **24**, 829.

Buttery, R. G., Guadagni, D. G. and Ling, L. C. (1976b). Geosmin, a musty off-flavor in dry beans. *J. Agr. Food Chem.* **224**, 419.

Buttery, R. G., Ling, L. C. and Bean, M. M. (1978). Coumarin off-odòr in wheat flour. *J. Agr. Food Chem.* **26**, 179.

Cant, P. A. E. (1979). Identification of alpha,alpha-dibromoanisole as the trace flavour contaminant of an abnormal commercial casein. *New Zealand J. Dairy Sci. and Tech.* **14**, 35.

Chang, S. S., Smouse, T. H., Krishnamurthy, R. G. and Reddy, R. B. (1966). Isolation and identification of 2-pentylfuran as contributing to the reversion flavor of soyabean oil. *Chemistry and Industry*, 1926.

Chang, S. S., Shen, G-H., Tang, J., Jin, Q. Z., Shi, H., Carlin, J. T. and Ho, C.-T. (1983). Isolation and identification of 2-pentenylfurans in the reversion flavor of soybean oil. *J. Amer. Oil Chem. Soc.* **60**, 553.

Cooper, D. M., Griffiths, N. M., Hobson-Frohock, A., Land, D. G. and Rowell, J. D. (1978). Fumigation of poultry food with methyl bromide: Effects on egg flavour, number and weight. *Brit. Poultry Sci.* **19**, 537.

Coxon, D. T., Peers, K. E. and Griffiths, N. M. (1986). Recent observations on the occurrence of fishy flavour in bacon. *J. Sci. Food Agric.* **37**, 867.

Crowell, E. A. and Guymon, J. F. (1975). Wine constituents arising from sorbic acid addition, and identification of 2-ethoxyhexa-3,5-diene as source of geranium-like off-odour. *Amer. J. Viticult.* **26**, 97.

CSIRO (1982). Taints in meat and other foods—chlorophenols and related compounds. *Meat Research Newsletter No. 82/5.* CSIRO Division of Food Research, Brisbane, Australia.

Curtis, R. F., Land, D. G., Griffiths, N. M., Gee, M. G., Robinson, D., Peel, J. L., Gee, J. M. (1972). 2,3,4,6-Tetrachloroanisole association with musty taint in chickens and microbial formation. *Nature* **235**, 223.

Curtis, R. F., Dennis, C., Gee, J. M., Gee, M. G., Griffiths, N. M., Land, D. G. and Robinson, D. (1974). Chloroanisoles as a cause of musty taint in chickens and their microbiological formation from chlorophenols in broiler house litters. *J. Sci. Food Agr.* **225**, 811.

Daise, R. L., Zottola, E. A. and Epley, R. J. (1986). Potato-like odor of retail beef cuts associated with species of *Pseudomonas. J. Food Protect.* **49**, 272.

Daley, J. D., Lloyd, G. T., Ramshaw, E. H. and Stark, W. (1986). Off-flavours related to the use of sorbic acid as a food preservative. *CSIRO Food Res. Q.* **46**, 59.

De Laveur, E. and Carpentier, J. (1968). The persistence of HCH on different supports. *Phytiat.-Phytopharm.* **17**, 41.

Dietz, F. and Traud, J. (1978). Geruchs- und Geschmacks-Schwellen-Konzentrationen von Phenolkorpern. *Gas- und Wassfach Wasser Abwass.* **119**, 318.

Dougherty, J. D., Campbell, R. D. and Morris, R. L. (1966). Actinomycetes: Isolation and identification of agent responsible for musty odors. *Science* **152**, 1372.

Dupuy, H. P., Flick, G. J., St Angelo, A. J. and Sumrell, G. (1986). Analysis of trace amounts of geosmin in water and fish. *J. Oil Chem. Soc.* **63**, 905.

Elsden, S. R., Hilton, M. G. and Waller, J. M. (1976), *Arch. Microbiol.* **107**, 283.

Fan, L. L., Tang, J.-Y. and Wohlman, A. (1983). Investigation of 1-decyne formation in cottonseed oil fried foods. *J. Amer. Oil Chem. Soc.* **60**, 1115.

Frijters, J. E. R. and Bemelmans, J. M. H. (1977). Flavor sensitivity for chloroanisoles in coagulated egg yolk. *J. Food Sci.* **42**, 1122.

Gee, M. G., Land, D. G. and Robinson, D. (1974). Simultaneous analysis of 2,3,4,6-tetrachloroanisole, pentachloroanisole and the corresponding chlorophenols in biological tissue. *J. Sci. Food Agric.* **25**, 829.

Gerber, N. N. (1968). Geosmin from microorganisms is *trans*-1,10-dimethyl-*trans*-9-decalol. *Tetrahedron Letters*, No. **25**, 2971.

Gerber, N. N. (1983). Volatile substances from actinomycetes: their role in the odor pollution of water. *Water Sci. Technol.* **15**, 115.

Goldenberg, N. and Matheson, H. R. (1975). Off-flavours in foods, a summary of experience 1948–74. *Chemistry and Industry* 551.

Griffiths, N. M. (1974). Sensory properties of the chloroanisoles. *Chemical Senses and Flavor* **1**, 187.

Griffiths, N. M. and Land, D. G. (1973). 6-Chloro-*o*-cresol taint in biscuits. *Chemistry and Industry*, 904.

Guadagni, D. G., Buttery, R. G. and Okano, S. (1963). Odour thresholds of some organic compounds associated with food flavours. *J. Sci. Food Agric.* **14**, 761.

Hall, G. and Andersson, J. (1983). Volatile fat oxidation products. Determination of odour thresholds and odour intensity functions by dynamic olfactometry. *Lebensm.-Wiss. u. Technol.* **16**, 354.

Harper, W. J. and Hall, C. W. (1976). *Dairy Technology and Engineering*. AVI, Westport, p. 125.

Harris, N. D., Karahdian, C. and Lindsay, R. C. (1986). Musty aroma compounds produced by selected molds and actinomycetes on agar and whole wheat bread. *Food Protect.* **49**, 964.

Harrison, G. (1970). *J. Inst. Brewing* **76**, 486.

Ho, C-T., Smagula, M. S. and Chang, S. S. (1978). The synthesis of 2-(1-pentenyl) furan and its relationship to the reversion flavor of soyabean oil. *J. Amer. Oil Chem. Soc.* **55**, 233.

Horwood, J. F., Lloyd, G. T., Ramshaw, E. H. and Stark, W. (1981). An off-flavour associated with the use of sorbic acid during Feta cheese maturation. *Aust. J. Dairy Technol.* **36**, 38.

Hsieh, T. C.-Y., Tanchotikul, U. and Matiella, J. E. (1988). Identification of geosmin as the major muddy off-Flavour of Louisiana brackish water clam (*Rangia cuneata*). *J. Food Sci.* **53**, 1228.

ISO (1992) Glossary of Terms relating to Sensory Analysis. ISO Standard 5492–1992.

Kellard, B., Busfield, D. M. and Kinderlerer, J. L. (1985). Volatile off-flavour compounds in desiccated coconut. *J. Sci. Food Agric.* **36**, 415.

Kimura, K., Nishimura, H., Iwata, I. and Mizutani, J. (1983). Deterioration mechanism of lemon flavor. 2. Formation mechanism of off-odor substances arising from citral. *J. Agric. Food Chem.* **31**, 801.

Larson, R. A. and Rockwell, A. L. (1979). *Env. Sci. Technol.* **13**, 325.

Lefebvre, A., Riboulet, J.-M., Boidron, J.-N. and Riberau-Gayon, P. (1983). *Sciences Aliments* **3**, 265.

Loos, M. A., Roberts, R. N. and Alexander, M. (1967). Formation of 2,4-dichlorophenol and 2,4-dichloroanisole from 2,4-dichlorophenoxy acetate by *Arthrobacter* sp. *Can. J. Microbiol.* **13**, 691.

Maarse, H. and ten Noever de Brauw, M. C. (1974). Another catty odour compound causing air pollution. *Chemistry and Industry*, 36.

Mackie, P. R., McGill, A. S. and Hardy, R. (1972). Diesel oil contamination of brown trout. *Environ. Pollut.* **3**, 9.

Marth, E. H., Constance, M., Capp, C. M., Hasenzahl, L., Jackson, H. W. and Hussong, H. V. (1966). Degradation of potassium sorbate by *Penicillium* species. *J. Dairy Sci.* **49**, 1197.

McGorrin, R. J., Pofahl, T. R. and Croasmun, W. R. (1987). Identification of the musty component from an off-odor packaging film. *Anal. Chem.* **59**, 1109A.

Medsker, L. L., Jenkins, D. and Thomas, J. F. (1968). Odorous compounds in natural waters. An earthy-smelling compound associated with blue-green algae and actinomycetes. *Env. Sci. Technol.* **2**, 461.

Meijboom, P. W. and Jongenotter, G. A. (1981). Flavour perceptibility of straight chain unsaturated aldehydes as a function of double bond position and geometry. *J. Amer. Oil Chem. Soc.* **558**, 680.

Meijboom, P. W. and Stroink, J. B. A. (1972). 2-*Trans*,4-*cis*,7-*cis*-decatrienal, the fishy off-flavour occurring in strongly autoxidised oils containing linolenic acid or *w*-3,6,9, etc. fatty acids. *J. Amer. Oil Chem. Soc.* **49**, 555.

Melnick, D., Luckman, F. H. and Gooding, C. M. (1954). Sorbic acid as a fungistatic agent in foods. *Food Res.* **19**, 44.

Miller, A., Scanlan, R. A., Lee, J. S., Libbey, L. M. and Morgan, M. E. (1973). Volatile compounds produced in sterile fish muscle (*Sebastes melanops*) by *Pseudomonas perolens*. *J. Applied Microbiol.* **225**, 257.

Milne, F. N. J. (1953). The effect of benzene hexachloride in poultry feed on meat and egg quality. *Queensland J. Agric. Sci.* **10**, 214.

Moll, C., Biermann, U. and Grosch, W. (1979). Occurrence and formation of bitter-tasting trihydroxy fatty acids in soybeans. *J. Agric. Food Chem.* **27**, 239.

Motohiro, T. and Inoue, N. (1973). *n*-Paraffins in polluted fish by crude oil from the Juliana wreck. *Bull. Fac. Fish Hokkaido Univ.* **23**, 204.

Motohiro, T. and Iseya, Z. (1976). Effect of water polluted by oil on aquatic animals. *Bull. Fac. Fish Hokkaido Univ.* **26**, 367.

Mottram, D. S., Patterson, R. L. S. and Warrilow, E. (1984). 2,6-dimethyl-3-methoxypyrazine; a microbiologically produced compound with an obnoxious musty odour. *Chemistry and Industry*, 448.

Newton, C. (1973). *Bristol Evening Post*, 4th October, 43.

Ogata, M. and Miyake, M. (1970). Offensive odour substance in fish in the sea along petrochemical industries. *Japan J. Public Health* **17**, 1125.

Ogata, M. and Miyake, M. (1973). Detection of causative compounds in petroleum oil for oily taint in fish. *Water Res.* **7**, 1493.

Palamand, S. R. (1974). *Brewers Digest*, 49, 58 and 90.

Park, R. J. (1969). Weed taints in dairy produce. I. Lepidium taint. *J. Dairy Res.* **36**, 31.

Park, R. J., Armitt, J. D. and Stark, D. P. (1969). Weed taints in dairy produce. II. *Coronopus* or land cress taint in milk. *J. Dairy Res.* **36**, 37–46.

Parks, O. W., Schwartz, D. P. and Keeney, M. (1964). Identification of *o*-aminoacetophenone as a flavour compound in stale dry milk. *Nature*, 185.

Parks, O. W., Wong, N. P., Allen, C. A. and Schwartz, D. P. (1969). 6-*Trans*-nonenal. An off-flavour component of foam spray dried milk powder. *J. Dairy Sci.* **52**, 953.

Parliment, T. H., Clinton, W. and Scarpellino, R. (1973). *Trans*-2-nonenal: A coffee component with novel organoleptic properties. *J. Agr. Food Chem.* **21**, 485.

Parr, L. J., Gee, M. G., Land, D. G., Robinson, D. and Curtis, R. F. (1974). Chlorophenols from wood preservatives in broiler house litter. *J. Sci. Food Agric.* **25**, 835.

Patterson, R. L. S. (1968). Catty odours in food: their production in meat stores from mesityl oxide in paint solvents. *Chemistry and Industry*, 548.

Patterson, R. L. S. (1969). Catty odours in food: confirmation of identity of 4-mercapto-4-methyl-pentan-2-one by GC–MS. *Chemistry and Industry*, 48.

Patterson, R. L. S. (1972). Disinfectant taint in poultry. *Chemistry and Industry*, 609.

Peacock, V. E. and Kuneman, D. W. (1985). Inhibition of the formation of alpha,*p*-dimethylstyrene and *p*-cymen-8-ol in a carbonated citral containing beverage system. *J. Agric. Food Chem.* **33**, 330.

Pearce, T. J. P., Peacock, J. M., Aylward, F. and Haisman, D. R. (1967). Catty odours in food: Reactions between hydrogen sulphide and unsaturated ketones. *Chemistry and Industry*, 1562. Notes on muddy odour: III. Variability of sensory response to 2-methylisoborneol.

Persson, P.-E. (1979). *Aqua Fennica* **9**, 48–52.

Reineccius, G. (1991). Off-flavours in food. *CRC Critical Reviews in Food Science and Nutrition*, 381.

Saxby, M. J., Stephens, M. A. and Chaytor, J. P. (1983). Detection and prevention of chlorophenol taints in liquid milk. *Leatherhead Food RA Research Report, No.416*.

Simpson, R. F., Amon, J. M. and Daw, A. J. (1986). Off-flavour in wine caused by guaiacol. *Food Technology Australia* **38**, 31.

Sloot, D. and Harkes, P. D. (1975). Volatile trace components in Gouda cheese. *J. Agr. Food Chem.* **23**, 356.

Spadone, J.-C., Takeoka, G. and Liardon, R. (1990). Analytical investigation of Rio off-flavour in green coffee *J. Agric. Food Chem.* **38**, 226.

Spencer, R. (1969). *BFMIRA Layman's Guide to Catty Taints in Foods*.

Strauss, C. R. and Heresztyn, T. (1984). 2-Acetyltetrahydropyridines—a cause of the mousy taint in wine. *Chemistry and Industry*, 109.

Suyama, K., Yeow, T. and Nakai, S. (1983). Vitamin A oxidation products responsible for haylike flavor production in non fat dry milk. *J. Agric. Food Chem.* **31**, 22.

Swoboda, P. A. T. and Peers, K. E. (1977a). Volatile odorous compounds responsible for metallic, fishy taint formed in butterfat by selective oxidation. *J. Sci. Food Agric.* **28**, 1010.

Swoboda, P. A. T. and Peers, K. E. (1977b). Metallic odour caused by vinyl ketones formed in the oxidation of butterfat. The identification of octa-1,*cis*-5-dien-3-one. *J. Sci. Food Agric.* **28**, 1019.

Thurston, P. A. (1986). The phenolic off-flavour test: A method for confirming the presence of wild yeasts. *J. Inst. Brew.* **92**, 9.

Tindale, C. R. and Whitfield, F. B. (1989). Production of chlorophenols by the reaction of fibre-board and timber components with chlorine-based cleaning agents. *Chemistry and Industry*, 835–836.

Tomkins, R. G., Mapson, L. W., Allen, R. J. L., Wager, H. G. and Baker, J. (1944). Drying of vegetables. III. Storage of dried vegetables. *J. Soc. Chem. Ind.* **63**, 225.

Watson, A. and Patterson, R. L. S. (1982). Tainting of pork meat by 1,4-dichlorobenzene. *J. Sci. Food Agric.* **33**, 103.

Wellnitz-Ruen, W., Reineccius, G. A. and Thomas, E. L. (1982). Analysis of the fruity off-flavour in milk using headspace concentration capillary column gas chromatography. *J. Agr. Food Chem.* **30**, 512.

Wenn, R. V., Macdonell, G. E. and Wheeler, R. E. (1987). A detailed mechanism for the formation of chlorophenolic taint in draught beer dispense systems: The rationale for taint persistence and ultimate removal. *Proc. 21st Congress European Brewery Convention*, Madrid.

Whitfield, F. B. and Freeman, D. J. (1983). Off-flavours in crustaceans caught in Australian coastal waters. *Wat. Sci. Technol.* **15**, 85.

Whitfield, F. B., Freeman, D. S. and Bannister, P. A. (1981). Dimethyl trisulphide: an important off-flavour component in royal red prawn (*Hymenopenaeus sibogae*). *Chemistry and Industry*, 692–693.

Whitfield, F. B., Last, J. H. and Tindale, C. R. (1982). Skatole, indole and *p*-cresol: Components in off-flavoured frozen french fries. *Chemistry and Industry*, 662.

Whitfield, F. B., Tindale, C. R. and Last, J. H. (1983). 2-Methylisoborneol: A cause of off-flavour in canned champignons. *Chemistry and Industry*, 316.

Whitfield, F. B., Ly Nguyen, T. H., Shaw, K. J., Last, J. H., Tindale, C. R. and Stanley, G. (1985). Contamination of dried fruit by 2,4,6-trichloroanisole and 2,3,4,6-tetrachloroanisole adsorbed from packaging materials. *Chemistry and Industry*, 661.

Whitfield, F. B., Shaw, K. J. and Ly Nguyen, T. H. (1986). Simultaneous determination of 2,4,6-trichloroanisole, 2,3,4,6-tetrachloroanisole and pentachloroanisole in foods and packaging materials by high resolution gas chromatography–multiple ion mass spectrometry. *J. Sci. Food Agric.* **37**, 85.

Whitfield, F. B., McBride, R. L. and Ly Nguyen, T. H. (1987). Flavour perception of chloroanisoles in water and selected processed foods. *J. Sci. Food Agr.* **40**, 357.

Whitfield, F. B., Last, J. H., Shaw, K. J. and Tindale, C. R. (1988). 2,6-Dibromophenol: The cause of an iodoform-like off-flavour in some Australian crustacea. *J. Sci. Food Agric.* **46**, 29.

Whitfield, F. B., Ly Nguyen, T. H. and Last, J. H. (1991a). Effect of relative humidity and chlorophenol content on the fungal conversion of chlorphenols to chloroanisoles in fibreboard cartons containing dried fruit. *J. Sci. Food Agric.* **54**, 595.

Whitfield, F. B., Shaw, K. J., Gibson, A. M. and Mugford, D. C. (1991b). An earthy off-flavour in wheat flour: geosmin produced by *Streptomyces griseus. Chemistry and Industry*, 841.

Witter, L. D. (1961). *J. Dairy Sci.* **44**, 983.

Wood, N. F. and Snoeyink, V. L. (1977), 2-Methylisoborneol improved synthesis and quantitative gas chromatographic method for trace concentrations producing odor in water. *J. Chromatog.* **132**, 405.

Würdig, G., Schlotter, H.-A. and Klein, E. (1974). Uber die Ursachen des sogenannten Geranientones. *Allgemeine Deuts. Weinfachzeit* **110**, 578.

Yurkowski, M. and Tabachek, J.-O. L. (1974). *J. Fish Res. Board Canada* **31**, 1851.

3 Analysis of taints and off-flavours

H. MAARSE

3.1 Introduction

The analysis of taints and off-flavours is only partly different from that of volatile compounds contributing positively to flavour. The differences are caused by the fact that in general many compounds contribute to the flavour of a product but only a few to taints and off-flavours.

Like flavour compounds, off-flavour compounds often occur in very low concentrations and may have extremely low threshold values. The selection and identification is therefore very demanding, and investigators in this field should have at their disposal sophisticated instrumental equipment. It is not sufficient simply to know the chemical name(s) of the compound(s) causing the problem; questions such as the following ones should be answered (Nijssen, 1991).

- Have the off-flavour compounds been formed from food constituents?
- Have they migrated into the product?
- Is the off-flavour the result of a serious disturbance of the balance of the various compounds constituting the flavour of a food product?

Answers to these questions are essential to enable the prevention of similar off-flavour problems in the future.

Many reviews and chapters and some books have been devoted to the analysis of flavours (e.g. Maarse and Belz, 1981; Cronin, 1982; Schreier, 1988; Maarse, 1991). Is there, then, any need to pay special attention to the analysis of off-flavours? In the author's opinion there is, for the following reasons.

- The preparation of concentrates need not be very refined because only one or two components are of interest. Instability and/or inefficient isolation of compounds is not important as long as these off-flavour compounds are recovered unchanged and in good yields.
- In a way, off-flavour compounds are often character-impact compounds. They are characteristic of the particular flavour of an off-flavoured or tainted product.

In this chapter, the general approach to the analysis of volatile compounds is first described, indicating the differences in procedures between flavour and off-flavour studies. Methods of isolation, concentration, separation and ident-ification are discussed. The importance of sensory analysis in guiding the investigations in all consecutive steps is emphasized and several examples are presented to elucidate the procedures of off-flavour studies. These examples concern off-flavours originating from product constituents, as well as foreign compounds originating from, for example, packaging material.

3.2 Preparation of the concentrates

3.2.1 Introduction

The methods and techniques used for the preparation of concentrates of vo-latile compounds have not changed greatly during the past decade. In this section, therefore, only some general remarks on methods and techniques will be made, and some new developments will be indicated. For more compre-hensive treatment of this subject the reader is referred to the literature (e.g. Bemelmans, 1981; Schaefer, 1981).

3.2.2 Extraction

One of the methods most frequently used to isolate volatile compounds from non-liquid food products is the combined steam distillation extraction (SDE) procedure according to Likens and Nickerson (Likens and Nickerson, 1964; Bemelmans, 1981). The SDE technique has also found many applications in off-flavour studies (e.g. Saxby, 1973).

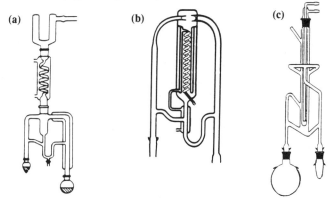

Figure 3.1 Various modifications of the Likens and Nickerson apparatus for simultaneous steam distillation and extraction: (a) modification after Maarse and Kepner (1970) with vacuum jacket to minimize premature condensation and a dry-ice condensor to reduce volatilization losses; (b) modification after Schultz *et al.* (1977)—can be used at reduced pressure; (c) modification after Godefroot *et al.* (1981)—a scaled-down version. Parts (a) and (b) reprinted from *J. Agric. Food Chem.* **18**, 1095–1101 and **25**, 446–449, respectively, with permission. Copyright (1970 and 1977) American Chemical Society. (c) reprinted from *J. Chromatography* (1981) **203**, 325–335 by permission of Elsevier Science Publishers B.V.

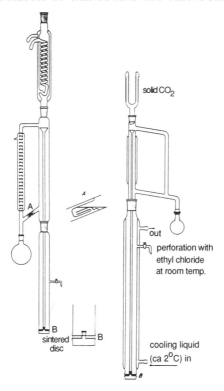

solid CO_2

out

perforation with
ethyl chloride
at room temp.

A

B
sintered
disc

B

cooling liquid
(ca 2^oC) in

Figure 3.2 Modifications of Kutscher–Steudel extractors according to Weurman (1969). Reprinted from *J. Agric. Food Chem.* **17**, 370–384, with permission. Copyright (1969) American Chemical Society.

Many (small) alterations have been made to the original apparatus as used by Likens and Nickerson. Some of these are depicted in Figure 3.1. The most important change was made by Schultz *et al.* (1977), who constructed an apparatus that can be used at reduced pressure (Figure 3.1(b)). This modification is used when the relevant compounds are not thermally stable. The micro-SDE apparatus described by Godefroot *et al.* (1981) is also worth mentioning. It is a scaled-down version of the original SDE apparatus. Núñez *et al.* (1984) used it for the isolation of the volatile constituents of grapefruit juice. The aroma of the concentrate was very similar to that of the juice, more so than that of the concentrates obtained with the original SDE apparatus or by solvent extraction. The micro-SDE apparatus was also used by Heil and Lindsay (1988) for the quantitative analysis of alkylphenols and aromatic thiols in fish (see also section 3.6.5).

Another way to isolate volatile compounds from a solid product is direct extraction in a Soxhlet apparatus (Bemelmans, 1981). This method was used by Paasivirta *et al.* (1987) for the study of off-flavours in fish. Direct extraction is also often used for liquid products, either manually, using a separator funnel, by vigorous magnetic stirring (Wigilius *et al.*, 1987), or in a continuous extraction apparatus such as the one shown in Figure 3.2. The extraction solvent should be purified by distillation before use to minimize the interference during subsequent analysis.

3.2.3 Adsorption

3.2.3.1 Headspace analysis. In some studies of off-flavours, headspace volatiles are collected on adsorbents such as Tenax (e.g. Veijanen *et al.*, 1983; Whitfield and Shaw, 1985), Chromosorb 105 (e.g. Murray and Whitfield, 1975) and charcoal (e.g. Grob and Zürchner, 1976; Cronin, 1982; Borén *et al.*, 1985). Further analysis of the collected compounds is possible by heat desorption directly on a gas chromatographic column (see section 3.4.2.1), or by extraction with a solvent.

The advantage of the headspace technique is that there is a lower risk of compounds being decomposed thermally. Enzymic activity should be prevented in some cases by the addition of, for example, salt solutions (e.g. Whitfield *et al.*, 1982; Buttery *et al.*, 1987). In this way flavour concentrates can be obtained that are more representative of the product under investigation (see also section 3.6.6). It is, however, not always possible to obtain a sufficient quantity of the concentrate to allow unequivocal identification of trace components (Whitfield and Shaw, 1985).

For the study of off-flavours in water, Borén *et al.* (1982, 1985) promote the use of an open modification of the standard water stripping system, using charcoal as the adsorbent, as constructed by Grob and Zürchner (1976). The

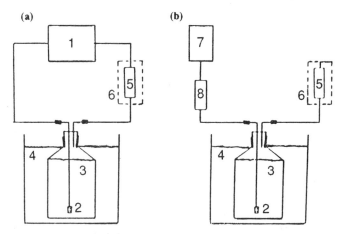

Figure 3.3 Closed (a) and open (b) stripping systems (from Borén *et al.*, 1982). 1 = pump; 2 = gas inlet; 3 = water sample; 4 = thermostatic water-bath; 5 = activated carbon adsorbent (analytical filter); 6 = oven; 7 = nitrogen gas cylinder; 8 = activated carbon adsorbent (gas cleaning filter). Reprinted from *J. Chromatography* (1982) **252**, 139–146 by permission of Elsevier Science Publishers B.V.

Table 3.1 Effect of extraction solvents on the recovery of adsorbed compounds on charcoal (adapted from Borén *et al.*, 1985)

Compound	Recovery (%)				
	Carbon disulphide	Diethyl ether	Methylene dichloride	Acetone–carbon disulphide (10:90)	Methanol–benzene–carbondisulphide (5:30:65)
Butyl acetate	55	62	63	61	65
Chlorobenzene	60	9	39	56	62
1-Clorohexane	57	61	60	61	60
1-Hexanol	43	57	67	62	66
Anisole	62	7	48	62	62
Phenol	3	12	4	52	52
1,4-Dichlorobenzene	60	2	20	60	60
n-Decane	62	53	52	63	59
Acetophenone	44	2	38	55	60
1-Chlorooctane	63	52	63	63	63
Guaiacol	36	4	39	52	58
1-Octanol	45	47	62	63	57
2,6-Dimethylphenol	29	6	26	54	53
Naphthalene	27	n.d.	3	27	46
1-Chlorodecane	65	37	63	64	65
1-Decanol	49	28	68	59	67
Methyl decanoate	66	33	68	65	65
2-Methoxynaphthalene	18	n.d.	2	21	35
1-Chlorodecane	67	45	62	64	63
Methyl dodecanoate	71	14	67	62	62
1-Chlorotetradecane	62	6	67	60	56
Anthracene	n.d.	n.d.	n.d.	1	4
1-Chlorohexadecane	23	n.d.	49	49	53
1-Chlorooctadecane	34	1	28	37	45
Mean	47	22	44	53	56

n.d. = not detected

original 'Grob system' and the adapted version are depicted in Figure 3.3. Borén *et al.* (1985) claim the advantages of the open system to be:

- better recoveries for compounds of low volatility and high polarity
- extremely low blank levels
- simplicity of operation

Borén *et al.* (1985) used extraction to desorb the compounds from the charcoal. They compared the efficiency of seven solvents and four solvent mixtures for the extraction of adsorbed compounds from charcoal filters. These desorption studies were performed by adding to the filter a mixture consisting of 50 ng each of 24 compounds widely differing in polarity. Some of the results are presented in Table 3.1. The mixture methanol–benzene–carbon disulphide (5:30:65) was found to be the most effective. When the solvent peaks of this mixture interfere with the subsequent gas chromatographic (GC) analysis, Borén *et al.* (1985) recommend the use of carbon disulphide with 10% acetone.

3.2.3.2 Direct adsorption from water. For the adsorption of traces of volatile components (<1 μg/kg) in water, Wigilius *et al.* (1987) used XAD-2 as the adsorbent. They compared this procedure with headspace stripping and direct extraction with dichloromethane (see section 3.2.4). Adsorption on XAD-2 was shown to be a suitable method in contrast to earlier reports. After optimization of the method the limit of detection of the XAD–GC–FID (FID = flame ionisation detection) method was shown to be generally below 20 ng/l for a sample volume of 10 l. However, compounds such as long-chain hydrocarbons and polar phenols are poorly adsorbed. An advantage of the XAD-2 technique is that large volumes can be processed.

3.2.4 Comparison of techniques

Lundgren *et al.* (1988) compared four methods to concentrate off-flavour compounds from water: (i) stripping with addition of salt; (ii) stripping without addition of salt; (iii) XAD-2 adsorption; and (iv) dichloromethane extraction. All methods were generally found to be satisfactory. In some off-flavour studies, however, better results were obtained with one method than with others. For example, the original odour of a certain water sample could be re-created from extracts obtained by dichloromethane extraction, but not from the corresponding stripping extracts. Sniffing the effluent from the GC showed there to be a large difference between the extracts. In the first extract a much larger number of medium- to strong-smelling compounds were observed.

The stripping technique is less time-consuming and very convenient to use; in addition, the absence of high-molecular-weight organic compounds in the stripping extracts permits the use of the on-column injection technique, without shortening the life of the GC capillary column. According to Lundgren *et al.* (1988), the stripping technique should therefore be preferred to the other techniques if, in a given situation, it has been shown that this technique gives a satisfactory enrichment of off-flavour compounds.

Wigilius *et al.* (1987) compared the XAD-2 method with liquid–liquid extraction and found the two techniques to be complementary. In the XAD extracts, compounds were found that were absent in the concentrates obtained by dichloromethane extraction, and vice versa.

3.3 The role of sensory analysis

3.3.1 Introduction

The only way to select the compounds responsible for an off-flavour is the use of sensory techniques. This implies that studies of off-flavours can only be started when persons are available who can actually detect it. In daily practice this means that panel members who are trained to detect and describe odours are asked to informally compare an off-flavoured product with a reference product. Those who are able to detect the off-flavour are used during the subsequent study of the relevant off-flavoured product. Sometimes a representative of the producer of the product, who is familiar with the off-flavour, is asked to co-operate in the study to ensure that it is indeed the compounds that caused the complaints that are searched for.

3.3.2 Control of concentration procedures

The second reason for needing the panel members is to control whether the isolation and concentration procedures have yielded a concentrate with the relevant off-flavour.

When collected headspace samples are heat-desorbed in cooled glass capillaries (see section 3.4.2.1) the contents of these capillaries can be used for sensory assessment (e.g. Jetten, 1985; Whitfield and Shaw, 1985). When the concentrate is a solvent extract these sensory tests are sometimes difficult because of interference by residual solvent in the concentrate. The use of paper odour strips can partly overcome this problem, but the technique requires the availability of persons with experience in this type of assessment (van Gemert, 1981).

Lundgren *et al.* (1989), studying off-flavours of water, used another method. They rapidly injected 2 μl of a dichloromethane extract in 200 ml of odourless water. The small droplets thus produced were left to dissolve slowly in the water without stirring. After 1 h the water sample was transferred to an Erlenmeyer flask, covered with a watch-glass and heated to 60°C in a water-bath for approximately 15 min. Finally, the odour quality was characterized and the threshold value of the solution was determined with triangle tests at 60°C. From the results obtained it could be established whether the extraction had been successful. The extract-to-water ratio ($1:10^5$) was close to the highest ratio permitting sensory evaluation of the extract without interference by the smell of the solvent. Non-polar volatile compounds are lost partly or almost

Table 3.2 Relative losses from water during dissolution and heating (60°C, 30 min) of 2μl of a dichloromethane extract (5 ng/μl of each compound in 200 ml of water). From Lundgren *et al.* (1989).

Compound	Boiling point (°C)	Relative loss (%)
n-Decane	174	90
Acetophenone	202	0
1-Chlorooctane	180	75
1-Octanol	194	0
Benzyl acetate	215	0
Naphthalene	218	25
2,6-Dichloroanisole	220	8
1,2,3,5-Tetrachlorobenzene	246	43
Methyl decanoate	224	58
2,3,6-Trichloroanisole	240	0
Diphenylether	258	0
2-Methoxynaphthalene	274	0

totally during this procedure (Table 3.2), therefore this method is less useful for these compounds.

3.3.3 Selection of off-flavour compounds

The most important contribution of the panel members is the selection of compounds that are supposedly contributing to the off-flavour. This is done by GC-sniffing, a technique that has become very popular among flavour analysts (e.g. van Gemert, 1981). The effluent of a GC column is split 1 to 10 to the detector and the sniffing port, respectively (Figure 3.4). Humidified air is added to the effluent to prevent drying out of the mucous membrane.

Sävenhed *et al.* (1985) did not use a split at the end of the column, but injected twice—once with the FID as detector and once with sensory detection. They found very similar retention times and concluded that their sniffing device could be used to determine the retention times of odorous compounds with high accuracy. They also tried a two-column system described by Veijanen *et al.* (1983). A fused silica SP 2100 column of 50 m × 0.2 mm was cut into two 25 m columns. Sävenhed *et al.* could not reproduce the results of Veijanen *et al.*, who reported that the retention times on their two columns of 25 m were equal. The reason might well be that Veijanen *et al.* used apolar stationary phases (OV 1 and SE 30), which are coated more homogeneously and show no differences in polarity.

The advantages of the two-column system claimed by Veijanen *et al.*, or of the system of Sävenhed *et al.* (1985) over a system with a well-constructed splitter at the end of the column are not obvious. Only 10% of the separated components are lost by splitting.

Some training is needed to be able to smell and describe compounds in the effluent of a capillary column. At the beginning of the chromatogram, compounds are leaving the column every 5–10 s, so in this period, particularly, fast 'translations' of sensory impressions into descriptions have to be made.

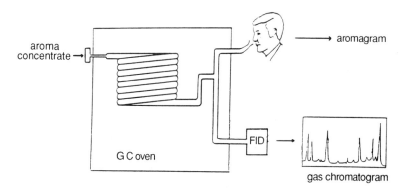

Figure 3.4   Gas chromatographic detection by sniffing.

The odour descriptions are noted on a form with a preprinted time scale. Preferably, the assessor should not see the chromatogram.

Sniffing should last no more than about 30 min. Longer periods are not effective due to olfactory fatigue. Sävenhed *et al.* (1985) used two observers for one run, each observer's period lasting 6–8 min. For health reasons, some investigators hesitate to use, or do not use at all, the sniffing technique (Saxby, 1991). It has been reported that smelling of chlorophenols may cause nasal cancer (Hardell *et al.*, 1983).

GC-sniffing results in an *aromagram*, a gas chromatogram in which all odours perceived are indicated on a time basis. GC-sniffing should preferably be carried out by at least two persons, who smell both the eluent of the off-flavoured product, and that of a reference product without the off-flavour. These persons should be able to observe and recognize the off-flavour. The results are compared and those peaks that may contribute to the off-flavour are selected.

Another method to select relevant compounds has been developed by Grosch and coworkers (Ulrich and Grosch, 1987), who called it 'aroma extract dilution analysis'. An aroma extract is stepwise diluted with a solvent, and the original extract and each dilution are analysed using the GC-sniffing method. The highest dilution at which a substance is still smelled is called its flavour dilution (FD) factor. An FD-factor of 20 means that the concentration of the odorous compound in the aroma extract is 20 times its odour threshold value as perceived by GC-sniffing. The FD-factors are determined both for the off-flavoured product, and for the reference product. It is supposed that compounds in the off-flavoured product having a FD-factor much higher than in the reference product, and compounds with a high FD-factor that are not present in the reference product, play a role in the off-flavour (see section 3.6.4). This method has the advantage compared with the determination of aroma values (ratio between the concentration of a compound in a product and its odour threshold value) that it can be used without knowing odour thresholds values. Both methods neglect the interaction between compounds—assessing all compounds separately—and, for that reason, have been subject to criticism (Maarse, 1991).

3.3.4 Confirmation of the contribution of identified compounds to the off-flavour

When the relevant compounds have been identified and quantified (see section 3.5.2.2), it is sometimes necessary to confirm the contribution of these compounds to the off-flavour. No confirmation is needed when one compound is found to have the same characteristic flavour as the off-flavoured product. In this final check the compounds are added to the reference product and the panel members are asked to establish whether the flavour is similar or equal to that of the off-flavoured product.

Preparation of a homogeneous mixture of the compound in the reference product is simple when the product under investigation is a liquid. In the case of fruit and vegetables eaten raw, a purée or juice can be prepared. When food products such as dried or heat-processed foods are involved, however, it is more difficult.

Whitfield and Shaw (1985) describe a procedure they have used successfully with dried foods. They used a 5 l cylindrical sealed glass jar, in which the food, together with filter paper impregnated with a known amount of the compound under investigation, was agitated by placing it on a pair of mechanically driven rollers. The quantity of the compound incorporated in the food can be calculated by re-weighing the filter paper after the agitation period. This period was 12 or 24 h depending on the volatility of the compound. For processed foods the same authors suggest to first grind, mince or blend the fresh product with a measured quantity of the off-flavour compound. Small portions are then rapidly cooked, preferably in a microwave oven, for a few minutes just before sensory assessment.

3.4 Separation

3.4.1 Introduction

In off-flavour studies a large majority of the separations are done by high-resolution gas chromatography (HRGC). Less frequently, high-performance liquid chromatography is applied (HPLC). It is outside the scope of this chapter to extensively describe these techniques; instead the reader is referred to the literature (e.g. Jennings, 1981; van Straten, 1981). Here, only a number of relatively new developments, which are specifically relevant to the study of off-flavours, are described.

3.4.2 High-resolution gas chromatography

3.4.2.1 Analysis of collected headspace samples. As mentioned in section 3.2.3.1, collected compounds can be desorbed by solvents, or can be directly injected onto a GC column by heat. For heat desorption many types of apparatus have been constructed (e.g. Murray *et al.*, 1972; Clark and Cronin, 1975; Schaefer, 1981), including commercially available equipment. To prevent peak-broadening, the thermally desorbed volatiles are condensed on a cooled capillary and then transferred to the GC column by heating this capillary. It is also possible to use these capillaries to check by sensory evaluation whether indeed the compound(s) causing the off-flavour has (have) been collected (see also section 3.3.2).

3.4.3 Two-dimensional gas chromatography

An elegant way to improve the separative power of a GC system is to use two columns in series, a pre-column and an analytical column. Interesting fractions of the first column can be introduced on-line onto the second column and further separated into individual compounds.

Two different column-switching systems are depicted in Figure 3.5. The main difference, apart from the price, is that in Figure 3.5(a) two ovens are needed and in Figure 3.5(b) only one. The second system can be adapted by introducing a split, either between the pre-column and the detector, or between the analytical column and the detector (Maarse, 1991). This enables the selection of those compounds that contribute to an off-flavour.

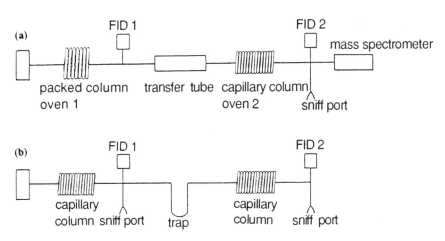

Figure 3.5 Two column-switching systems for multidimensional GC (from Maarse and van den Berg, 1989). (a) Schematic drawing of the Siemens system (Model Sichromat 2) for total transfer from packed or capillary columns, with intermediate trapping and a device for parallel MS and GC-sniffing or MS–FID registration; (b) schematic drawing of the Chrompack MUSIC system for total transfer from pre-columns to analytical columns, with intermediate trapping and a device for parallel sniffing and FID registration, or trapping and FID registration. Reprinted by permission of Ellis Horwood, Chichester.

3.4.4 Trapping of compounds

Although direct injection of an off-flavour concentrate is sometimes possible, generally pre-concentration of the relevant fraction in the chromatogram by trapping that fraction in a cooled capillary is necessary. Lundgren *et al.* (1989) determined the recoveries of 14 compounds in a test mixture using the set-up depicted in Figure 3.6. When the PTFE trap was pretreated with 3 µl of dichloromethane they found recoveries between 68 and 83%.

Jetten (1985) developed a method that enabled the injection of the content of the trap (40 cm × 0.32 mm internal diameter (i.d.)) fused silica capillary

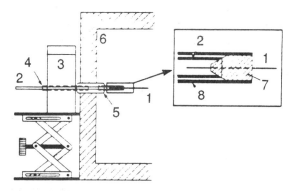

Figure 3.6 Device for cold trapping of different fractions of the GC effluent (from Lundgren *et al.*, 1989). 1 = GC column; 2 = PTFE tubing for fraction collection; 3 = liquid nitrogen container; 4 = metal tubing through the liquid nitrogen; 5 = connecting PTFE block; 6 = GC oven; 7 = PTFE block; 8 = PTFE tubing. Reprinted from *J. Chromatography* (1989) **482**, 23–34 by permission of Elsevier Science Publishers B.V.

coated with a chemically bonded stationary phase) directly on a gas chromatograph coupled to a mass spectrometer (GC–MS). The capillary can be mounted in the gas chromatograph between the injector and the capillary column using a zero dead-volume union. Injection on a column with another stationary phase, to enable further separation followed by GC-sniffing, is also possible. Recoveries between 80 and 97% were obtained. This method has the advantage over that of Lundgren *et al.* (1989) in that no solvents are used to remove the compound(s) from the trap. In this way optimal sensitivity is obtained. The relatively low yields of the first method are due to incomplete recovery of solvent from the trap (Lundgren *et al.*, 1989). Another advantage of using fused silica capillaries is their flexibility.

3.4.5 High-performance liquid chromatography

HPLC has not found many applications in off-flavour studies. One of the main reasons is that it is impossible with the usual eluents to evaluate by sensory analysis the contribution of separated compounds/fractions to an off-flavour of a product. The eluents either are toxic, or totally obscure the observation of flavour compounds in a fraction. Consequently, an eluent should be chosen that:

   (i)   is not toxic;
 (ii)   has no or only a very weak flavour; and
(iii)   still enables sufficient separation.

Such a system has been used to fractionate a wood extract. A gradient of ethanol and a phosphoric acid–water mixture was used as eluent on a RP-18 column (Maarse and van den Berg, 1989). HPLC or LC–GC combinations can be used to isolate relevant fractions and thus concentrate off-flavour compounds when the identity of the compounds is known. HPLC is also used to quantify known off-flavour compounds, such as chlorophenols. The sensitivity of the detector should be sufficiently low to enable the measurement of concentrations below that of the odour threshold value of the compound under investigation (Whitfield and Shaw, 1985).

3.5 Identification

3.5.1 Introduction

Most workers in the field of the analysis of flavours and off-flavours realize that the identity of a compound cannot be decided upon on the basis of the measurement of one or two retention times on a chromatographic column. Additional properties have to be determined, such as mass and/or infrared (IR) and/or nuclear magnetic resonance (NMR) spectra (Belz, 1981; Maarse, 1991). In off-flavour studies the comparison of the flavour properties of an unknown and a reference compound could also be important. This means that reference compounds should be available to measure retention data (preferably Kovats Indices (KI)), and spectra and flavour properties. Except in particular cases, when it is crucial to know with certainty that the identity of a compound is correct, it is not possible in daily routine to keep up this extremely high standard.

Considering mass and IR spectra as indispensable pieces of information, the reliability of identification can be roughly classified (Table 3.3).

Table 3.3 Rough classification of the reliability of identifications using some major identification methods (from Belz, 1981)

Spectroscopic evidence	Confirmation by chromatography[a, b]		
	Number of retention values		
	None	One	Two[c]
IR without reference spectrum	−	−	+
MS without reference spectrum	−	+	+
IR with reference spectrum	+, ++[d]	++	++
MS with reference spectrum	+	++	++
IR and MS with both reference spectra	++	++	++

[a] GC, TLC or LC with reference values
[b] − = not reliable; + = moderately reliable; ++ = very reliable
[c] GC, polar and apolar stationary phase; thin-layer chromatography (TLC) and liquid chromatography (LC), two different systems
[d] ++ = first members of a homologous series; + = higher members of a homologous series

3.5.2 Mass spectrometry

3.5.2.1 Introduction. The GC–MS combination is by far the most popular technique for the identification of volatile compounds in foods and beverages.

Many reviews have been published on this subject (e.g. ten Noever de Brauw, 1981; ten Noever de Brauw and van Ingen, 1981; Schreier, 1988; Sharpe and Chappell, 1990). In this section, only the more recent applications in off-flavour studies will be highlighted.

3.5.2.2 Selected-ion-monitoring mass spectrometry (SIM-MS). The SIM-MS technique has successfully been used in off-flavour studies. In this technique the mass spectrometer is equipped with a peak selector, which instructs the instrument to record selected ions characteristic of the compound of interest. It is not possible to record mass fragmentograms after the analysis, as can be done by total ion mass spectrometry (ten Noever de Brauw and van Ingen, 1981). This implies that it is only used when one is looking for a compound of known identity. Some advantages of the SIM technique over full scanning are (Sharpe and Chappel, 1990):

- SIM can improve sensitivity since scanning time can be focused specifically upon ions of interest, and is not spread over a wide range. This allows for a better signal-to-noise ratio. A 1000-fold increase in sensitivity can be attained, so reaching the picogram range (Schreier, 1988).
- SIM can improve accuracy and reproducibility for quantitation purposes.
- With GC–MS, SIM can filter out interfering peaks which may co-elute on the GC column, but only if the ion being monitored is absent from the spectrum of the interfering substance.

Whitfield *et al.* (1986) have used SIM-MS for the quantitative analysis of chloroanisoles in foods and packaging materials. The detection limits of these compounds were found to be $0.01\,\mu g/kg$ which is sufficiently low (the taste threshold of 2,4,6-trichloroanisole in dried fruit is $0.2\,\mu g/kg$, for example). Based on 20 g of initial sample, Heil and Lindsay (1988) report a quantitative detection with SIM-MS of alkylphenols and aromatic thiols down to a concentration of $0.5\,\mu g/kg$ (see also section 3.6.5).

3.6 Examples of off-flavour studies

3.6.1 Introduction

To give an impression of the way off-flavour studies are carried out in practice, some of them are described in the following sections. These examples are partly chosen from the many investigations carried out in the author's laboratory and partly taken from the literature. In this selection an attempt has been made to show a variety of techniques. This might give an insight into the dependence of the choice of the analytical procedure on the type of off-flavour, and on the matrix. The availability of equipment is also important.

3.6.2 Onion- and almond-like off-flavour of packaged cheese

3.6.2.1 Introduction. In the early 1980s the author's laboratory was requested to investigate the cause of an off-flavour of tainted cheese (Nijssen *et al.*, 1987). The cheese was packed as blocks of 15 g in cellophane, coated with polyvinylidene chloride (PVDC). The cheese had an offensive flavour, which was described as onion- and almond-like. It was supposed that the off-flavour was caused by the packaging material, because it had not previously been reported in the dairy industry. The taint developed slowly after the cheese had been packed, and the contamination took place only occasionally.

3.6.2.2 Methods. A concentrate of the volatiles of sound and tainted cheese samples was prepared by simultaneous steam distillation extraction according to Likens and Nickerson (see section 3.2.2). These concentrates were analysed on a CP Sil 5 CB column (25 m × 0.3 mm i.d.) and detected by FID and GC-sniffing (see section 3.3.3).

To enable the identification of the most volatile compounds, 20 ml of the headspace over ten blocks of cheese in a 800 ml jar were collected on a Tenax–GC trap (see section 3.2.3.1). The volatiles were heat-desorbed, and analysed by GC–MS.

To check on the presence of taint compounds and their precursors a static headspace method was used. The headspace was taken from a 25 ml vial containing 10 g of cheese at a temperature of 50°C, and was analysed on a Carbowax capillary column (25 m × 0.3 mm i.d.).

3.6.2.3 Results and discussion. By GC-sniffing, one position was found in the chromatogram where there was an odour quite similar to the almond-like taint. The compound eluting at this retention time was identified by GC–MS and by comparing retention data of the unknown and a reference compound, as 2-methyl-2-pentenal. This compound can be formed by aldol condensation of two molecules of propanal. Addition of propanal to sound cheese led to the formation of 2-methyl-2-pentenal after a one-week incubation period at 4°C (see Figure 3.7(b)). Propanal was not found in sound and tainted cheese, nor in printed and unprinted packaging material. It might, however, be an intermediate in the formation of 2-methyl-2-pentenal, its precursor being propanol. This compound was found in tainted cheese, but not in sound cheese (see Figure 3.8).

The disinfectant for the cutting and packaging machines was investigated as a possible source of propanol. According to the label it contained 11% w/w propanol-1 and 3% w/w propanol-2. The following model experiment was carried out. The disinfectant was added to slices of sound cheese, which were subsequently wrapped separately in packing material. After a one-week incubation period no 2-methyl-2-pentenal was found (Figure 3.7(c)). A possible

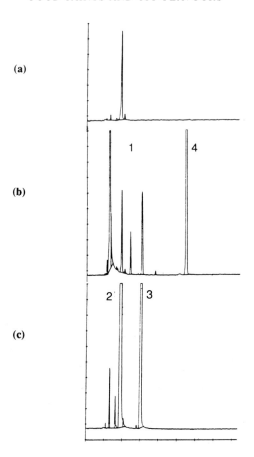

Figure 3.7 Headspace chromatogram of cheese at 50°C (from Nijssen *et al.*, 1987): (a) sound cheese; (b) propanal added, plus one-week incubation at 4°C; (c) disinfectant added, plus one-week incubation at 4°C. 1 = propanal; 2 = ethanol; 3 = propanol; 4 = 2-methyl-2-pentenal. Reprinted from *Flavour Science and Technology* (eds Martens, M., Dalen, G. A. and Russwurm, H.) by permission of John Wiley & Sons, Ltd.

explanation for this result is that propanal can only be formed from propanol by oxidation. An oxidant should be present and the reaction rate might be increased by higher temperatures (sealing?) and decreased pH.

The onion-like taint might be caused by the product formed by the addition of hydrogen sulphide to 2-methyl-2-pentenal. Hydrogen sulphide is readily available in cheese. A preliminary experiment showed that these two compounds could indeed give rise to an onion-like flavour.

3.6.2.4 Conclusion. The almond-like taint observed in a batch of cheese blocks packed in cellophane is caused by 2-methyl-2-pentenal. The onion-like note may be caused by the product formed by the addition of hydrogen

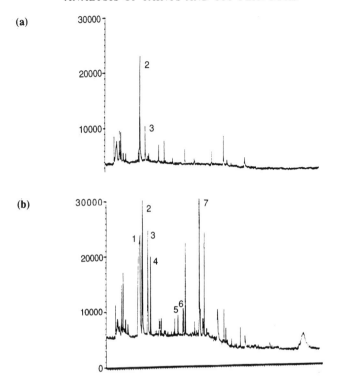

Figure 3.8 MS total-ion chromatogram of cheese volatiles obtained by Tenax trapping (from Nijssen *et al.*, 1987): (a) sound cheese; (b) tainted cheese. 1 = propanol; 2 = butanone-2; 3 = butanol-2; 4 = tetrahydrofuran; 5 = 2-methyl-2-butenal; 6 = 2-methylpentanal; 7 = 2-methyl-2-pentenal. Reprinted from *Flavour Science and Technology* (eds Martens, M., Dalen, G. A. and Russwurm, H.) by permission of John Wiley & Sons, Ltd.

sulphide to this compound. It is most likely that 2-methyl-2-pentenal is formed from a propanol-containing disinfectant. Propanol is first oxidized to propanal, and 2-methyl-2-pentenal is formed via aldol condensation.

3.6.3 Determination of maximal allowable levels of chlorophenols

3.6.3.1 Introduction. Chloroanisoles give rise to taint problems in many food products and are, together with chlorophenols, responsible for the industry's most costly losses caused by this type of problem, principally expressed as damage to reputation and destruction of the contaminated product (Whitfield and Shaw, 1985). Prevention of these taint problems is therefore very important.

Chloroanisoles can be formed from chlorophenols present in packaging materials, such as cardboard and wooden pallets. Consequently, one of the ways to prevent taint problems caused by chloroanisoles is to require that the concentrations of their precursors, chlorophenols, in the packaging material are below maximum allowable levels (Maarse *et al.*, 1988). A procedure to determine these levels was developed at TNO, and an example of its application is described in the following section.

3.6.3.2 Method. Suppose that a food product is stored in a container for several weeks. The container also holds wooden pallets and cardboard boxes. First, the odour and taste threshold values of 2,4,6-trichloroanisole (TCA)— the isomer with the lowest threshold value—have to be determined in the relevant food product. Following this the diffusion rate of TCA from the surrounding atmosphere through the packaging material into the food product has to be determined in a storage experiment. This can be done by placing the food product in an airtight container, e.g. a tin can. A known amount of TCA is injected into the can through the lid and its concentration in the food product is measured at predetermined time intervals. This must be done with several TCA concentrations in the can. The storage time at a certain concentration of TCA in air, at which the level of TCA in the food product exceeds the threshold value, can be read from a graph (Figure 3.9).

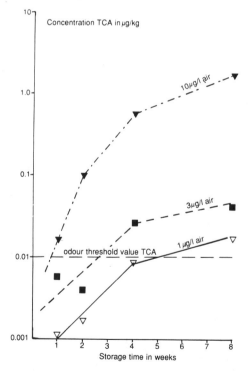

Figure 3.9 Concentration of TCA in a product in relation to the concentration of TCA in the surrounding air during a storage experiment (from Maarse *et al.*, 1988). Reprinted by permission of Akademie Verlag, Berlin.

To be able to estimate the maximum allowable level of chlorophenols in the packaging material, the following assumptions are made:

(i) the total amount of the trichlorophenol is converted into TCA within a relatively short period of time;

(ii) the total amount of TCA is transferred to, and evenly distributed in the surrounding atmosphere; and

(iii) the container is leak-tight.

3.6.3.3 Results. A typical example of such an experiment is shown in Figure 3.9. For three concentrations of TCA in the atmosphere the concentration of TCA in the food product is given as a function of storage time. Suppose that the odour threshold value of TCA in the food product is $0.01 \, \mu g/kg$. The level of TCA in the atmosphere must then be maintained below $1 \, \mu g/l$ as can be read from Figure 3.9. The empty space in the container is about 50 000 l, so the total amount of chlorophenols is allowed to be 50 mg. If the container is filled with 250 kg of wood from pallets, the concentration of chlorophenols in the wood must be below $0.2 \, mg/kg$. Concentrations higher than this level are often found. In this example the diffusion is rather low.

3.6.3.4 Conclusion. The result of this experiment leads to the following conclusions (Maarse *et al.*, 1988):

(i) the occurrence of off-flavours due to chloroanisoles could be limited if maximum allowable levels of chlorophenols in relation to various applications were known and adhered to in practice; and

(ii) these levels can be determined by the procedure described.

3.6.4 A study of an off-flavour of beer by the aroma extract dilution analysis

3.6.4.1 Introduction. The change of the flavour of beer during storage has been the subject of many publications (e.g. Meilgaard and Peppard, 1986). An interesting new approach to the study of this beer off-flavour problem has been presented by Schieberle (1992), who used the aroma extract dilution analysis, resulting in flavour dilution (FD) factors for those compounds that are the most important odorants (see section 3.3.3). The method used and the results obtained by Schieberle are described in the following sections.

3.6.4.2 Method. A sample of pale lager beer (500 ml), to which oxygen was added, was stored at 40°C for two weeks. After this period the beer had an off-flavour described as sweet, honey-like, somewhat resembling the flavour of ribes. The volatile compounds were isolated by extraction with diethyl ether, followed by sublimation *in vacuo*. The extract was fractionated into neutral and acidic compounds by treatment with sodium bicarbonate. The

FD-factors of this extract were determined and compared with those of an extract of the neutral compounds of fresh beer, which was obtained in the same way.

3.6.4.3 Results.   The results of the determination of the FD-factors are given in Table 3.4. A number of odour-contributing compounds not detectable by sensory analysis in the fresh beer were found in stored beer. Schieberle (1992) supposes that phenylacetaldehyde, having an intensely sweet, honey-like odour, contributes to the sweeter overall flavour, while 3-methyl-3-mercaptobutyl formate contributes to the ribes flavour of the stored beer.

Table 3.4   FD-factors of primary odorants (FD-factor > 32) identified in a stored beer (two weeks; 40°C), compared with those of the same compounds in fresh beer (from Schieberle, 1992)

Compound[a]	Odour quality	FD-factor	
		Stored	Fresh
Phenylacetaldehyde	Sweet, honey-like	256	< 1
3-Methyl-3-mercaptobutyl formate[b]	Catty, ribes-like	32	< 1
Unknown	Sweet, aniseed-like	64	< 1
(E.E)-2,4-Nonadienal	Fatty, waxy	64	< 1
(E)-2-Nonenal	Green, tallowy	64	16
4-Vinyl-2-methoxyphenol	Spicy, clove-like	128	1024

[a] The compounds were identified by comparing their mass spectra (EI, CI), their retention times on two stationary phases, and their odour quality (as determined by GC–sniffing) with those of reference compounds
[b] The MS signals were too weak for an unequivocal interpretation. The compound was identified on the basis of the other criteria mentioned in the preceding footnote.

3.6.4.4 Conclusion.   Measurement of FD-factors is a new and valuable approach to the study of off-flavours. Phenylacetaldehyde and 3-methyl-3-mercaptobutyl formate might be used as indicators for the development of an off-flavour in pale lager beers (Schieberle, 1992).

3.6.5 Quantitative analysis of tainting compounds in fish

3.6.5.1 Introduction.   Many studies have reported tainting of fish by polluted water (e.g. Reineccius, 1979; Bemelmans, 1983; Hiatt, 1983; Nijssen, 1991). Compounds that cause the taints are often present in very low quantities and have (very) low threshold values. One example of these studies is presented in the following sections. It was chosen because it describes the use of the micro-SDE apparatus as developed by Godefroot *et al.* (1981) for the preparation of the flavour concentrates (see also section 3.2.2), and SIM-MS analysis for quantification (see also section 3.5.2.2).

3.6.5.2 Method.   Heil and Lindsay (1988) developed a method to quantify flavour-tainting alkylphenols and thiophenols from fish tissue. For the isolation they used the micro version of the SDE apparatus according to Godefroot *et al.* (1981) (see also section 3.2.2). Fish samples were minced in a 300 ml Warren Blender for 30 s. In a 100 ml round bottom flask, 20 g of sample were weighed and 100 µl of an ethanolic solution containing 0.5 mg/l of each alkylphenol and thiophenol were added. Prior to SDE the flask was agitated vigorously by hand-shaking for 5 min. Subsequently, 40 ml saturated sodium chloride, 400 mg cysteine (to prevent oxidation of the thiols), and HCl to pH 6 were added. t-Butyl methyl ether (3 ml + 1.0 µg 2,4,6-trimethylphenol as internal standard) was used as the solvent and the distillation time was 1.5 h The extract was concentrated to about 20 µl. The concentrates were analysed on a Carbowax 20 M fused silica capillary column (60 m × 0.32 mm i.d.) interfaced with a mass spectrometer in the selected ion monitoring mode (see also section 3.5.2.2).

3.6.5.3 Results. The ratios of ion signal intensity of each alkyl- and thiophenol, and that of the internal standard were measured at three concentration levels corresponding to 50, 5 and 0.5 µg/kg of these compounds. The last concentration is the approximate flavour-tainting threshold value for the most

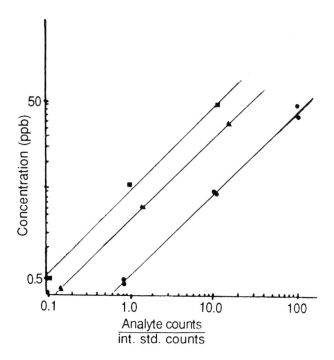

Figure 3.10 Log–log calibration curves for alkylphenols and thiophenols, obtained by GC-SIM-MS (from Heil and Lindsay, 1988): (●) thiophenols; (■) isopropylphenols; (▲) 2,4-3,5-djisopropyl-phenol.

Table 3.5 Recovery of alkyl- and thiophenols from rainbow trout and ions monitored for each compound (from Heil and Lindsay, 1988)

Compound	Ions monitored		Recovery[a]
	Primary	Secondary	(%)
2,4,6-Trimethylphenol	121	136	100
2-Isopropylphenol	121	136	100
3-Isopropylphenol and 4-isopropylphenol	121	136	101
2,4-Diisopropylphenol	163	178	95
2,5-Diisopropylphenol	163	178	96
2,6-Diisopropylphenol	163	178	81
3,5-Diisopropylphenol	163	178	96
Carvacrol	135	150	96
Thymol	135	150	103
Thiophenol	110	109	55
Thiocresol	91	124	85

[a] Standard deviations for % recovery were 2% or less for each compound

potent alkylphenol in fish. Three separate curves were obtained (Figure 3.10) showing linear responses for all of the compounds over the concentration ranges tested.

The recovery was determined at the 50 µg/kg level, analysing all samples in triplicate. The results are given in Table 3.5. The recovery of 2,6-diisopropylphenol is lower than that of the other alkylphenols, most probably because of the shielding effect of the two ortho-substituted isopropyl groups on the aromatic hydroxyl moiety, influencing the extraction efficiency. The recovery of the thiophenols is lower because of their sensitivity to oxidation–reduction

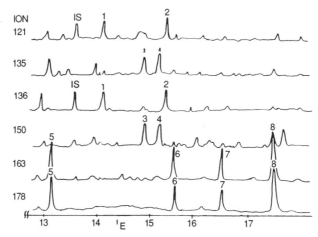

Figure 3.11 Ion chromatograms of alkylphenols in fish tissue (from Heil and Lindsay, 1988). IS = internal standard; 1 = 2-isopropylphenol; 2 = 3- and 4-isopropylphenol; 3 = thymol; 4 = carvacrol; 5 = 2,6-diisopropylphenol; 6 = 2,4-diisopropylphenol; 7 = 2,5-diisopropylphenol; 8 = 3,5-diisopropylphenol. The retention times, based on standard esters, are given on the horizontal axis.

conditions. Without addition of cysteine, the recovery of thiophenol and thio-cresol was 31 and 60%, respectively. Ion chromatograms of alkylphenols in fish tissues are shown in Figure 3.11.

3.6.5.4 Conclusion.   The analytical method described is sufficiently sensitive to allow for the routine quantification of tainting alkyl- and thiophenols at µg/kg levels in fish.

3.6.6 An important off-flavour problem in sand lobsters

3.6.6.1 Introduction.   Whitfield *et al.* (e.g. 1981, 1982 and 1985) have studied a number of off-flavours of crustacean sea-foods. One of their studies concerned the identification of the compound responsible for an intense garlic flavour of the sand lobster (*Ibacus peroni* and *Ibacus ultricrenatus*).

3.6.6.2 Method.   Uncooked off-flavoured crustaceans were chosen for the analyses, for the following reasons:

  (i)   the number of volatile compounds in these samples was less than in cooked material; and
  (ii) the concentration of the off-flavour compounds was higher in the un-cooked than in the cooked samples.

To prevent the formation of artefacts, 100 g of shelled crustaceans were ho-mogenized in 150 ml of a saturated aqueous solution of sodium chloride. Headspace (38 l) over this homogenate was collected on a Chromosorb 105 trap at 40°C. The traps consisted of a bed of Chromosorb 105 (50/60 mesh, 100 mg) packed into stainless steel tubes (90 mm × 3.2 mm i.d.). They were preconditioned in a nitrogen stream, initially at 225°C for 24 h and thereafter, following each use, at 180°C for 30 min (Murray and Whitfield, 1975). The collected volatiles were transferred to a cooled capillary (see section 3.4.2.1) and then injected on to a glass capillary column (60 m × 0.6 mm i.d.) coated with Carbowax 20 M. The retention time of the compound causing the off-fla-vour was determined by GC-sniffing. Identification was carried out by ana-lysing the trapped compound by GC–MS, and by proton NMR.

3.6.6.3 Results.   The compound with an intense garlic aroma, detected by GC-sniffing, gave mass and NMR spectra that agreed with that of bis-(methyl-thio)-methane. The addition of this compound to samples of fresh minced prawn flesh of normal flavour in concentrations of 1–5 µg/kg, followed by cooking for 3 min at 100°C, gave a material with a distinctive garlic flavour.

The source of the sulphur compound is the animal's 'head' section, which includes the stomach and the digestive gland. The tail, being the edible part, is free of the off-flavour compound. On cooking the animals whole, a significant proportion of the bis-(methylthio)-methane migrates from the head to the tail flesh.

3.6.6.4 Conclusion. Whitfield and coworkers showed that the headspace method in combination with GC-sniffing is a very useful method to study off-flavour problems. Artefact formation by enzymic action can be prevented by addition of sodium chloride.

3.7 Conclusions

It has been shown that several methods and techniques can be used in the study of off-flavours and taints. In practice it is impossible to have all these techniques operational in industrial or institute's laboratories. Each laboratory will choose those techniques that are useful to solve the majority of the off-flavour problems. This might well imply that the most appropiate method cannot always be used to solve a particular off-flavour problem.

References

Belz, R. (1981). Identification—introduction. In *Isolation, Separation and Identification of Volatile Compounds in Aroma Research*. Eds H. Maarse and R. Belz. Akademie Verlag, Berlin and Reidel, Dordrecht, pp. 152–155.

Bemelmans, J. M. H. (1981). Isolation and concentration from the product phase. In *Isolation, Separation and Identification of Volatile Compounds in Aroma Research*. Eds H. Maarse and R. Belz. Akademie Verlag, Berlin and Reidel, Dordrecht, pp. 4–37.

Bemelmans, J. M. H. (1983). Investigation in an iodine-like taste in herring from the Baltic sea. *Wat. Sci. Technol.* **15**, 105–113.

Borén, H., Grimvall, A. and Sävenhed, R. (1982). Modified stripping technique for the analysis of trace organics in water. *J. Chromatogr.* **252**, 139–146.

Borén, H., Grimvall, A., Palmborg, J., Sävenhed, R. and Wigilius, B. (1985). Optimization of the open stripping system for the analysis of trace organics in water. *J. Chromatogr.* **348**, 67–78.

Buttery, R. G., Teranishi, R. and Ling, L. C. (1987). Fresh tomato volatiles: a quantitative study. *J. Agric. Food Chem.* **35**, 540–544.

Clark, R. G. and Cronin, D. A. (1975). The use of activated charcoal for the concentration and analysis of headspace vapours containing food aroma volatiles. *J. Sci. Food Agric.* **26**, 1615–1624.

Cronin, D. A. (1982). Techniques of analysis of flavours. Chemical methods including sample preparation. In *Food Flavours. Part A. Introduction*. Eds I. D. Morton and A. J. MacLeod. Elsevier Scientific Publishing Company, Amsterdam, pp. 15–48.

van Gemert, L. J. (1981). Coordination of sensory and instrumental analysis. In *Isolation, Separation and Identification of Volatile Compounds in Aroma Research*. Eds H. Maarse and R. Belz. Akademie Verlag, Berlin and Reidel, Dordrecht, pp. 240–258.

Godefroot, M., Sandra, P. and Verzele, M. (1981). New method for quantitative essential oil analysis. *J. Chromatogr.* **203**, 325–335.

Grob, K. and Zürchner, F. (1976). Stripping of trace organic substances from water, equipment and procedure. *J. Chromatogr.* **117**, 285–294.

Hardell, L., Axelson, O. and Rappe, Ch. (1983). Nasal cancer and chlorophenols. *Lancet*, 1167.

Heil, T. P. and Lindsay, R. C. (1988). A method for quantitative analysis of flavor-tainting alkylphenols and aromatic thiols in fish. *J. Environ. Sci. Health* **B23**, 475–488.

Hiatt, M. H. (1983). Determination of volatile organic compounds in fish samples by vacuum distillation and fused silica gas chromatography–mass spectrometry. *Anal. Chem.* **55**, 505–516.

Jennings W. G. (ed.) (1981). *Application of Glass Capillary Gas Chromatography*. Chromatographic Science Series, vol. 15. Marcel Dekker, Inc., New York.

Jetten, J. (1985). Simple and versatile trapping and reinjection technique for capillary columns. *J. High Resolut. Chromatogr. Chromatogr. Commun.* **8**, 696–697.

Likens, S. T. and Nickerson, G.B. (1964). Detection of certain hop oil constituents in brewing products. *Am. Soc. Brew. Chem. Proc.*, 5–13.

Lundgren, B. V., Borén, H., Grimwall, A., Sävenhed, R. and Wigilius, B. (1988). The efficiency and relevance of different concentration methods for the analysis of off-flavours in water. *Wat. Sci. Technol.* **20** (8/9), 81–89.

Lundgren, B., Borén, H., Grimwall, A. and Sävenhed, R. (1989). Isolation of off-flavour compounds in water by chromatographic sniffing and preparative gas chromatography. *J. Chromatogr.* **482**, 23–34.

Maarse, H. (1991). Introduction. In *Volatile Compounds in Foods and Beverages*. Ed. H. Maarse. Marcel Dekker, Inc., New York, pp. 1–39.

Maarse, H. and Belz, R. (eds.) (1981). *Isolation, Separation and Identification of Volatile Compounds in Aroma Research*. Akademie Verlag, Berlin and Reidel, Dordrecht.

Maarse, H. and Kepner, R. E. (1970). Changes in composition of volatile terpenes in Douglas fir needles during maturation. *J. Agric. Food Chem.* **18**, 1095–1101.

Maarse, H. and van den Berg, F. (1989). Current issues in flavour research. In *Distilled Beverage Flavour: Recent Developments*. Eds J. R. Piggott and A. Paterson. Ellis Horwood, Chichester, pp. 1–15.

Maarse, H., Nijssen, L. M. and Angelino, S. A. G. F. (1988). Halogenated phenols and chloroanisoles: occurrence, formation and prevention. In *Characterization, Production and Application of Food Flavours*. Ed. M. Rothe. Akademie Verlag, Berlin, pp. 43–61.

Meilgaard, M. C. and Peppard, T. L. (1986). The flavour of beer. In *Food Flavours. Part B. The Flavour of Beverages*. Eds I. D. Morton and A. J. MacLeod. Elsevier Science Publishers B.V., Amsterdam, pp. 99–170.

Murray, K. E. and Whitfield, F. B. (1975). The occurrence of 3-alkyl-2-methoxypyrazines in raw vegetables. *J. Sci. Food Agric.* **26**, 973–986.

Murray, K. E., Shipton, J. and Whitfield, F. B. (1972). The chemistry of food flavour. I Volatile constituents of passionfruit, *Passiflora edulis*. *Austr. J. Chem.* **25**, 1921–1933.

Nijssen, L. M. (1991). Off-flavours. In *Volatile Compounds in Foods and Beverages*. Ed. H. Maarse. Marcel Dekker, Inc., New York, pp. 689–735.

Nijssen, L. M., Jetten, J. and Badings, H. T. (1987). An unexpected off-flavour in small blocks of packaged cheese. In *Flavour Science and Technology*. Eds M. Martens, G. A. Dalen and H. Russwurm Jr. John Wiley & Sons Ltd., Chichester, pp. 127–132.

ten Noever de Brauw, M. C. (1981). The horizons of identification and analysis with mass spectrometry. In *Flavour '81*. Ed. P. Schreier. W. de Gruyter & Co, Berlin, pp. 253–286.

ten Noever de Brauw, M. C. and van Ingen, C. (1981). Mass spectrometry. In *Isolation, Separation and Identification of Volatile Compounds in Aroma Research*. Eds H. Maarse and R. Belz. Akademie Verlag, Berlin and Reidel, Dordrecht, pp. 155–171.

Núñez, A. J., Bemelmans, J. M. H. and Maarse, H. (1984). Isolation methods for the volatile components of grapefruit juice. Distillation and solvent extraction. *Chromatographia* **18**, 153–158.

Paasivirta, J., Klein, P., Knuutila, M., Knuutinen, J., Lahtiperä, M., Paukku, R., Veijanen, A., Welling, L., Vuorinen, M. and Vuorinen, P. J. (1987). Chlorinated anisoles and veratroles in fish. Model compounds, instrumental and sensory determinations. *Chemosphere* **16**, 1231–1241.

Reineccius, G. A. (1979). Off-flavours in meat and fish—a review. *J. Food Sci.* **44**, 12–21.

Sävenhed, R., Bóren, H. and Grimwall, A. (1985). Stripping analysis and chromatographic sniffing for the source identification of odorous compounds in drinking water. *J. Chromatogr.* **328**, 219–231.

Saxby, M. J. (1973). Taints in foodstuffs. Detection and identification. *FMF Rev.* **8** (11), 19, 20, 24 and 26.

Saxby, M. J. (1992). Personal communication.

Schaefer, J. (1981). Isolation and concentration from the vapour phase. In *Isolation, Separation and Identification of Volatile Compounds in Aroma Research*. Eds H. Maarse and R. Belz. Akademie Verlag, Berlin and Reidel, Dordrecht, pp. 37–59.

Schieberle, P. (1992). Important odorants of fresh and stored beer. In *Proc. 3rd Wartburg Aroma*

Symposium. Eds M. Rothe. and H.-P. Kruse. Deutsches Institut für Ernärungsforschung, Potsdam–Rehbrücke, pp. 137–145.

Schreier, P. (1988). On-line coupled HRGC techniques for flavour analysis. In *Characterization, Production and Application of Food Flavours.* Ed. M. Rothe. Akademie Verlag, Berlin, pp. 23–42.

Schultz, T. H., Flath, R. A., Mon, T. R., Eggling, S. B. and Teranishi, R. (1977). Isolation of volatile components from a model system. *J. Agric. Food Chem.* **25**, 446–449.

Sharpe, F. R. and Chappell, C. G. (1990). An introduction to mass spectrometry and its application in the analysis of beer, wine, whisky and food. *J. Inst. Brew.* **96**, 381–393.

van Straten, S. (1981). Gas chromatography. In *Isolation, Separation and Identification of Volatile Compounds in Aroma Research.* Eds H. Maarse and R. Belz. Akademie Verlag, Berlin and Reidel, Dordrecht, pp. 59–94.

Ulrich, F. and Grosch, W. (1987). Identification of the most intensive volatile flavour compounds formed during autoxidation of linoleic acid. *Z. Lebensm. Unters. Forsch.* **184**, 277–282.

Veijanen, A., Lahtiperä, M., Paukku, R., Kääriäinen, H. and Paasivirta, J. (1983). Recent development in analytical methods for identification of off-flavour compounds. *Wat. Sci Technol.* **15**, 161–168.

Weurman, C. (1969). Isolation and concentration of volatiles in food odour research. *J. Agric. Food Chem.* **17**, 370–384.

Whitfield, F. B. and Shaw, K. J. (1985). Analysis of food off-flavours. In *Progress in Flavour Research.* Ed. J. Adda. Elsevier Science Publishers B. V., Amsterdam, pp. 221–238.

Whitfield, F. B., Freeman, D. J. and Bannister, P. A. (1981). Dimethyltrisulphide: an important off-flavour component in the royal red prawn (*Hymenopenaeus sibogae*). *Chemistry and Industry (London)*, 692–693.

Whitfield, F. B., Freeman, D. J., Last, J. H., Bannister, P. A. and Kennett, B. H. (1982). Oct-1-en-3-ol and (5Z)-octa-1,5-dien-3-ol compounds important in the flavour of prawns and sand-lobsters. *Aust. J. Chem.* **35**, 373–383.

Whitfield, F. B., Shaw, K. J. and Nguyen, T. H. L. (1986). Simultaneous determination of 2,4,6-trichloroanisole, 2,3,4,6-tetrachloroanisole and pentachloroanisole in foods and packaging materials by high-resolution gas chromatography–multiple ion monitoring mass spectrometry. *J. Sci. Food Agric.* **37**, 85–96.

Wigilius, B., Borén, H., Carlberg, G. E., Grimwall, A., Lundgren, B. V. and Sävenhed, R. (1987). Systematic approach to adsorption on XAD-2 resin for the concentration and analysis of trace organics in water below the µg/kg level. *J. Chromatogr.* **391**, 169–182.

4 Off-flavors in raw and potable water

I. H. SUFFET, J. HO and J. MALLEVIALLE

4.1 Introduction

Today, in North America and Western Europe, the quality and availability of drinking water is generally good. However, customer complaints concerning taste and odor are a continuing concern of the water supply industry because they are associated with poor drinking water quality by consumers. Many tastes and odors are caused by low concentration levels of organic compounds. Some of these compounds are naturally occurring, while others are of industrial origin. Of primary concern is the identification of organoleptic compounds in water supplies to enable their efficient removal. At present this is primarily achieved through knowledge of the compounds present in the water. However, there is a general lack of identification of compounds that cause tastes and odors, especially if the compounds are present in complex mixtures at or below individual odor threshold concentrations. This chapter will evaluate the present state of knowledge of off-flavors in raw and potable water.

4.2 Off-flavors in drinking water and their relation to drinking water standards

Drinking water standards have been developed with the philosophy that the best raw water source is for drinking water. As the population increases, less than optimum new sources of drinking water have to be considered. Consequently, more extensive evaluations of water quality have become necessary to choose the best available site and to determine the optimum sequence of drinking water unit operations.

Drinking water quality standards are separated into microbiological, radionuclide, inorganic and organic chemical areas (Pontius, 1990). The usual situation for the United States, Canada and Western Europe is:

  (i) radionuclides are not a problem;
 (ii) the water source is not vulnerable to direct sewage discharge; and
(iii) inorganic or organic chemicals that would exceed the standards are not usually present without an accidental spill condition.

The water treatment plant design is then determined by considerations of disinfection, turbidity, trihalomethanes (THMs) of $< 100\,\mu g/l$, and secondary drinking water standards of color, taste and odor (EPA, 1977).

The EPA (1977) odor standard is measured by the threshold odor number (TON). The TON is the number of times a water must be diluted with odor-free water before the odor is just barely perceptible. The TON value should be $\leqslant 3$ to ensure public acceptance. The Council of European Communities (1980) recommends 'dilution numbers' as 0 for both taste and odor, but specifies maximum admissible dilutions of 2 at 12°C and 3 at 25°C for both taste and odor. The World Health Organization (1981) goal is that neither the taste nor odor of water shall be objectionable to 90% of the population.

The water quality characteristics that are used to describe organic chemicals in water are measures of:

(i)   naturally occurring organic matter (e.g. UV absorption and dissolved organic carbon (DOC));
(ii) disinfectant use (e.g. chlorine demand and ozone utilized);
(iii) hazardous potential (e.g. trihalomethane formation potential (THMFP));
(iv) hazardous chemicals (e.g. chloroform); and
(v)   aesthetics (e.g. color, taste and odor)

River water that is used for drinking can contain organic matter as high as 10 mg/l as DOC. It is estimated that 90% of the DOC in drinking water supplies is high-molecular-weight (> 1000 daltons) natural humic material. The remaining 10% of the DOC is characterized as low-molecular-weight soluble organics (Thurman, 1986). Many chemical identifications of the low-molecular-weight fraction, which is non-polar, hydrophobic and volatile (boiling point < 400°C) have been made. It is this fraction that contains compounds that cause taste and odor problems in drinking water.

The aesthetic quality of drinking water as a secondary drinking water standard and as an indirect measure of organic matter indicates that off-flavors are a major concern of potable water treatment. The causes of tastes and odors in water supply and their treatment are evaluated in this chapter.

4.3 Classifying tastes and odors

A new classification scheme based on the flavor profile analysis (FPA) sensory method is being used for characterizing drinking water (Brady et al., 1988). The FPA relies on reference standards to identify and estimate the magnitude of a taste or odor. The 17th edition of Standard Methods for the Examination of Water and Wastewater (APHA, 1989) presents three methods for evaluating taste and odor: (1) the flavor threshold test (FTT), also called threshold odor number; (2) the flavor rating scale (FRS); and (3) the flavor profile analysis (FPA).

The FPA method is based on using trained panelists to examine water samples. Panelists identify taste and odor qualities and rate the strength of each taste or odor separately. Each odor or flavor is assigned a quantitative rating on a seven-point scale, e.g.

)(threshold
2   very weak (very slight)
4   weak
6   weak to moderate (slight to moderate)
8   moderate
10 moderate to strong
12 strong

FPA is currently used at several large water utilities—the Metropolitan Water District of Southern California (MWDSC), the Philadelphia Water Department (PWD) and the Philadelphia Suburban Water Department (PSWD), in the United States, and the Lyonnaise des Eaux (LDE) in France. FPA was included for the first time in the 17th edition of Standard Methods for the Analysis of Water and Wastewater, although it is routinely used in the food industry.

Bruvold (1989) has succinctly described the differences between the TON and FPA methods and stressed their specific strengths and weaknesses. FPA does not require dilution of the sample while TON does. Since dilution changes the nature of the sample (Mallevialle and Suffet, 1987), it is impossible to directly compare results. Dilution may be a disadvantage of the TON method because the chemical which produces an objectionable odor in the undiluted sample may not be the chemical which controls the dilution ratio required to reach the threshold odor of the sample. There may be a relatively innocuous chemical in the sample, which requires more dilutions before it becomes undetectable. Consequently, the TON may be controlled by a chemical that does not offend consumers. In fact, the objectionable odor may be the result of an interaction among chemicals rather than one specific chemical. Dilution may change the nature of the interaction.

TON does not correlate well with public acceptability of drinking water (Mallevialle and Suffet, 1987). The correlation between FPA results and public acceptability has not been indicated by many studies. The MWDSC's experience is that customer complaints occur when a slight rating exists for objectionable odorant (Mallevialle and Suffet, 1987).

According to Bruvold (1989), the choice of the sensory technique depends on the objective. The best application of TON is for detection threshold determination or to determine if dilution of a water will reproduce a better flavor, while FPA is more appropriate for planning, monitoring and evaluating water treatment processes.

Both the TON and FPA methods provide a quantitative rating of the strength of the overall odor in the drinking water, but only FPA provides a description

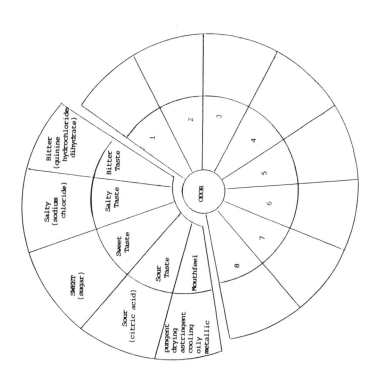

Odour wheel classifications with reference standards

Tier 1	Tier 2	Reference standards
1. Earthy, musty	Earthy Mildewy, damp basement Musty	Geosmin/2-methyl isoborneol 2,3-Diethylpyrazine 2-Isobutyl and 2-isopropyl 3-methoxypyrazines
2. Chlorinous	Chlorinous Bleach	Monochloramine Hypochlorous acid
3. Grassy, hay, woody	Fresh grass	Cis-3-hexen-1-ol
4. Marshy, swampy, septic, sulfurous	Decaying vegetation Septic Putrid, sickening Rancid oily	Dimethyl sulfide Dimethyl disulfide Butyric acid Heptanal
5. Fragrant, flowery, fruity	Cucumber Geranium Citrusy Sweet almond Vanilla/black cherry	Trans-2-cis-6-nonadienal Diphenyl ether d-Limonene Benzaldehyde Coumarin
6. Fishy, algae		
7. Medicinal, phenolic, alcoholic	Alcoholic Medicinal	Butanol Chlorophenols
8. Chemical, hydrocarbon, miscellaneous	Solvently Plastic Painty Glue Detergenty Organic chemical	Cumene, indene Methylmethacrylate Methylisobutyl ketone Toluene, styrene 1-Dodecanol Xylene

Figure 4.1 Taste and odor wheel classification for drinking water, with odor reference standards (Burlingame *et al.*, 1991). Reprinted from AWWA-WQTC, Orlando, Florida, by permission. Copyright © 1991, American Water Works Association.

of the type of odor or taste. A description of the type of odor may provide clues as to the cause of the odor. For example, an earthy/musty odor may be associated with 2-methyl isoborneol (MIB) or geosmin, which are known to be microbial metabolites with earthy/musty characters.

Classification of odors and odor-causing agents is difficult because numerous substances (e.g. algae or chemical compounds) can impart the same odor to water (Amoore, 1962). For example, metabolites of an alga such as *Synura* produce an odor that progresses from cucumber to fishy as the abundance of alga increases, and chemicals change their odor character at low and high concentrations, e.g. MIB changes from musty to camphor. Work has begun to assign descriptors and to determine the specific chemicals that cause odors or off-flavors commonly found in drinking water (Mallevialle and Suffet, 1987).

The Philadelphia Water Department began developing a list of odor reference standards in 1984 (Bartels *et al.*, 1986). The standards represented odors typically occurring or having the potential to occur in the untreated or finished drinking water. The selection of compounds to be used as standards was divided into categories.

I.   Knowns: Standards of single chemicals that have occurred or can occur in the untreated or finished drinking water. Examples include free chlorine (chlorinous), geosmin (earthy), MIB (musty), *trans*-2-*cis*-6-nonadienal (cucumber) and chlorinated phenols (medicinal).

II. Representatives: Standards of single chemicals that may not be actually present in water but have odor qualities similar to those found in water, e.g. *cis*-3-hexen-1-ol (grassy).

III. Substitutes: Standards made from natural materials producing an odor that occurs in water and can be created in a consistent manner. Examples include decayed vegetation and septic odors made from aged solutions of grass.

Studies using FPA have shown that different water utilities are agreeing on a common set of classes of odor description for untreated and finished drinking waters, as cross-panel comparisons indicate that trained panelists from various utilities can agree on descriptors (Bartels *et al.*, 1987). This led to the evolving development of a common classification scheme (Figure 4.1) as a taste and odor wheel for drinking water (Mallevialle and Suffet, 1987; Suffet *et al.*, 1988; Burlingame *et al.*, 1991). The wheel highlights those odors that have been documented to cause off-tastes and odors in water supplies.

The taste and odor wheel for drinking water is an evolving concept. Refinement of the odor wheel will be achieved with further development of odor reference standards, and the active participation of the water supply industry that uses the FPA method. Only known and representative odors are shown on Figure 4.1 (Burlingame *et al.*, 1991).

4.4 The causes of tastes and odors in water supplies

Taste and odor problems are naturally produced, industrially-made, produced as disinfection by-products during water treatment disinfection, or produced by substances that leach from water pipes or storage facilities (Bartels *et al.*, 1986). Table 4.1 lists compounds that can cause tastes and odors in drinking water. Naturally occurring tastes and odors are produced by microscopic organisms, notably algae and bacteria, in both surface water and groundwater supplies, as well as during finished water storage and water distribution. These tastes and odors include geosmin and 2-methyl isoborneol, with the threshold odor concentrations in the ng/l range. Table 4.1 outlines these, and other microbiological causes of tastes and odors in drinking water.

Table 4.1 Compounds causing tastes and odors in water

Compound	Source	Odor	Reference
Geosmin	Actinomycetes Blue-green algae Cyanobacteria	Earthy	Gerber and Lechevalier (1965); Gerber (1967, 1968, 1979); Saffermann *et al.* (1967); Medsker *et al.* (1968); Rosen *et al.* (1970); Henley (1970) Piet *et al.* (1972); Kikuchi *et al.* (1973a, b); Narayan and Núñez (1974); Tabachek and Yurkowski (1976); Tsuchiya *et al.* (1978, 1981); Persson (1979a); Izaguirre *et al.* (1982); Berglind *et al.* (1983); Wood *et al.* (1983); Burlingame *et al.* (1986); Wu and Juttner, (1988a, b); Matsumoto and Tsuchiya (1988); Izaguirre (1992); van Breemen *et al.* (1992)
2-Methyl isoborneol	Actinomycetes Blue-green algae Cyanobacteria	Musty	Rosen *et al.* (1968); Gerber (1969); Medsker *et al.* (1969); Collins *et al.* (1970); Tsuchiya *et al.* (1978); Izaguirre *et al.* (1982, 1983); Negoro *et al.* (1988); Wu and Juttner (1988b) Martin *et al.* (1991); Izaguirre (1992)
Cadinene-ol	Actinomycetes	Woody-earthy	Collins (1971); Gerber (1971)
Isopropyl methyl pyrazine	Actinomycetes	Potato-bin Musty	Buttery and Ling (1973); Gerber (1983)
Trans-2-cis-6- nonadienal	Algae	Cucumber	Burlingame *et al.* (1990)
Unknown	Algal decomposition by fungi and bacteria	Decaying vegetation	MacKenthum and Keup (1970)

Table 4.1 *(continued.)*

Compound	Source	Odor	Reference
n-Hexanal and *n*-heptanal	Flagellated algae Diatom	Fishy	Collins and Kalnins (1965a, b, 1966, 1967) Kikuchi *et al.* (1974, 1986)
Decadienal	Flagellated alga	Cod liver oil	Juttner (1981, 1983)
Hepta- and deca-dienals	Dinobryon alga	Fishy	Bayliss (1951); Juttner *et al.* (1986); Yano *et al.* (1988)
Mercaptans	Decomposed or living blue-green algae	Odorous Sulfur	Jenkins *et al.* (1967); Slater and Block (1983)
Dimethyl polysulphides	Bacteria	Swampy/fishy	Giger and Schaffner (1981); Wajon *et al.* (1985a, b, c); Krasner *et al.* (1986)
Chlorophenols	Phenol chlorination/ chloramination	Chlorophenolic	Burtschell *et al.* (1959); Lee (1967); Bryan *et al.* (1973); White (1980); Krasner and Barrett (1984); Krasner *et al.* (1986)
Aldehydes (low molecular weight)	Chlorination of amino acids	Swampy/ swimming pool	Hrudey *et al.* (1989)
Iodinated trihalomethanes	Chloramination	Medicinal	Bruchet *et al.* (1989); Gittelman and Yohe (1989)
Unknown	Algae (moderate to large amounts)	Fishy Grassy/septic	Palmer (1962, 1977, 1980)
Hydrogen sulphide	Anaerobic bacteria (reduce SO_4 to S^-)	Rotten egg	Hack (1981)
Aldehydes (higher molecular weight)	Ozonation	Fruity/fragrant	Anselme *et al.* (1985a, 1988); Suffet *et al.* (1986)
Phenolic antioxidants	Polyethylene pipes	Plastic/burnt plastic	Burman and Colbourne (1979); Anselme *et al.* (1985b, c)

Updated from Ibrahim *et al.* (1990) in *Chemical Water and Wastewater Treatment*, eds H. H. Hann and R. Klute by permission of Springer-Verlag

Groundwater off-flavors are caused either by chemicals contributed to an aquifer by the geologic formation, or by biological activity. High concentrations of minerals can alter the taste of water markedly, making it, for example, salty or astringent-tasting. Biologically induced tastes and odors are generated by reducing bacteria that occur in some groundwaters to form hydrogen sulfide, with a typical rotten-egg odor.

Surface water supplies produce off-flavors by:

(i) algae and other aquatic plants during growth and decay, e.g. algae and decaying vegetation odors;

(ii) thawing soils, which carry synthetic organic compounds to some surface water supplies, e.g. oily chemicals; and

(iii) chemicals in run-off, point-source discharges, and accidental spills from industries.

Table 4.1 outlines odors caused from organic compounds that originate from these sources. This is not a complete list, but is representative.

4.5 Cause-and-effect relationships in drinking water taste and odor problems

The taste and odor literature is filled with presumptive statements about the causes of taste and odor problems, which can mislead the water industry. For example, the numbers of algae or actinomycetes may not affect a taste and odor problem, and the cause of the earthy/musty odor in water is not necessarily geosmin or MIB. The chemical analysis of the water can only prove that geosmin or MIB is present, while the evaluation of the metabolites of the microorganisms can only prove they cause the problem. Rules of evidence must be satisfied before a cause-and-effect relationship can be established between the presence of an organism or chemical, and a taste or odor problem in the water. The following rules of evidence have been suggested to enable more precise communication of information in future studies (Mallevialle and Suffet, 1987).

(1) *Microbiological causes*
   (a) Presumptive test: identify the organism believed to be the cause of the problem and quantify its population density.
   (b) Confirmed test: (1) isolate the organism by culture technique and determine the odor produced by the culture; (2) isolate microbial products; and (3) isolate chemicals from the culture, and test by gas chromatography–mass spectroscopy for metabolites that cause that odor.
(2) *Chemical causes*
   (a) Presumptive test: develop a statistical correlation between the chemical compound in the water sample and the odors determined by sensory panel techniques.
   (b) Confirmed test: (1) determine by the sensory panel technique if the chemical in a reference water does cause the odor response. Develop a graphical plot of log concentration versus odor response (Weber–Fechner Law); and (2) develop a relationship between the chemical concentration in actual water samples and the odor determined by the sensory panel technique.

This procedure describes 'the scientific method,' regarding presumptive and confirmatory testing procedures to validate the cause of a taste or odor in drinking water.

4.6 Specific taste and odor-causing compounds

The taste and odor wheel shown in Figure 4.1 highlights those tastes and odors that have been confirmed to cause drinking water tastes and odors. This section of the chapter will describe the primary causes of taste and odor problems that are highlighted in Figure 4.1 and initially described in Table 4.1.

4.6.1 Geosmin

Gerber and Lechevalier (1965) were the first to give an accurate description of the major actinomycete metabolite responsible for producing the earthy odor in water supplies. They also reported that the compound, when treated with strong acid, changed into an odorless hydrocarbon that they named 'argosmine'. Gerber (1968) and Medsker et al. (1968) further elucidated the chemical structure of geosmin.

Rosen and co workers (1970) first reported the isolation of geosmin from natural water. Geosmin was isolated from the *Streptomyces* genus, as well as from strains of *Nocardia, Micromonospora, Microbispora, Oscillatoria, Aphanizomenon* and *Phormidium*. In addition, geosmin has been identified as a metabolite of many genera of algae in drinking water (see Table 4.1).

4.6.2 Methyl isoborneol (MIB)

Medsker et al. (1969) and Gerber (1969) identified with taste and odor episodes in natural waters, a musty-smelling compound, 2-methyl isoborneol, which was recovered from pure cultures of actinomycetes. Rosen et al. (1970) isolated MIB not only from pure actinomycete cultures but also from a natural water. Although a variety of other odorous compounds have been isolated from actinomycete cultures, geosmin and MIB have been most frequently identified with taste and odor episodes. In addition, blue-green algae produce MIB as well as geosmin (see Table 4.1). Strains of *Oscillatoria tenuis, Uroglena americana* and *Phormidium* have been identified as organisms that produce MIB.

4.6.3 Chlorine

Aqueous chlorine solutions include a variety of species such as hypochlorous acid (HOCl) and hypochlorite ion (OCl$^-$); when ammonia is present, the aqueous solution can contain several chloramine species, e.g. dichloramine (NHCl$_2$), monochloramine (NH$_2$Cl) and trichloramine (NCl$_3$). In general, hypochlorous acid and hypochlorite ions have the same odor descriptor—'bleach'. Hypochlorous acid has a pK$_a$ of 7.6 and ionizes to hypochlorite ion and hydronium ion. Hypochlorous acid (pH < 6) has an odor threshold level of 0.28 mg/l and a taste threshold level of 0.24 mg/l. Hypochlorite ion (pH > 9) has an odor threshold level of 0.36 mg/l and a taste threshold level of 0.30 mg/l (Krasner and Barrett, 1984) (Table 4.2). The chlorination process and its by-products formation are affected by pH. A previous study by Bryan *et al.* (1973) showed a slightly lower threshold intensity for the flavor of the hypochlorite ion. In general, both studies agree that hypochlorous acid has a lower threshold level than the hypochlorite ion, and both groups describe similar chlorinous flavors and aromas.

Table 4.2 Sensory threshold values (mg/l as Cl_2)

Compound	Descriptor	Aroma	Flavor
Hypochlorous acid	Bleach	0.28	0.24
Hypochlorite ion	Bleach	0.36	0.30
Monochloramine	Swimming pool	0.65	0.48
Dichloramine	Swimming pool	0.15	0.13
Trichloramine	Geranium	0.02	

Source: Krasner and Barrett (1984)

4.6.4 Chloramine

Figure 4.2 shows the breakpoint chlorination curve with different ratios of chlorine and ammonia. The dominant chlorine species in water prior to the breakpoint are inorganic chloramines (combined chlorine residual). Monochloramine, dichloramine and trichloramine are the three major inorganic chloramines. The threshold odor and taste levels of monochloramine are 0.65 mg/l and 0.48 mg/l, respectively (Table 4.2). According to data from the Metropolitan Water District of Southern California, monochloramine at concentrations of 0.5 to 1.5 mg/l has an intensity level of)((i.e. threshold) to 2.0 (very slight) on the seven-point flavor profile analysis scale. Only some panelists can perceive monochloramine at this level. In fact, monochloramine rarely causes taste and odor problems in drinking water unless its concentration exceeds 5 mg/l.

The odor and taste threshold levels of dichloramine are much lower than monochloramine (Table 4.2). The descriptors for dichloramine at higher concentration are either 'swimming pool' or 'bleach'. At concentrations from 0.1 to 0.5 mg/l in water, dichloramine has an odor intensity level of 4 (slight) to 8 (moderate), which is in the 'acceptable' range on the flavor profile scale. However, if the concentration of dichloramine reaches 0.9 to 1.3 mg/l, the odor will be described as moderate to very strong, which is offensive and not acceptable. The empirical threshold flavor limit of dichloramine in water is 0.5 mg/l (Krasner and Barrett, 1984).

The monochloramine and dichloramine concentrations prior to the breakpoint are mainly determined by chlorine:ammonia-nitrogen ratio in water. Once the chlorine dose is above the breakpoint, the free chlorine residual will

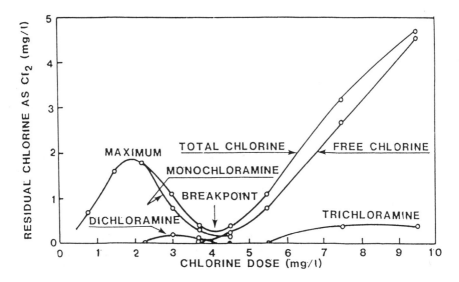

Figure 4.2 Chlorine residual as a function of chlorine dose (Krasner and Barrett, 1984). NH₃–dose 0.5 mg/l; temperature 25°C; pH 7.8→7.4; in-the-dark contact time 60 min. Reprinted from AWWA-WQTC, Denver, Colorado, by permission. Copyright © 1984, American Water Works Association.

increase rapidly and dominate the other chlorine species in water (Figure 4.2). The results from different studies conclude that a low pH solution is a preferred environment for the formation of more odorous chlorine residuals such as dichloramine. Consequently, water treatment plants should maintain a higher pH to control the chlorinous taste and odor at the end of the treatment process. For example, the Metropolitan Water District of Southern California usually maintained a pH of approximately 8 in the treated water (Krasner and Barret, 1984).

Trichloramine produces a geranium-like odor, which could be described as chlorinous and fragrant. The odor threshold concentration of trichloramine is 0.02 mg/l, which is much lower than the other chloramine species (Table 4.2). Fortunately, the rate of trichloramine formation is catalyzed by H^+. The rate of trichloramine formation increases as the pH decreases below pH 3. Such a low pH level is far below normal in water treatment processes. As a result, the taste and odor problem caused by trichloramine is insignificant under normal conditions.

4.6.5 Chlorination by-products

Some taste and odor problems that develop in water utilities are indirect consequences of chlorination. There are two important reaction mechanisms of chlorination: (i) oxidation; and (ii) substitution. In general, oxidation is the dominant mechanism in water. The substitution reaction rate is low unless the reacting chemical is activated by some other means. The disinfection by-products from chlorine and trace amounts of organic compounds present in water can be classified into three major groups: (i) organic nitrogen and aldehydes; (ii) phenols and chlorophenols; and (iii) halomethanes, containing iodine.

4.6.5.1 Aldehydes. Amino acids are responsible for a large part of the surface water chlorine demand. The occurrence of amino acids is associated with biological activity. The breakpoint curves for most amino acids are similar to that of ammonia (NH_3) described in the chloramination process. Amino acids are important precursors for the aldehydes and nitriles as a result of oxidation during drinking water treatment. For example, phenylalanine and glycine react with hypochlorite to produce phenylacetaldehyde and formaldehyde, respectively. A comparison of aldehyde generation by chlorine, chloramine and chlorine dioxide shows the lowest production of aldehydes by chlorine dioxide, and the highest by chlorine after 10 h of reaction time (Hrudey *et al.*, 1989, 1990). This result is consistent with the conclusion that chloramination is a better alternative than chlorination for taste and odor control.

The odor characteristics of aldehydes are complicated since a single aldehyde may have many descriptors to an untrained consumer panel. The most common descriptors for a variety of aldehydes are chlorine, earthy, stale, disinfectant, bitter, ammonia, organic, muddy, moldy and bleach-like (Hrudey *et al.*, 1988). The μg/l odor threshold concentrations of some low-molecular-weight aldehydes, which are more prevalent in treated water than raw water, are summarized in Table 4.3 (Hrudey *et al.*, 1988). The specific odor of each of these aldehydes alone and in mixtures needs to be defined by a sensory panel. Following this, the relationship between sensory and chemical analysis can be correlated. Subsequently, appropriate odor and chemical removal strategies can be developed.

Table 4.3 Aldehydes detected in drinking water after chlorination (swampy and swimming pool) (Hrudey *et al.*, 1988)

Compound	Threshold level (ug/l)	Reference
2-Methyl propanal	0.9	Guadagni *et al.* (1963)
	1.0	Guadagni *et al.* (1972)
	2.3	Amoore *et al.* (1976)
3-Methyl butanal	0.15	Guadagni *et al.* (1963)
	0.2	Guadagni *et al.* (1972)
	2.0	Amoore *et al.* (1976)
2-Methyl butanal	12.5	Amoore *et al.* (1976)
Phenylacetaldehyde	4.0	Buttery *et al.* (1971)

4.6.5.2 Phenols and chlorophenols. Phenols are important taste and odor compounds, which react with chlorine to produce chlorophenols. The majority of phenols in a source for drinking water come from urban run-offs and industrial wastes.

Phenol has a typical medicinal taste and odor. The odor threshold concentration is more than 1000 ppb (Burtschell *et al.*, 1959) (Table 4.4); however the 2-chlorophenol, 2,4-dichlorophenol and 2,6-dichlorophenol have threshold odor and taste concentrations lower than 10 ppb. The reaction pathways and products of chlorination of phenols are affected by factors such as chlorine:phenol ratio, chlorine concentration, temperature, ammonia concentration and pH value in a water treatment plant. The chlorine:phenol ratio is the major determinant of the phenolic by-products of the chlorination process. The maximum taste level is developed at a 2:1 chlorine:phenol ratio because the odorous 2,6-dichlorophenol is the dominant chlorination by-product at this ratio (Table 4.4). When the chlorine:phenol ratio increases to 4:1, and the chlorine concentration increases to 10 ppm, none of these taste and odor compounds can be detected. Consequently, the chlorine concentration and chlorine:phenol ratio are critical in determining the aromatic characteristic of chlorination by-products (Ettinger and Ruchhoft, 1951; Lee, 1967).

Table 4.4 Taste and odor threshold concentrations (Burttschell *et al.*, 1959)

Component	Geometric mean thresholds (ppb)	
	Taste	Odor
Phenol	> 1000	> 1000
2-Chlorophenol	4	2
4-Chlorophenol	> 1000	250
2,4-Dichlorophenol	8	2
2,6-Dichlorophenol	2	3
2,4,6-Trichlorophenol	> 1000	> 1000

All tests were made at room temperature, about 25°C

The formation of odorous chlorophenol is highly dependent on pH. There is no significant development of chlorophenolic taste at a pH less than 7. The optimum pH value for development of chlorophenols is between pH 8 and 9. When ammonia is present in the phenol solution, it consumes free chlorine and lowers the free chlorine residual level. Consequently, ammonia can inhibit the formation of chlorophenols and reduce the taste and odor problems associated with chlorophenols. Since the production rate of chlorophenols is reduced by the chloramination process, the best alternative to control chlorophenols and their odors is to substitute chlorine with chloramine as the disinfectant. However, once chlorophenols do form, breakpoint chlorination is necessary to destroy the chlorophenols at a chlorine:phenol ratio of 4:1 or higher.

4.6.5.3 Iodomethanes. The formation of iodomethanes in drinking water is related to the organic content of raw water and the chlorination process. Free chlorine can react with and oxidize both organic and inorganic compounds in water.

Trace amounts of bromide and iodide at concentration levels of 0.1 mg/l in the water supplies can be converted into bromine and iodine by chlorination. The natural humic material in water reacts by the haloform reaction to produce noxious brominated and iodinated haloforms. The odor and taste threshold concentrations of iodoform are 20 ng/l (Bruchet *et al.*, 1989) and 5 µg/l (Hansson *et al.*, 1987), respectively. As a result, the presence of iodinated haloforms at concentrations between 0.02 ppb and 10 ppb will cause medicinal taste and odor problems in drinking water (Bruchet *et al.*, 1989; Gittleman and Yohe, 1989).

Prechlorination before ammoniation, to produce chloramines during the water treatment process, favors the predominance of bromoform and chloro-bromomethane in the treated water; only trace amounts of iodoforms will appear. On the other hand, if ammonia is added first or at the same time as chlorine, then the predominant species is iodoform, which causes medicinal odor in water (Hansson *et al.*, 1987). One study showed that 8 µg/l of iodoform was produced by the sequence of NH_3/Cl_2, compared with less than 1 µg/l of iodoform, which was produced by the sequence of Cl_2/NH_3 (Hansson *et al.*, 1987).

4.7 Treatment of specific odorous chemicals

Table 4.5 shows a set of specific odorous chemicals that have been studied by water treatment processes. Table 4.6 shows the percent reductions of the organoleptic compounds from conventional water treatment unit process (Baker *et al.*, 1986). It illustrates that compared with all the disinfection methods, powdered activated carbon (PAC) has the highest chemicals removal efficiency. Chlorine is only efficient in removing diphenylamine (49%). On the other hand, chloramines are not efficient in removing any of these chemicals. Table 4.7 shows similar results for 1-hexanal, 1-heptanal, MIB and geosmin (Glaze *et al.*, 1990). Dimethyl trisulfide (a swampy odorant) is well removed by all of the oxidants (Wajon *et al.*, 1988). The unsaturated deca-dienal (a cucumber odorant) found in water supplies (Burlingame *et al.*, 1990) is removed by oxidation by chlorine or monochloramine.

Table 4.5 Test compounds and treatment reagents used

Compound	Odor type	Reference	Aqueous ppb concentration		OTC (µg/l)	OTC reference
			Expts. #1 and 2	Expt. #3		
Toluene	Model glue	APHA (1989)	10	20	170 1000	Mallevialle and Suffet (1987)[b] Zoeteman (1971)[b]
Ethylbenzene			10	20	200 140 100 29	Cherkinski (1961)[b] Rosen (1963)[b] Zoeteman (1971)[b] Amoore and Hautala (1983)
1,4-Dichlorobenzene			10	20	11 30 3	Amoore and Hautala (1983) Kolle (1972)[b] De Grunt (1975)[b]
2-Ethyl-1-hexanol	Sweet (medicinal, mint, menthol) Shoe polish	Suffet (1992)	50	100	27 1280 198	Rosen (1963)[b] Lillard (1975)[b] Suffet (1992)
2-Isopropyl-3-methoxypyrazine	Potato bin, earthy-musty	Seifert et al. (1970)	15	30	0.002	Seifert et al. (1970)
Nonanal	Fruity/sweet		25	50	0.5 0.08	Forss (1962)[b] Buttery (1971)[b]
2-Methyl isoborneol	Earthy-musty Camphorous	Persson (1979a) Persson (1980)	50	50	0.02 0.029	Zoeteman (1973)[b] Persson (1979a)
Trans-2-nonen-1-al	Cucumber with skin Woody, fatty, oily and cucumber	APHA (1989) Olhoff and Thomas (1971)[a]	25	50		
1,2,4-Trichlorobenzene			10	20	5	Kolle (1972)[b]

Compound	Odour description	Reference				Reference
Naphthalene	Mothballs	Amoore and Hautala (1983)	10	10	21 500 5	Amoore and Hautala (1983) Holluta (1960)[b] Zoetman (1971)[b]
2-Isobutyl-3-methoxypyrazine	Green bell pepper, earthy-musty	Buttery et al. (1969)	10	20	0.002 16	Buttery. (1969)[b] Takken (1975)[b]
2,3,6-Trichloroanisole	Musty	Guadagni and Buttery (1978)	10	20	0.007	Guadagni and Buttery (1978)
Geosmin	Earthy-musty	Zoeteman and Piet (1974)	10	10	0.01	Zoeteman and Piet (1974)
Diethylphthalate		APHA, (1989)	25	50		
Diphenylamine	Fishy Pleasant, floral	Harper et al. (1968)	10	20		

Treatment reagents

(1) Alum—20 mg/l as $Al_2(SO_4)_3$
(2) Chlorine—12 mg/l as Cl
(3) Chloramines—12 mg/l as Cl
(4) Chlorine dioxide—3 mg/l as Cl
(5) Potassium permanganate—2 mg/l as $KMnO_4$
(6) Powdered activated carbon—50 ml/l Aqua Nuchar

[a] McGugan, W. A. (1980) *Description of Flavor Chemicals. Food Research Reports*, Vol. 4, No.1
[b] Van Gemert, L. J. and Nettenbreijer, A. H. (1977). Compilation of odour threshold values in air and water. National Institute for Water Supply. Voorburg, Netherlands, Central Institute for Nutrition and Food Research, TNO, Zeist, Netherlands

Note: Mallevialle and Suffet (1987) are compilation and not original researchers.
Reprinted from Baker et al. (1988) in *Chemical Water and Wastewater Treatment*, eds Hahn, H. H. and Klute, R. by permission of Springer-Verlag

Table 4.6 Percent reductions of compounds from conventional water treatment unit processes (Baker et al., 1986)

Compound	Filtered blank	Coagulation	Chlorine	Chloramines	Chlorine dioxide	Permanganate	PAC
Toluene	10.0 ± 8.5	13.7 (n = 2)	16.7 ± 5.8	13.9 (n = 1)	19.5 ± 14.9	10.3 ± 26.7	95.2 ± 8.0
Ethylbenzene	9.1 ± 2.4	8.6 ± 1.7	13.9 ± 11.5	3.6 ± 5.7	3.8 ± 6.0	7.4 ± 6.0	95.1 ± 4.4
1,4-Dichlorobenzene	7.9 ± 2.0	7.1 ± 6.0	11.8 ± 13.3	1.4 ± 9.0	− 0.4 ± 4.5	6.9 ± 4.6	100.0
2-Ethyl-1-hexanol	− 2.0 ± 9.1	− 1.0 ± 9.1	4.3 ± 19.2	0.9 ± 7.0	− 3.2 ± 10.7	1.5 ± 18.1	73.9 ± 26.8
2-Isopropyl-3-methoxypyrazine	1.0 ± 1.5	1.8 ± 3.7	3.5 ± 2.9	2.2 ± 8.4	3.5 ± 2.0	9.1 ± 2.7	86.1 ± 12.8
Nonanal	−	−	−	−	−	−	−
2-Methyl isoborneol	4.0 ± 0.6	4.5 ± 1.2	3.7 ± 2.0	4.2 ± 4.2	3.6 ± 1.6	1.8 ± 5.1	53.7 ± 1.8
Trans-2-nonen-1-al	−	−	−	−	−	−	−
1,2,4-Trichlorobenzene	10.3 ± 3.6	− 6.9 ± 10.0	15.6 ± 17.2	− 2.7 ± 11.1	− 3.0 ± 11.0	10.3 ± 8.6	100.0
Naphthalene	7.3 ± 2.2	4.7 ± 7.0	6.8 ± 10.0	− 0.6 ± 6.6	1.3 ± 8.2	4.3 ± 8.6	100.0
2-Isobutyl-3-methoxypyrazine	5.0 ± 3.5	2.3 ± 6.5	8.5 ± 7.0	2.2 ± 3.3	8.9 ± 10.2	3.1 ± 3.4	93.1 ± 6.7
2,3,6-Trichloroanisole	− 3.7 ± 1.1	1.2 ± 9.9	7.8 ± 8.8	− 5.2 ± 7.1	− 3.4 ± 8.6	3.7 ± 6.5	100.0
Geosmin	5.5 ± 3.8	0.6 ± 10.9	10.0 ± 13.2	− 4.7 ± 11.3	− 3.2 ± 10.3	1.1 ± 5.5	79.2 ± 5.2
Diethylphthalate	− 3.7 ± 1.1	− 2.5 (n = 2)	− 7.6 ± 5.9	− 2.0 ± 3.2	1.3 ± 1.1	− 14.7 ± 21.3	100.0
Diphenylamine	− 5.2 (n = 2)	1.6 ± 25.6	49.0 ± 10.5	− 4.7 ± 37.1	36.3 ± 27.1	49.6 ± 28.3	37.5 ± 23.7

Nonanal and nonenal were unstable in the water sampled, therefore no results are available for these compounds

Table 4.7   Percent removal of model taste and odor compounds result from oxidation (Glaze *et. al.*, 1990)

Model	HOCl	ClO$_2$	NH$_2$Cl	KMnO$_4$	H$_2$O$_2$
1-Hexanal	5	− 28	− 79	33	43
1-Heptanal	− 79	− 83	− 60	49	47
Dimethyl trisulfide	> 99	> 99	> 99	82	90
2,4-Decadienal	54	35	57	95	52
MIB	10	2	15	13	29
Geosmin	16	17	27	15	31

Reaction time 120 min, oxidant dose 3 mg/l; negative percent-removal values indicate formation of the taste and odor compound on oxidation.

In a study of swampy odors at three water treatment plants in Australia, Wajon *et al.* (1988) showed an 'inverse relationship' between the concentration of dimethyl trisulfide (DMTS) and free chlorine in the water. This was probably due to DMTS being oxidized to non-odorous methyl sulfonyl chloride. The oxidation of DMTS by chlorine, and the implication presented that chloramines are ineffective was observed by Krasner and Barrett (1984) when the Metropolitan Water District of Southern California changed from chlorine to chloramines as their primary disinfectant. However, the DMTS was reduced during a fishy episode and not a swampy episode. The field data did not show the same thing as the lab data (Table 4.7), as monochloramine was not as efficient as chlorine for the reduction of DMTS.

Many known chemical odorants including geosmin (earthy), MIB (musty) and 2,3,6-trichloroanisole (musty, cork) cannot be removed efficiently by chlorination or chloramination as shown in Table 4.6. Free chlorine is needed before final chloramination to control swampy odors (DMTS) and the medicinal odor of iodoforms in drinking water treatment. Chlorophenols are minimized by using chloramination. However, once chlorophenols are formed, chlorine is needed to completely oxidize them and remove the medicinal odor of chlorophenols.

4.8 Taste and odor treatment of off-odors by chlorination and chloramination

It is important to define causal relationships between chemical constituents and specific odors. Sometimes, however, more than 100 organic compounds can exist in one raw water sample. Each of these chemicals may have an odor descriptor, and no single compound may explain the odor problem in a drinking water sample. Since the identification of all these chemicals by the current analytical methods such as GC/MS is very time consuming, one control strategy is to investigate only the specific odor and its removal efficiency. The off-odors removal efficiency by chlorination and chloramination have been studied in water treatment plants. This section will evaluate known chlorination and chloramination studies, based on actual odor removal as measured by a sensory panel.

Marshy and swampy odors, which are associated with anoxic waters, contain sulfur compounds, such as dimethy trisulfide, and have been investigated in Australia by Wajon *et al.* (1988). Sensory panels showed that chlorination efficiently removes the swampy odor from a groundwater of an unconfined aquifer. At a 2 mg/l chlorine dose the swampy odor disappears within 1 h. A swampy odor also occurred when the Metropolitan Water District of Southern California switched from chlorination to chloramination. A 2 mg/l dose of chlorine also appeared to eliminate the problem, confirming the work of Wajon *et al.* (1988).

Fishy odors have been related to natural processes of algae and microorganisms (Mallevialle and Suffet, 1987). High algal concentrations can produce fishy odors. However, *Ceratium* and *Volvox*—two genera of algae—can have fishy odors in moderate concentration. Although fishy odors are normally associated with a nitrogenous functional group, no chemical compounds have been confirmed to be the cause of a fishy odor episode (Suffet and Mallevialle, 1987). In general, fishy odor can be removed by chlorination. Chloramination appears ineffective against fishy odors. A fishy odor incident occurred when the Metropolitan Water District of Southern California switched over to chloramination. The odor was described as a fish, cod-liver oil odor. The odor occurred during periods of high levels of diatoms and algae in the raw water. The use of free chlorine before chloramination eliminated the problem. At the same time, the chlorination process eliminated the presence of DMTS in the water. Although DMTS has been associated with swampy water (Wajon *et al.*, 1988) here it was correlated with fishy odors.

Medicinal odor has different origins. Both phenols and haloforms can create the medicinal odor by reaction with chlorine and chloramines. The sequence of addition of chlorine and ammonia at the chloramination process can influence the intensity of medicinal odor. In the case of iodoforms, if ammonia is added before chlorine, the medicinal odor is stronger than if chlorine is added first, because the major product is iodoform instead of bromoform (Hansson *et al.*, 1987). Since most of the haloform precursors are consumed by the chlorination process, chlorination is more efficient in removing medicinal odor than is chloramination. Increasing the reaction time with chlorine before ammonia addition can also reduce the occurrence of medicinal odors.

Another study was performed using flavor profile analysis by sensory panels on three different water treatment plants: Neshaminy Water Treatment Plant (Philadelphia Suburban Water Company (PSWC)), Baxter Water Treatment Plant (Philadelphia Water Department (PWD)), and Morsang Water Treatment Plant (France). The common organoleptic descriptions of influent and effluent water samples at the three plants are listed in Tables 4.8, 4.9 and 4.10 (Bartels *et al.*, 1989). The disinfection practice at the Philadelphia Suburban Water Company treatment plant included chloramination. Treatment processes at the

Table 4.8 Common flavors and aromas at the PSWC Neshaminy Plant–Chloramination (Bartels *et al.*, 1989)

Sample location	Major response	Percent occurrence		Intensity (scale 0–12)
		Inc. notes*	Ex. notes*	
Effluent aroma	Chlorinous	47	34	3.6
	Musty	46	22	2.0
	Swimming pool	53	43	4.4
Effluent flavor	Chlorinous	84	72	2.2
	Metallic	56	38	2.1
	Astringent	56	28	2.0
	Bitter	47	22	2.7
Influent aroma	Sewage	84	83	4.3
	Creeky	47	20	3.5
	Musty	27	10	3.7

* 'Other notes' occur when less than 50% of the panel detect a flavor or aroma. No intensity is assigned to 'other notes', thus the intensity average is based on the occurrences of excluding notes (ex. notes) rather than including notes (inc. notes)

Table 4.9 Common flavors and aromas at the PWD Baxter Plant—Chlorination (Bartels *et al.*, 1989)

Sample location	Major response	Percent occurrence		Intensity (scale 0–12)
		Inc. notes*	Ex. notes*	
Effluent aroma	Chlorinous	97	91	3.3
	Musty	69	46	3.1
	Earthy	52	35	3.1
Effluent flavor	Chlorinous	89	54	2.7
	Musty	66	37	2.2
	Earthy	29	19	2.3
Influent aroma	Earthy	81	43	4.4
	Septic	67	52	4.0
	Decayed vegetation	68	62	5.0
	Musty	38	17	3.6
	Fishy	32	13	5.0
	Vegetation	24	22	4.3

* 'Other notes' occur when less than 50% of the panel detect a flavor or aroma. There is no intensity assigned to 'other notes', the numerical average is based on the occurrences of excluding notes (ex. notes) rather than including notes (inc. notes)

Baxter Water Treatment plant in Philadelphia included chlorination. The Morsang treatment plant used pre-ozonation and final chlorination.

The results showed that earthy, chlorinous and swimming pool odors were dominant odor descriptors in the effluent of the PSWC plant (Table 4.8). The influent odors such as sewage and creeky were completely removed. However, the occurrence of a musty odor increased after the treatment process (27% to 46%), although the intensity level of musty odor decreased from 3.7 to 2.0 on the 12-point flavour profile analysis scale. At the PWD Baxter Water Treatment Plant (Table 4.9), the septic, decaying vegetation, vegetation and fishy

Table 4.10 Common flavors and aromas at the Lyonnaise des Eaux Morsang plant (Bartels *et al.*, 1989)

Sample location	Major response	Percent occurrence Inc. notes*	Ex. notes*	Intensity (scale 0–12)
Effluent aroma	Chlorinous	–	88	4.7
	Musty	–	81	2.5
	Swimming pool	–	24	4.3
Effluent flavor	Chlorinous	–	94	4.6
	Musty	–	18	2.0
	Astringent	–	18	2.0
Influent aroma	Muddy	–	82	6.7
	Fishy	–	71	6.5
	Septic	–	29	9.0
	Musty	–	38	3.5
Influent flavor	Muddy	–	71	6.2
	Musty	–	59	6.9
	Earthy	–	47	5.3

* 'Other notes' occur when less than 50% of the panel detect a flavor or aroma. There is no intensity assigned to other notes, thus the numerical average is based on the occurrences of excluding notes (ex. notes) rather than including notes (inc. notes)

odors were effectively removed by chlorination. On the other hand, the earthy and musty odors were not effectively removed by chlorination. The occurrence of earthy odor decreased from 81% to 52% occurrence, and the odor intensity level also decreased from 4.4 to 3.1. However, the occurrence of musty odor increased from 38% to 69% with the musty odor intensity level dropping slightly after treatment (Table 4.9). At both the PSWC and PWD treatment plants, the earthy and musty odors were not related to MIB and geosmin, as shown by closed-loop stripping analysis and GC/MS. Table 4.10 indicates that the results at the PWD plant and at Morsang in France were similar, although ozone was the primary disinfectant, and chlorine the final disinfectant. The musty odor increased after treatment by ozone and chlorine.

4.9 Taste and odor treatment of off-odor by ozonation

Ozonation is well known for its ability to remove unpleasant tastes and odors (McGuire and Gaston, 1988). Table 4.11 summarizes examples in the literature which demonstrate the effectiveness of ozone for taste and odor control. However, ozone cannot be considered the answer to all taste and odor problems. Lalezary *et al.* (1986) studied pre-stripped organic free water that was spiked with geosmin and MIB, and compared alternate oxidants for the destruction of the specific taste- and odor-causing compounds. It was found that low doses of ozone (2 mg/l) could remove 20% of the geosmin. However, the removal efficiency decreased at higher ozone dosages. Glaze *et al.* (1990)

Table 4.11 Ozonation for taste and odor control (from McGuire and Gaston, 1988)

Site	Results of ozonation	Reference
Worcester, MA, USA	Produced odor-free water	Prendiville and Thompson (1988)
Hackensack, NJ, USA	Changed musty/grassy to non-objectionable sweet odor	Weng et al. (1989)
Bay City, MI, USA	Eliminated objectionable tastes and odors	Dekam (1983)
Monroe, MI, USA	Eliminated objectionable tastes and odors	LePage (1981, 1985)
Sherbrooke, Quebec, Canada	Effectively eliminated raw water tastes and odors	Soule and Medlar (1980)
Osaka, Japan	O_3 at 2 mg/l oxidized 80–90% of MIB. 100% geosmin removed by O_3/GAC	Tasumi (1987)
Morsang, Paris, France	O_3 at 2.5 mg/l and 10 min contact time reduced all raw water tastes and odors	Suffet et al. (1986) Baker et al. (1986) Anselme et al. (1988)
Germany	Pre-ozonation showed slight improvement, while O_3/GAC was effective	Sontheimer et al. (1988)

GAC = granular activated carbon

and Duguet et al. (1989) point out that this discrepancy in results may be due to the type of water used in the two studies. In natural waters ozone reacts by two competing pathways: (i) direct oxidation by the O_3 molecule; and (ii) indirect oxidation by the much more powerful hydroxyl free radical $OH^\bullet$. Since natural waters contain many substances that can initiate formation of $OH^\bullet$ radicals compared to purified water, it is reasonable to expect higher percent removals in studies conducted on natural waters.

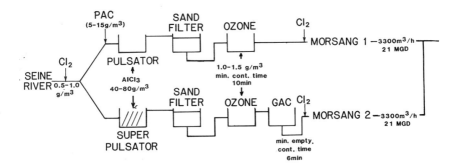

Figure 4.3 Morsang Water Treatment Plant (Anselme et al., 1988). Reprinted from *AWWA Journal*, Vol. 80, No. 10 (October 1988), Denver, Colorado, by permission. Copyright © 1988, American Water Works Association.

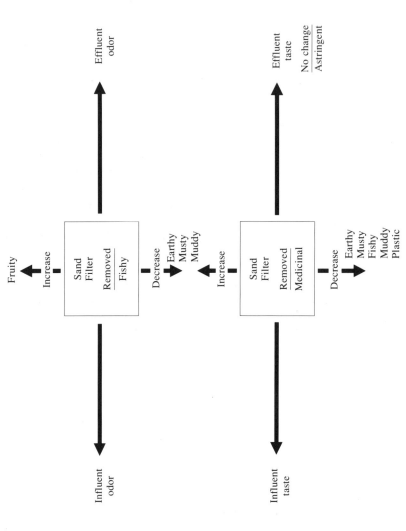

Figure 4.4 Morsang Water Treatment Plant—odor intensity changes caused by ozonation (Adapted from Anselme *et al.*, 1985a).

Figure 4.3 is a diagram of the Morsang Water Treatment Plant near Paris, France, which treats Seine River water (Anselme *et al.*, 1988). River Seine water enters the plant and is split into two lines after breakpoint chlorination (1.5–3.0 mg chlorine/l). In line 1, 5–15 g powdered activated carbon/m^3 water is normally injected for taste and odor control. The water in line 1 then enters a coagulation process (Pulsator*). The prechlorinated water in line 2 flows directly to the alum coagulation process (Super Pulsator). The water in each line then undergoes rapid sand filtration. The chlorine residual after sand filtration is < 0.1 mg/l. The odors found in raw water, except for chlorine and musty-earthy odors, decrease in percentage frequency of occurrence and intensity during these treatment processes. After sand filtration an ozone dosage of 2.5–2.7 mg/l is applied (10 min contact time). An ozone residual of < 0.2 mg/l leaves the ozonation process, and residual chlorine reacts with the ozone.

Figure 4.4 shows the odors present in the raw water at the Morsang plant as measured by the FPA method. All odors were reduced to < 50% frequency of occurrence by ozonation, and fishy odors were completely eliminated. Fruity odors, produced during ozonation, increased to 40% frequency of occurrence. Figure 4.5 presents FPA intensity values for earthy-musty odors during the time period of the ozonation process in line 2. The intensities of all descriptors were reduced to below the slight level on the FPA scale (intensity 4.5), except for an astringent taste. Figure 4.6 shows that a medicinal taste was also not removed by the treatment. Figure 4.7 shows that the fruity odor increased from a negligible intensity to just perceptible in the spring; increased in spring, summer and into the fall to about a monthly average of 2; and then

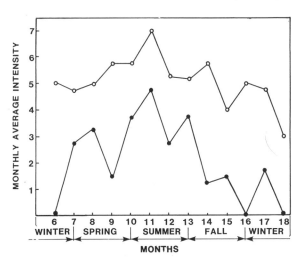

Figure 4.5 Removal of earthy-musty odor by the ozonation water treatment process (Anselme *et al.*, 1988): (○) ozone inlet; (●) ozone outlet. Reprinted from *AWWA Journal*, Vol. 80, No. 10 (October 1988), Denver, Colorado, by permission. Copyright © 1988, American Water Works Association.

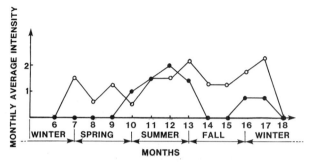

Figure 4.6 Removal of plastic odor by the ozonation water treatment process (Anselme *et al.*, 1988): (○) ozone inlet; (●) ozone outlet. Reprinted from *AWWA Journal*, Vol. 80, No. 10 (October 1988), Denver, Colorado, by permission. Copyright © 1988, American Water Works Association.

dramatically increased to a high of 4–7 in subsequent months. The increase in fruity odors is suspected to be due to the formation of aldehydes during ozonation. This hypothesis was tested using a statistical correlation (see Figure 4.8).

Figure 4.4 summarizes the taste and odor changes caused by ozonation in lines 1 and 2. Three interesting points were observed. First, most tastes and odors decreased in frequency of detection. The intensities of all qualities decreased slightly during treatment up to rapid sand filtration. Second, the

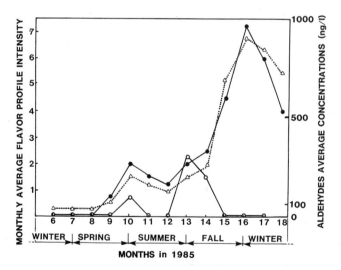

Figure 4.7 Relationship between intensity of fruity odor and concentration of total aldehydes (Anselme *et al.*, 1988): (○) fruity odor in ozone inlet; (●) fruity odor in ozone outlet; (Δ) aldehydes in ozone outlet. FPA scale: 1 = threshold; 4 = slight; 8 = moderate; 12 = strong. Reprinted from *AWWA Journal*, Vol. 80, No. 10 (October 1988), Denver, Colorado, by permission. Copyright © 1988, American Water Works Association.

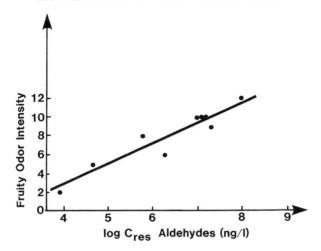

Figure 4.8 Correlation between intensity of fruity odor and log of aldehyde concentration (Anselme *et al.*, 1988). Linear regression $l = f(\log C_{res})$. $R^2 = 0.98$. Reprinted from *AWWA Journal*, Vol. 80, No. 10 (October 1988), Denver, Colorado, by permission. Copyright © 1988, American Water Works Association.

ozone unit process was effective for removing all tastes and odors, except for astringent and plastic tastes. Ozonation also produced a fruity odor. Third, although no geosmin or MIB was found during the study, other causes of earthy and musty tastes and odors increased in frequency of detection as the water progressed through prechlorination and settling. At the present time there is no explanation for the fact that prechlorination appears to increase the frequency with which musty and earthy problems occur.

4.10 Taste and odor treatment by carbon adsorption

Carbon adsorption can be very effective for taste and odor control. Both granular activated carbon (GAC) and powdered activated carbon (PAC) have been successfully used to treat tastes and odors in many water treatment facilities (Baker *et al.*, 1986). GAC is usually more economical than PAC if more than 25 mg/l PAC has to be applied for more than three months per year (Sontheimer *et al.*, 1988). In most cases, a PAC dose of 5 to 10 mg/l is sufficient for taste and odor control (Sontheimer *et al.*, 1988). These dosages are similar to those used in the American Water System (Dixon, 1989). In a survey of 36 American plants, Dixon (1989) found that the average PAC dosage ranged from 1.5–14 mg/l. Several American plants also use GAC, in a sand replacement filter adsorber mode at depths of 15 to > 30 inches, for taste and odor control. Graese *et al.* (1987) reported on a case study experience at Stockton, California, where geosmin and MIB were identified as the principal odor-causing compounds (dirty, musty and moldy tastes and odors). PAC and potassium permanganate were found to be inadequate in controlling these tastes and odors. GAC was installed in a sand replacement filter adsorber mode, and taste and odor complaints were reduced 60–70% for the two year period following installation. During the same period, geosmin and MIB removals ranged from 60–70% and 45–50%, respectively. Herzing *et al.* (1977) also studied GAC adsorption of geosmin and MIB, and found that both compounds were readily adsorbed. In addition, they found that the presence of humic substances significantly reduced the amount of adsorption, but that MIB and geosmin were more strongly adsorbed than the humics. Suffet (1980) reviewed data in a National Academy of Science Report, which showed that the musty odors of MIB and geosmin can be removed by GAC for years due to favourable competition with humic·substances. In pilot plant work conducted by Lalezary-Craig *et al.* (1988) geosmin and MIB concentrations of 66 ng/l were reduced to acceptable levels (< 7 ng/l) by PAC dosages as low as 10 mg/l in distilled water. They also found that chlorine, chloramines, and the presence of humic substances adversely affected adsorption.

4.11 Summary

In summary, a drinking water taste and odor wheel has been used to help describe the present state of knowledge of off-flavors in raw and potable water. The primary causes of naturally occurring tastes and odors are microorganisms that produce geosmin and 2-methyl isoborneol. The disinfectants chlorine, chloramine and ozone, when present in drinking water, produce their own off-flavours. In the United States the chlorine disinfectants are the most common taste and odor problem. The disinfectants by-products—e.g. low-molecular-weight aldehydes from amino acid oxidation, chlorophenols from phenol, iodomethanes from iodide, and humic materials—are another important source of tastes and odors in drinking water. The water treatment processes of disinfection and activated carbon adsorption are the primary methods used to remove tastes and odors. Disinfection processes are chemical specific whereas activated carbon is non-specific. Activated carbon is the optimum process, but also the most expensive one. Present research is unravelling the causes of tastes and odors, e.g. cucumber from microorganisms, and plastic from plastic pipes. However, the causes of some odors, e.g. decaying vegetation and fishy, are still not known nor understood.

References

Amoore, J. E. (1962). The stereochemical theory of olfaction. *Prod. Sci. Toilet Goods Assn.* **37**, 1.
Amoore, J. E. and Hautala, E. (1983). Odor as an aid to chemical safety. Thresholds compared with threshold limit values and volatilities for 214 industrial chemicals in air and water dilution. *J. Appl. Toxicol.* **3**(6), 272.

Amoore, J. E., Forrester, L. J. and Pelosi, P. (1976). Specific anosmia to isobutyraldehyde: the malty primary odor. *Chemical Senses and Flavour* **2**, 17–25.

Anselme, C., Mallevialle, J. and Suffet, I. H. (1985a) Removal of tastes and odors by ozone-granular activated carbon water treatment processes. *Proceedings of the 7th Ozone World Congress*, Incl. Ozone Association, Tokyo, Japan.

Anselme, C., N'Guyen, K., Bruchet, A and Mallevialle, J. (1985b). Can polyethylene pipes impart odors in drinking water? *Envir. Technol. Letters* **6**, 477.

Anselme, C., N'Guyen, K., Bruchet, A. and Mallevialle, J. (1985c). Characterization of low molecular weight products desorbed from polyethylene tubings. *Sci. Total Envir.* **47**, 371.

Anselme, C., Suffet, I. H. and Mallevialle, J. (1988). Effects of ozonation on tastes and odors. *J. AWWA* **80**, 45.

APHA (1985). Standard methods for the examination of water and wastewater. AWWA and WPCF. Washington, DC (16th ed.).

APHA (1989). Standard methods for the examination of water and wastewater. AWWA and WPCF. Washington, DC (17th ed.). *Method 2170. Flavor Profile Analysis.*

Baker, R. J., Suffet, I. H., Anselme, C. and Mallevialle, J. (1986). Evaluation of water treatment methods for removal of taste and odor causing compounds from drinking water. *Proceedings Water Quality Technology Conference*, Portland, OR, November.

Bartels, J. H. M., Burlingame, G. A. and Suffet, I. H. (1986). Flavor profile analysis: taste and odor control of the future. *J. AWWA* **78** (3), 50.

Bartels, J. H. M., Brady, B. M. and Suffet, I. H. (eds) (1989). Taste and odor in drinking water supplies Phase I and II, Vol.1. *AWWA Research Foundation Report*. American Water Works Association, Denver, CO.

Bayliss, J. R. (1951). Fishy odor in water caused by *Dinobryon. Pure Water* **3**, 128.

Berglind, L., Johnson, I. J., Ormerod, K. and Skulberg, O. M. (1983). *Oscillatoria brevis* (Kutz.) Gom. and some other, especially odouriferous benthic cyanophytes in Norwegian inland waters. *Wat. Sci. Technol.* **15** (6/7), 241.

Brady, B., Bartels, J. H. M., Mallevialle, J. and Suffet, I. H. (1988). Sensory analysis of drinking water. *Water Quality Bulletin* **13** (2–3), 61.

van Breemen, L. W. C. A., Dits, J. S. and Ketelaars, H. A. M. (1992). Production and reduction of geosmin and 2-methylisoborneol during storage of river water in deep reservoirs. *Wat. Sci. Technol.*, in press.

Bruchet, A., N'Guyen, K., Mallevialle, J. and Anselme, C. (1989). Identification and behaviour of iodinated haloform medicinal odor. AWWA Seminar Proceedings *The Identification and Treatment of Taste and Odor Compounds*, AWWA Annual Conference, Los Angeles, CA, pp.125–141.

Bruvold, W. H. (1989). A critical review of methods used for the sensory evaluation of water quality. *Critical Reviews of Environmental Control* **19** (4), 291.

Bryan, P., Kuzcinski, L., Sawyer, M. and Feng, R. (1973). Taste threshold of halogens in water. *J. AWWA* **65** (5), 363.

Burlingame, G. A., Dann, R. M. and Brock, G. L. (1986). A case study of geosmin in Philadelphia's waters. *J. AWWA* **78**, 56.

Burlingame, G. A., Muldowney, J., Taylor, I., Hinton, M. and Maddrey, R. (1990). *A Case Study of Cucumber Taste and Odor in Philadelphia Water*. Philadelphia Water Dept. Report.

Burlingame, G. A., Bartels, J. H. M., Khiari, D. and Suffet, I. H. (1991). Odor reference standards: the universal language. *Proceedings of the AWWA Water Quality Technology Conference*, Orlando, Florida.

Burttschell, R. L., Rosen, A. A., Middleton, F. M. and Ettinger, M. B. (1959). Chlorine derivatives of phenol causing taste and odor. *J. AWWA* **51**, (2), 205.

Buttery, R. G. and Ling, L. C. (1973). Earthy aroma of potatoes. *J. Agric. Food Chem.* **21**, 745.

Buttery, R. G., Seifert, R. M., Guadagni, D. G. and Ling, L. C. (1969). Characterization of some volatile constituents of bell pepper, *J. Agr. Food Chem.* **17**, 1322.

Buttery, R. G., Seifert, R. M., Guadagni, D. G. and Ling, I. C. (1971). Characterization of additional volatile components of tomato. *J. Agr. Food Chem.* **19**, 524.

Collins, R. P. (1971). *Characterization of Taste and Odors in Water Supplies*. USEPA, Office Res. Monitor., Wat. Poll. Ctrl. Res. Series, Project 16 040-DGM-08/71, Washington, DC.

Collins, R. P. and Kalnins, K. (1965a). Volatile constituents produced by the alga *Synura petersenii*. II. Alcohols, esters, and acids. *Int. J. Air Wat. Poll.* **9**, 501.

Collins, R. P. and Kalnins, K. (1965b). Volatile constituents produced by the alga *Synura petersenii*. I. The carbonyl fraction. *Lloydia* **28**, 48.

Collins, R. P. and Kalnins, K. (1966). Carbonyl compounds produced by *Cryptomanas ovata* var. *Palustris. J. Protozoa* **13**, 435.

Collins, R. P. and Kalnins, K. (1967). The fatty acids of *Synura petersenii*. *Lloydia* **30**, 437.

Collins, R. P., Knaak, L.E. and Soboslai, J. W. (1970). Production of geosmin and 2-exo-hydroxy-2-methylborane by *Streptomyces odorifer. Lloydia* **33**, 199.

Dekam, J. A. (1983). Ozonation experiences at the Bay Metropolitan Water Treatment Plant. *The 6th World Ozone Congress*, Washington, DC, May.

Dixon, K. L. (1989). American Water Works Service Co. Letter to Jeffry Adams, US EPA, March 31.

Duguet, J. P., Bruchet, A. and Mallevialle, J. (1989). Geosmin and 2-methylisoborneol removal using ozone or ozone/hydrogen peroxide coupling. *Proceedings of the Ninth Ozone World Congress*, New York.

EPA (Environmental Protection Agency) (1977). National secondary drinking water regulation. *Federal Register* CFR 42, 17144.

Ettinger, M. B. and Ruchhoft, C. C. (1951). Effect of stepwise chlorination on taste and odor producing of some phenolic compounds. *J. Am. Water Works Assoc.* **43**, 561.

European Communities (1980). Council Directive of 15 July 1980 relating to the quality of water intended for human consumption. *Official Journal of the European Communities* No. L 229/11.

van Gemert, L. J. and Nettenbreijer, A. H. (1977). *Compilation of Odor Threshold Values in Air and Water*. National Institute for Water Supply, Vooburg, Netherlands, Central Institute for Nutrition and Food Research TNO, Zeist, Netherlands.

Gerber, N. N. (1967). Geosmin, an earthy smelling substance isolated from actinomycetes. *Biotechnol. Bioeng.* **9**, 321.

Gerber, N. N. (1968). Geosmin, from microorganisms, is *trans*-1,10-dimethyl-trans-9-decalol. *Tetrahedron Letters* **25**, 2971.

Gerber, N. N. (1969). A volatile metabolite of actinomycetes, 2-methylisoborneol. *J. Antibiot.* **22**, 508.

Gerber, N. N. (1971). Sesquiterpenoids from actinomycetes: cadin-4-ene-1-ol. *Phytochemistry* **10**, 185.

Gerber, N. N. (1979). Volatile substances from actinomycetes: their role in the odor pollution of water. *CRC Critical Reviews in Microbiol.* **7**, 191.

Gerber, N. N. (1983). Volatile substances from actinomycetes: their role in the odor pollution of water. *Wat. Sci. Technol.* **15** (6/7), 115.

Gerber, N. N. and Lechevalier, H. A. (1965). Geosmin, an earthy smelling substance isolated from actinomycetes. *Appl. Microbiol.* **13**, 935.

Giger, W. and Schaffner, C. (1981). Groundwater pollution by volatile organic chemicals. *Env. Sci. Technol.* **17**, 7.

Gittelman, T. S. and Yohe, T. L. (1989). Treatment of the iodinated haloform medicinal odors in drinking water. AWWA Seminar Proceedings *The Identification and Treatment of Taste and Odor Compounds*, AWWA Annual Conference, Los Angeles, CA, pp. 105–123.

Glaze, W. H., Raymond, S., Chauncey, W., Ruth, H. C., Zarnoch, J. J., Aieta, E. M., Tate, C. H. and McGuire, M. (1990). Evaluating oxidants for the removal of model taste and odor compounds from municipal water supplies. *J. AWWA* **82** (5), 79–84.

Graese, S. L., Snoeyink, V. L. and Lee, R. G. (1987). GAC filter-adsorbers. *AWWA Research Foundation Report*, June.

Guadagni, D. G. and Buttery, R. G. (1978). Odor threshold of 2,3,6-trichloroanisole in water. *J. Food Sci.* **43**, 1346.

Guadagni, D. G., Buttery, R. G. and Okano, S. (1963). Odour thresholds of some organic compounds associated with food flavours. *J. Sci. Food Agric.* **14**, 761.

Guadagni, D. G., Buttery, R. G. and Turnbaugh, J. G. (1972). Odour thresholds and similarity ratings of some potato chip components. *J. Sci. Food Agric.* **23**, 1435–1444.

Hack, D. J. (1981). Common customer complaint: my hot water stinks. *Op. Flow*, 3, 7 and 9.

Hansson, C. R., Henderson, M. J., Jack, P. and Taylor, R. D. (1987). Iodoform taste complaints in chloramination. *Water Research* **21** (10), 1265–1271.

Harper, R., Bate-Smith, E. C. and Land, D. G. (1968) *Odor Description and Odor Classification*. J. & A. Churchill Ltd., London.

Henley, D. E. (1970). Odorous metabolite and other selected studies of cyanophyta. *Doctoral dissertation*, North Texas State University, Denton, Texas.

Herzing, D. R., Snoeyink, V. L. and Wood, N. F. (1977). Activated carbon adsorption of the odorous coumpounds 2-methylisoborneol and geosmin. *J. AWWA* **69** (4), 223.

Hrudey, S. E., Gac, A. and Daignault, S. A. (1988). Potent odor causing chemicals arising from drinking water disinfection. *Water Sci. Technol.* **20** (8/9), 55.

Hrudey, S. E., Daignault, S. A., Gac, A., Poole, D., Walker, G. and Birkholz, D. A. (1989). Swampy or swimming pool aldehyde odours caused by chlorination and chloramination. AWWA Seminar Proceedings *The Identification and Treatment of Taste and Odor Compounds*, AWWA Annual Conference, Los Angeles, CA, pp. 143–156.

Hrudey, S. E., Huck, P. M. and Roodselaar, A. V. (1990). The causes of and possible remedies for offensive odor and related water quality problems in Edmonton's drinking water. *City of Edmonton Quarter Report*, Vol. 1, Main Report.

Ibrahim, E. A., Becker, W. C., Capangpangan, M. C. and Suffet, I. H. (1990). Optimization of sequential unit operations for removal of organics from drinking water. In *Chemical Water and Wastewater Treatment*. Eds H. H. Hahn and R. Klute. Springer-Verlag, Berlin, and Heidelberg, pp. 305–332.

Izaguirre, G. (1992). A copper-tolerant *Phormidium* sp. from Lake Mathews, California, that produces 2-methylisoborneol and geosmin. *Wat. Sci. Technol.*, in press.

Izaguirre, G., Hwang, C. J., Krasner, S.W. and McGuire, M. J. (1982). Geosmin and 2-methylisoborneol from cyanobacteria in three water supply systems. *Appl. Envir. Microbiol.* **43**, 708.

Izaguirre, G., Hwang, C. J., Krasner, S. W. and McGuire, M. J. (1983). Production of 2-methylisoborneol by two benthic cyanophyta. *Wat. Sci. Technol.* **15** (6/7), 211.

Jenkins, D., Medsker, L. L. and Thomas, J. F. (1967). Odorous compounds in natural waters: some sulfur compounds associated with blue-green algae. *Envir. Sci. Technol.* **1**, 731.

Juttner, F. (1981). Detection of lipid degradation products in the water of a reservoir during a bloom of *Synura uvella. Appl. Envir. Microbiol.* **41**, 100.

Juttner, F. (1983). Volatile odorous excretion products of algae and their occurrence in the natural aquatic environment. *Wat. Sci. Technol.* **15** (6/7), 247.

Juttner, F., Hoflacher, B. and Wurster, K. (1986). Seasonal analysis of volatile organic biogenic substances (VOBS) in freshwater phytoplankton populations dominated by *Dinobryon, Microcystis*, and *Aphanizomenon. J. Phycol.* **22**, 169.

Kikuchi, T. *et al.* (1972). Odorous components of the diatom, *Syneda rumpens kutz*, isolated from the water in Lake Biwa. Identification of *n*-hexanal. *Yakugaku Zasshi* **92** (12), 1567.

Kikuchi, T., Mimura, T., Itoh, Y., Moriwaki, Y., Negoro, K., Masada, Y. and Inoue, T. (1973a). Odorous metabolites of actinomycetes biwako-C and -D Strain isolated from the bottom deposits of Lake Biwa. Identification of geosmin, methylisoborneol and furfural. *Chem. Pharm. Bull., Tokyo*, **21**, 2341.

Kikuchi, T., Mimura, T., Harimaya, K., Yano, H., Arimoto, T., Masada, Y. and Inoue, T. (1973b). Odorous metabolites of blue-green algae *Schizothrix mueller nageli* collected in the southern basin of Lake Biwa. Identification of geosmin. *Chem. Pharm. Bull., Tokyo* **21**, 2342.

Kikuchi, T. *et al.* (1974). Metabolites of a diatom, *Syneda rumpens kutz*, isolated from the water in Lake Biwa. Identification of odorous compounds, *n*-hexanal, and *n*-heptanal, and analysis of fatty acids. *Chem. Pharm. Bull., Tokyo* **22** (4), 945.

Krasner, S. W. and Barrett, S. E. (1984). Aroma and flavor characteristics of free chlorine and chloramines. *Proceedings of the AWWA Water Quality Technology Conference*, Denver, CO.

Krasner, S. W., Barrett, S. E., Dale, M. S. and Hwang, C. J. (1986). Free chlorine versus monochloramine in controlling off-tastes and -odors in drinking water. *Proceedings of the AWWA Annual Conference*, Denver, CO.

Lalezary, S., Pirbazari, M. and McGuire, M. J. (1986). Oxidation of five earthy-musty taste and odor compounds. *J. AWWA* **78** (3), 62.

Lalezary-Craig, S., Pirbazari, M., Dale, M. S., Tanaka, T. S. and McGuire, M. J. (1988). Optimizing the removal of geosmin and 2-methylisoborneol by powdered activated carbon. *J. AWWA* **80**, 73.

Lee, G. F. (1967). Kinetics of reactions between chlorine and phenolic compounds. In *Principles and Applications of Water Chemistry*. Eds. S. D. Faust and J.V. Hunter. John Wiley and Sons, Inc., New York.

LePage, W. L. (1981). The anatomy of an ozone plant. *J. AWWA* **73**, 105.

LePage, W. L. (1985). A treatment plant operator assesses ozonation. *J. AWWA* **77**, 44.

MacKenthun, K. M. and Keup, L. E. (1970). Biological problems encountered in water supplies. *J. AWWA* **62** (8), 520.

Mallevialle, J. and Suffet, I. H. (1987). Identification and treatment of tastes and odors in drinking water. *AWWA Research Foundation Report*. American Water Works Association, Denver, CO.

Martin, J. F., Izaguirre, G. and Waterstrat, P. (1991). A planktonic *Oscillatoria* species from Mississippi catfish ponds that produces the off-flavor compound 2-methylisoborneol. *Wat. Res.* **25**, 1447–1451.

Matsumoto, A. and Tsuchiya Y. (1988). Earthy-musty odor-producing cyanophytes isolated from five water areas in Tokyo. *Water Sci. Technol.* **20**, 179–183.

McGugan, W. A. (1980). *Descriptions of Flavor Chemicals*. Food Research Reports, vol. 4., No.1.

McGuire, M. J. and Gaston, J.M. (1988). Overview of the technology for controlling off-flavors in drinking water. *Waterworld News*, May/June.

Medsker, L. L., Jenkins, D. and Thomas, J. F. (1968). Odorous compounds in natural waters. An earthy-smelling compound associated with blue-green algae and actinomycetes. *Envir. Sci. Technol.* **2**, 461.

Medsker, L. L., Jenkins, D., Thomas, J. F. and Koch, C. (1969). Odorous compounds in natural waters. 2-exo-hydroxy-2-methybornate, the major odorous compound produced by actinomycetes. *Envir. Sci. Technol.* **3**, 476.

Narayan, L. V. and Núñez, W. J. (1974). Biological control: isolation and bacterial oxidation of the taste and odor compound geosmin. *J. AWWA* **66** (9), 532.

Negoro, T., Ando, M. and Ichikawa, N. (1988). Blue-green algae in Lake Biwa which produce earthy-musty odors. *Wat. Sci. Technol.* **20** (8/9), 117–123.

Palmer, C. M. (1962). *Algae and Water Supplies*. Public Health Service Publication 657. US Department of Health, Education and Welfare, Washington, DC.

Palmer, C. M. (1977). *Algae and Water Pollution*. Munic. Envir. Res. Lab., Ofc. Res. & Devel., USEPA, Cincinnati, OH.

Palmer, C. M. (1980). Taste and odor algae. In *Algae and Water Pollution*. Castle House, London.

Persson, P. E. (1979a). The source of muddy odor in bream (*Abramis brama*) from the Porvoo Sea area (Gulf of Finland). *J. Fish. Res. Board Can.* **36**, 883.

Persson, P. E. (1979b). Notes on muddy odour: III. Variability of sensory response to 2-methylisoborneol. *Aqua Fennica* **9**, 48–52.

Persson, P. E. (1980). On the odor of 2-methylisoborneol. *J. Agri. Food Chem.* **28**, 1344.

Piet, G. J., Zoeteman, B.C.J. and Kraayeveld, A. J. A. (1972). Earthy smelling substances in surface waters of the Netherlands. *Wat. Trmt. Exam.* **21**, 281.

Pontius, F. W. (1990). Complying with the new drinking water quality regulations. *J. AWWA* **82**, 3.

Prendville, P. C. and Thompson, J. C. (1988). Ozone-direct filtration at Worcester and Springfield, Massachusetts. *The International Ozone Association Spring Technical Conference*, Monroe, MI, April.

Rosen, A. A., Safferman, R. S., Mashni, C. I. and Romano, A. H. (1968). Identity of odorous substances produced by *Streptomyces griseoluteus*. *Appl. Micrbiol.* **16**, 178.

Rosen, A. A., Mashni, C. I. and Sufferman, R. S. (1970). Recent developments in the chemistry of odor in water: the cause of earthy/musty odor. *Wat. Trmt. Exam.* **19**, 106.

Safferman, R. S., Rosen, A. A., Mashni, C. I. and Morris, M. E. (1967) Earthy-smelling substance from a blue-green alga. *Envir. Sci. Technol.* **1**, 429.

Seifert, R. M., Buttery, R. G., Guadagni, D. G., Black, D. R. and Harris, J. G. (1970). Synthesis of some 2-methoxy-3-alkylpyrazines with strong bell pepper-like odors. *J. Agric. Food Chem.* **18**, 246.

Slater, G. P. and Block, V. C. (1983). Volatile compounds of the cyanophyceae—a review. *Wat. Sci. Technol.* **15** (6/7), 181.

Sontheimer, H., Crittenden, J. C. and Summers, S. (1988). *Activated Carbon for Water Treatment*. DVGW-Forschungsstelle, distributed by AWWA Research Foundation, Denver, CO.

Soule, B. E. and Medlar, S. J. (1980). $O_3 + Cl_2$ produces taste-free No THM, safe piped product. *Water and Sewage Works*, March.

Suffet, I. H. (1980). An evaluation of activated carbon for drinking water treatment: a National Academy of Science Report. *J. AWWA* **72**, 41.

Suffet, I. H. (1992). Private communication.

Suffet, I. H., Anselme, C. and Mallevialle, J. (1986). Removal of tastes and odors by ozonation. AWWA Seminar Proceedings *Ozonation, Recent Advances and Research Needs*, Denver, CO, June.

Suffet, I. H., Baker, R. J. and Yohe, T. L. (1988). Pretreatment of drinking water to control organic contaminants and taste and odor pretreatment. *In Chemical Water and Wastewater Treatment.* Eds Hahn H. H. and Klute R. Springer-Verlag, Berlin and Heidelberg, pp. 15–40.

Tabachek, J. L. and Yurkowski, M. (1976). Isolation and identification of blue-green algae producing muddy odor metabolites, geosmin and 2-methylisoborneol, in saline lakes in Manitoba *J. Fish. Res. Board Can.* **33**, 25.

Tasumi, S. (1987). Removal of trihalomethanes and musty odorous compounds from drinking water by ozonation/granular activated carbon treatment. *Summary AWWA Research News*, November.

Thurman, E. M. (1986). Dissolved organic compounds in normal waters. In *Organic Carcinogens in Drinking Water, Detection, Treatment and Risk Assessment.* Eds N. M. Ram, E. J. Calabrese and R. E. Christman. J. Wiley and Sons, New York, pp. 55–92.

Tsuchiya, Y., Matsumoto, A. and Okamoto, T. (1978). Volatile metabolites produced by actinomycetes isolated from Lake Tairo at Miyakejima. *Yakugaku Zasshi* **89**, 454.

Tsuchiya, Y., Matsumoto, A. and Okamoto, T. (1981). Identification of volatile metabolites produced by blue-green algae, *Oscillatoria splendida, O. amoena, O. germinata* and *Aphanizomenon* sp. *Yakugaku Zasshi* **101**, 852.

Wajon, J. E., Alexander, R. and Kagi, R. I. (1985a). Determination of trace levels of dimethyl polysulphides by capillary gas chromatography. *J. Chromatog.* **319**, 187.

Wajon, J. E., Kagi, R. I. and Alexander, R. (1985b). *The Occurrence and Control of Swampy Odour in the Water Supply of Perth, Western Australia.* Report submitted to the Water Authority of Western Australia by the School of Appl. Chem., Western Australian Inst. Technol., Bentley, Western Australia, November.

Wajon, J. E., Alexander, R., Kagi, R. I. and Kavanagh, B. (1985c). Dimethyl trisulfide and objectional odours in potable water. *Chemosphere* **14** (1), 85.

Weng, C., Hoven, D. L. and Schwartz, B. J. (1986). Ozonation: an economic choice for water treatment, *J. AWWA* **78**, 83.

Wood, S., Williams, S. T. and White, W. R. (1983). Microbes as a source of earthy flavours in potable waters—A Review. *Intl. Biodeterioration Bull.* **19** (3/4).

World Health Organization (1981). Guidelines for drinking water quality, consultation on aesthetic and organoleptic aspects. *ICP/RCE 209*, p. 4. Copenhagen.

Wu, J. T. and Juttner, F. (1988a). Effect of environmental factors on geosmin production by *Fischerella muscicola. Wat. Sci. Technol.* **20** (8/9), 143–148.

Wu, J. T. and Juttner, F. (1988b). Differential partitioning of geosmin and 2-methylisoborneol between cellular constituents in *Oscillatoria tennis, Archiv. Microbiol.* **150**, 580–583.

Yano, H., Nakahara, M. and Ito, H. (1988). Water blooms of *Uroglena americana* and the identification of odorous compounds. *Wat. Sci. Technol.* **20** (8/9), 75–80.

Zoeteman, B. C. J. and Piet, G. J. (1974). Cause and identification of taste and odour compounds in water. *Sci. Total Environ.* **3**, 103–115.

5 Undesirable flavors in dairy products

I. J. JEON

5.1 Introduction

A dairy product should provide the pleasing, characteristic flavor that consumers expect. Occasionally, however, one may experience an unpleasing flavor and aftertaste associated with an individual product. For example, normal milk has a bland but characteristic milk flavor that is pleasing and slightly sweet. In some milk, however, the flavor may be uncharacteristic and often undesirable. Undesirable flavors in dairy products, or any food, are due to the presence of chemical compounds that impart an odor and flavor uncharacteristic of the products.

Comprehensive reviews are available on off-flavors from various sources in milk (Bassette *et al.*, 1986; Shipe, 1980) and whey protein concentrates (Morr and Ha, 1991), and on those caused by microorganims in dairy products (Margalith, 1981). The objective of this chapter is to review off-flavors in various dairy products, and the associated chemical compounds that cause the off-flavor taints.

5.2 Sensory characteristics of off-flavors

In dairy products, a number of different off-flavors are recognized, although terms for describing them may not be the same from one country to another. The American Dairy Science Association (ADSA) has recognized the off-flavors in various dairy products as flavor defects, which are used as part of the judging criteria for product quality. The list of flavor defects for milk and three selected milk products is summarized in Table 5.1 and the list for four common fermented dairy products is shown in Table 5.2. The off-flavor terms listed in Tables 5.1 and 5.2 have evolved over many years and include a mixture of associative terms, such as barny and cowy, and more scientific terms, such as high acid and oxidized (Shipe, 1980). However, some traditional off-flavor descriptions are not included in the tables because of their recent elimination from the ADSA scorecards.

In 1978, the ADSA Committee on Flavor Nomenclature and Reference Standards recommended the adaptation of a classification of undesirable flavors in milk on the basis of their causes (Shipe *et al.*, 1978). The classification

Table 5.1 List of off-flavors in milk and some milk products (adapted from American Dairy Science Association scorecards)

Milk	Butter[a]	Ice cream[b]	Dry milk[c]
Acid	Acid	Acid	Acid
Barny	Bitter	Cooked	Astringent
Bitter	Cheesy	Lacks freshness	Bitter
Cooked	Feed	Metallic	Chalky
Cowy	Flat	Old ingredients	Cooked
Feed	Garlic/onion	Oxidized	Feed
Fruity/fermented	Metallic	Rancid	Fermented
Flat	Musty	Salty	Flat
Foreign	Oxidized	Storage	Foreign
Garlic/onion	Rancid	Whey	Gluey
Lacks freshness	Scorched		Metallic
Oxidized	Storage		Oxidized
Rancid	Unclean		Rancid
Salty	Whey		Salty
Unclean	Yeasty		Scorched
			Stale
			Unclean
			Weedy

[a] Sweet butter
[b] Vanilla, dairy ingredients as a source only
[c] Bodyfelt *et al.* (1988)

Table 5.2 List of off-flavors in some fermented dairy products (adapted from American Dairy Science Association scorecards)

Cottage cheese	Cheddar cheese	Cultured milk[a]	Yogurt
Bitter	Bitter	Astringent	Acetaldehyde
Cooked	Feed	Bitter	Bitter
Diacetyl	Fruity/fermented	Chalky	Cooked
Feed	Flat	Cheesy	Foreign
Fruity/fermented	Garlic/onion	Cooked	High acid
Flat	Heated	Fruity/fermented	Lacks freshness
Foreign	High acid	Foreign	Low acid
High acid	Moldy	Green	Oxidized
Lacks freshness	Rancid	High acid	Rancid
Malty	Sulfide	Flat	Unclean
Metallic	Unclean	Lacks freshness	Yeasty
Musty	Whey taint	Oxidized	
Oxidized	Yeasty	Rancid	
Rancid		Salty	
Unclean		Unclean	
Yeasty		Yeasty	

[a] Buttermilk, kefir, sour cream, etc. (Bodyfelt *et al.*, 1988)

is shown in Table 5.3. The purpose was to make the classification system more useful in quality control work by focusing attention on causes. Nevertheless, this classification can be applied to all dairy products because most off-flavors described for them are in the same categories as those in milk (Tables 5.1, 5.2 and 5.3). For convenience, therefore, this classification is utilized in the following discussions.

Table 5.3 Categories of off-flavors in milk (from Shipe *et al.*, 1978)

Causes	Descriptive or associative terms
Heated	Cooked, caramelized, scorched
Light-induced	Light, sunlight, activated
Lipolyzed	Rancid, butyric, bitter*, goaty
Microbial	Acid, bitter*, fruity, malty, putrid, unclean
Oxidized	Papery, cardboard, metallic, oily, fishy
Transmitted	Feed, weed, cowy, barny
Miscellaneous	Absorbed, astringent, bitter*, chalky, flat, chemical, foreign, lacks freshness, salty

* May arise from a number of different causes.

5.3 Transmitted off-flavors

Milk flavor can be tainted before the milk leaves the udder, by the transfer of substances from the cow's feed or environment. This transfer of substances may occur via the respiratory and/or digestive system and bloodstream. Most of the flavor taints encountered in freshly drawn milk are due to these transmitted substances. These types of off-flavor taints in milk are called transmitted flavors, and flavor descriptions include feed, weed, cowy and barny. Some of these flavors are mild in intensity and may not be considered objectionable by consumers. However, because the majority of the transmitted substances will remain in the milk during processing and some may get concentrated, they may cause objectionable off-flavors in finished products such as butter and cheese. The mechanism for the transmission of flavor compounds appears to occur both through lungs to the udder and through rumen walls to the bloodstream (Dougherty *et al.*, 1962; Shipe *et al.*, 1962). It seems that when very volatile materials from the feed, in the atmosphere surrounding the cow, or in the gas eructated from the cow's rumen are inhaled, they pass rapidly from lungs to udder, whereas volatile and non-volatile flavor materials pass more slowly from the digestive system to the udder.

5.3.1 Barny flavor

This associative term is used by flavor panelists to describe the milk flavor associated with a poorly ventilated barn. Obviously, the primary cause of barny flavor is inadequate ventilation. This flavor taint will occur if cows are forced continuously to inhale strong odors from damp, dirty barns. However, the nature of barny flavor has not been characterized nor distinguished clearly from cowy flavor (Shipe *et al.*, 1978). In addition, this flavor, as well as the cowy defect, has been eliminated from the ADSA scorecards. Its incidence has become rare in recent years because of the improved environment of dairy farms in North America.

5.3.2 Cowy flavor

The term 'cowy flavor' has been used by dairy researchers and dairymen to describe a poorly defined 'unclean' flavor usually associated with barn odors and poor ventilation (Bassette *et al.*, 1986). Usually, this flavor taint in milk has a distinct cow odor and a persistent unpleasant aftertaste. Earlier works suggested that cows suffering from ketosis or acetonemia produce milk with a 'cow-like' odor. The odor or the breath of affected cows was found to be similar to that of their milk. In severe cases, the odor was so strong that it was transmitted to the milk of neighboring cows when ventilation was not adequate. Patton *et al.* (1956) suggested that a high level of methyl sulfide also imparts a cowy flavor to milk. Toan *et al.* (1965) reported that when commercial milk was contaminated with *Enterobacter aerogenes*, a cow-like odor was produced by the production of methyl sulfide. As indicated earlier, cowy or barny flavor is uncommon in milk if cows are fed in a well-ventilated area. Therefore, it is not usually a major potential flavor taint in milk or any other dairy products.

5.3.3 Feed flavor

Some odorous feeds taint the milk if fed to cows shortly before milking. The tainted milk will have a sweet and aromatic taste and an odor that may be characteristic of the feed. Common feeds related to the feed flavors are silage, green forage, lush green pastures, etc. However, if these feeds are fed immediately after milking and are withheld for 4 to 5 h before milking, they may not produce a feed flavor in the milk (Shipe, 1980). Feed flavor problems are often associated with an abrupt change in feeding from dry winter rations to one including lush green pasture forage (Shipe *et al.*, 1978). The feed flavor in milk can be carried through the processing steps to appear in butter, dry milk and a variety of cheeses, although this is relatively uncommon.

The chemical compounds responsible for off-flavors in milk appear to be numerous. For example, Morgan and Pereira (1962) found a mixture of methyl sulfide, lower aldehydes, ketones, alcohols, and simple methyl, ethyl, and propyl esters in volatile constituents of grass and corn silage. Among these compounds, methyl sulfide and esters were considered as principal contributors to the off-flavors. The same researchers also reported that freshly cut alfalfa hay contained high concentrations of *trans*-2-hexenal, 3-hexenals and 3-hexenols, which would impart a green grassy flavor (Morgan and Pereira, 1963). When Shipe *et al.* (1962) introduced a mixture of acetone, 2-butanone, methyl sulfide, and *cis*-3-hexene-1-ol through the rumen and lungs of cows, the mixture imparted flavors closely resembling those that could be found in commercial milk. Methyl sulfide is a normal component of milk. However, an unusually high level of methyl sulfide in milk was found to impart a cowy or molasses-like flavor (Patton *et al.*, 1956; Toan *et al.*, 1965). Ammonia, and propyl-and hexylamines are present in trace amounts in fresh milk. However, feed-tainted milk was found to contain higher concentrations of these amines, which were implicated in the feed taints (Cole *et al.*, 1961). An excellent review on chemical compounds that are responsible for the feed flavor taints in milk can be found elsewhere (Bassette *et al.*, 1986).

5.3.4 Weed flavor

One of the most common and readily recognized weed flavors is garlic/onion flavor, which is caused by cows consuming wild garlic or onion. This flavor taint can be readily detectable in milk, butter and cheese because of its characteristic pungent odor and persistent taste. This highly objectionable flavor defect may be expected in the spring and autumn in areas where pastures are infested with weeds (Nelson and Trout, 1981). In addition to garlic and onion, many weeds will taint milk if eaten by cows, especially a short time before milking. The character and intensity of the weed flavor will depend on the kind of weed and the elapse of time between feeding and milking. Weed flavor is frequently associated with bitterness. The different weeds that will taint milk differ with localities and are generally seasonal.

According to a review by Heath and Reineccius (1986), the distinctive flavor of garlic and onion is due to S-alk(en)yl derivatives of L-cysteine sulfoxide, which are rapidly hydrolyzed by the enzyme alliinase to give an unstable sulfenic acid derivative together with pyruvic acid and ammonia. The sulfenic acid compound either breaks down and rearranges to form the relatively more stable thiopropanal S-oxide, which has lachrymatory properties, or reacts with other compounds to produce a complex mixture of di-or trisulfides, which ultimately characterize the product. The characteristic aroma of garlic and onion is due primarily to a mixture of allyl disulfide, allyl thiosulfonate, and allyl trisulfide. The flavor difference between garlic and onion is suggested to be due to the qualitative and quantitative differences in the precursors present, which may be degraded differently by the enzyme.

The off-flavor produced when cows eat plants of peppergrass (*Lepidium* spp.) has been attributed to skatole and indole, with skatole being the major

Table 5.4 Chemical compounds in some weeds that are suspected of transmitting off-flavors to milk via cows (from Roberts, 1959)

Name	Botanical classification	Suspected chemical compounds
Land cress	*Coronopus didymus (Senebiera didyma)*	Benzyl mercaptan
Penny cress, French weed, or stinkweed	*Thlaspe arvense*	Allyl isothiocyanate
Penny royal	*Mentha pulegium*	Pulegone
Pepper grass	*Lepidium virginicum*	Indole
Onion and garlic	*Allium cepa and Allium sativum*	Di-n-propyl sulfide, isopropyl mercaptan, propionaldehyde

contributor (Park, 1969). The primary cause of the flavor in milk produced by cows eating land cress (*Coronopus didymus*) is benzyl methyl sulfide (Park *et al.*, 1969). This compound is considered to be a metabolite of benzyl thiocyanate. Benzyl isothiocyanate, benzyl cyanide, indole, and skatole also can be found in traces in land cress-tainted milk. Table 5.4 lists some weeds that produce off-flavor taints in milk, together with their causative compounds.

5.4 Lipolyzed flavor

One of the common and important off-flavors in milk and dairy products is lipolyzed flavor, which results from hydrolytic cleavage of fatty acids from milk fat triglycerides by the enzyme lipase. Historically, several terms have been used to describe this flavor defect. The term 'rancid' has been used most commonly by people in the dairy industry. However, to avoid a confusion with the flavor defect associated with lipid oxidation in other segments of the food industry, lipolyzed flavor has been recommended to denote the lipase-induced flavor defects in fluid milk (Shipe *et al.*, 1978). Hydrolytic rancidity is also commonly used to describe the lipolyzed flavour (Weihrauch, 1988). As shown in Tables 5.1 and 5.2, the term 'rancid' is used extensively for all dairy products to describe the flavor defect resulting from the lipolysis of triglycerides. The descriptive terms 'goaty', 'soapy', 'butyric', and 'bitter' have also been used to describe the lipolyzed flavor (Shipe *et al.*, 1978; Bassette *et al.*, 1986). The term 'bitter', however, may be ambiguous, because bitter flavors can occur from protein degradation. Lipases secreted by microbial contaminants in milk reportedly can produce flavor defects that are usually accompanied by bitterness resulting from concurrent protein degradation (Shipe *et al.*, 1978). Nevertheless, tainting of raw milk by lipolysis of milk fats could be a serious problem both for the acceptability of market milks to consumers, and for other milk products that are made from the lipolyzed milk.

5.4.1 *Lipoprotein lipase*

Most milk contains sufficient amounts of lipase to cause the flavor defect. However, the enzyme known as a lipoprotein lipase is not normally active as the milk leaves the cow, because the substrate (triglycerides) and enzymes are well partitioned and a multiplicity of factors affect enzyme activity (Weihrauch, 1988). The factors that affect the activity of lipase include (i) the stage of lactation; (ii) temperature manipulation; and (iii) handling, storing and processing procedures of milk. Literature on the activation of milk lipase is voluminous; a few excellent reviews can be found elsewhere (Walstra and Jenness, 1984; Bassette *et al.*, 1986; Weihrauch, 1988). Heat treatment is the most important practical means of inactivating the lipoprotein lipase in milk. Although a number of discrepancies have been reported on the heat inactivation of the milk lipase, heating milk slightly above pasteurization temperature (76.7°C, 16 s) appears to be sufficient to protect it from lipolyzed flavor problems (Shipe and Senyk, 1981). However, heat treatments or any other regular manufacturing steps would not remove the rancid off-flavor if it were formed in raw milk. Consequently as indicated earlier, tainting may become a problem not only for fluid milk but also for all other dairy products produced from it.

5.4.2 Microbial lipases

Heat-resistant lipases from psychrotrophic bacteria, predominantly *Pseudomonas* species, may also contribute to the rancid flavor in milk and other dairy products. The microflora developing in raw milk in holding tanks under refrigeration may produce extracellular lipases and proteases that may survive ordinary pasteurization and sterilization temperatures (Weihrauch, 1988). However, lipolytic spoilage of pasteurized milk by the enzymes from psychrotrophs is rarely reported, because it requires microbial counts exceeding 10^6 per ml in milk to cause noticeable lipolysis (Muir *et al.*, 1978).

The heat-resistant microbial lipases may also cause serious flavor problems in fermented products such as cheese. This is because microbial lipases may have different specificities for releasing fatty acids from triglycerides. Law *et al.* (1976) reported that strong lipolytic rancidity was observed in cheese when butyric acid and medium-chain fatty acids were released from milk fat by *Pseudomonas fluorescens* lipase. In fact, lipases may be specific or nonspecific with regard to the liberation of fatty acids from milk fat molecules (Kwak *et al.*, 1989). Lipoprotein lipase is nonspecific; it releases fatty acids in nearly the same proportion as those in milk (Nelson, 1972). On the other hand, microbial lipases may or may not be specific, depending on the source of the enzyme. Kwak *et al.* (1989) demonstrated patterns of the preferential release of short-chain fatty acids from milk fat by various sources of lipase (Figure 5.1). According to their study, lipases of ruminant-animal origin showed an extremely high ratio of butyric acid but low ratios of caprylic and capric acids (Group 1, Figure 5.1). *Pseudomonas fluorescens* and all porcine lipases showed no extreme ratios in the fatty acids liberated (Group 2, Figure 5.1). Lipases from molds and *Chromobacterium viscosum* produced lower ratios of butyric acid than other groups but a higher ratio of capric acid (Group 3, Figure 5.1). The difference in specificity makes it difficult to measure chemically the intensity of rancid flavor in milk and other dairy products. This would be particularly true with a product containing a significant amount of lipases that are relatively specific for a certain type of fatty acid. Consequently total fatty acid measurements such as acid degree value may not be good indicators for rancid flavor in milk and dairy products that are particularly high in microbial lipases (Duncan *et al.*, 1991), because they cannot distinguish the type of fatty acids released from milk fats.

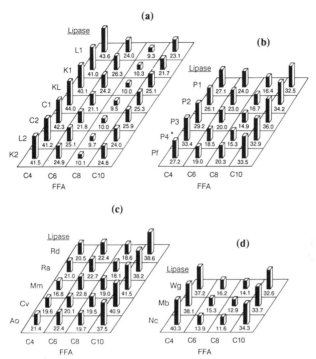

Figure 5.1 Preferential release of short-chain free fatty acids (FFA) in cheese slurries by various sources of lipases according to the ratios in percent grouped by principal component analysis. The lipases were from: (a) Group 1—lamb (L1, L2), kid (K1, K2), kid and lamb (KL), and calf (C1, C2); (b) Group 2—pig (P1–4) and *Pseudomonas fluorescens* (Pf); (c) Group 3—*Rhizopus delema* (Rd), *Rhizopus arrhizus* (Ra), *Mucor miehei* (Mm), *Chromobacterium viscosum* (Cv), and *Aspergillus oryzae* (Ao); (d) Group 4—wheat germ (Wg), *Mucor bacillus* (Mb), and natural cheese (Nc) as a control. Redrawn and adapted from Kwak *et al.* (1989).

5.4.3 Flavor characteristics of lipolytic products

Chemical compounds associated with rancid flavor are relatively well known, and are primarily, if not exclusively, volatile free fatty acids of butyric, caproic, caprylic, capric and lauric acids (Shipe, 1980). Flavor characteristics of these fatty acids, as well as their flavor threshold values, are summarized in Table 5.5. Studies suggest that no single fatty acid is a predominant contributor to the rancid flavor in milk, although it is associated with short-chain fatty acids and, rarely, long-chain (C14 to C18) fatty acids (Scanlan *et al.*, 1965; Bills *et al.*, 1969). Sodium salts of capric and lauric acids appear to be associated with the 'soapy' taste of rancidity (Bills *et al.*, 1969). However, the relationship between rancid flavor in cheese, and short-chain free fatty acids is less clear in the literature (Lin and Jeon, 1987). Normal Cheddar cheese contains a considerable amount of free fatty acids, and literature suggests that short-chain fatty acids play a significant role in the aged flavor of Cheddar cheese (Kristofferson and Gould, 1958; Patton, 1963; Ohren and Tuckey,

1965). Generally, young Cheddar cheese contains low levels of free fatty acids, whereas aged, desirably flavored, Cheddar cheese has intermediate concentrations. Literature also suggests that excess free fatty acids cause the rancid off-flavor in cheese (Bills and Day, 1964; Woo and Lindsay, 1982). Ohren and Tuckey (1965) reported that the concentrations of specific free fatty acids had an important influence on the flavor of cheese, because samples of cheeses that had abnormally high amounts of C10, C12, and C14 acids had unclean and rancid flavors. In cheeses having the most desirable flavor, the concentrations of C10, C12, and C14 acids were always much lower than in those cheeses having fermented, unclean and rancid flavours. Woo and Lindsay (1982) reported that large amounts of free fatty acids, especially C10 and C12 acids, caused soapy flavor in cheese.

Table 5.5 Flavor characteristics and flavor threshold values of short-chain free fatty acids in milk

Fatty acids	Flavor characteristic (in milk)[1]	Flavor threshold (ppm)[2] Milk	Oil
Butyric	Butyric	25.0	0.6
Caproic	Cowy (goaty)	14.0	2.5
Caprylic	Cowy, goaty	–	350
Capric	Rancid, unclean, bitter, soapy	7.0	200
Lauric	Rancid, unclean, bitter, soapy	8.0	700

1 Al-Shabibi *et al.* (1964)
2 Kinsella (1969)

5.5 Microbial flavors

A variety of flavor taints may result in both milk and other dairy products from an accumulation of the metabolic products of contaminating microorganisms (Margalith, 1981; Bassette *et al.*, 1986). The types of flavor defects include acid, bitter, fruity (or fermented), malty, putrid and unclean (Table 5.3). The nature and intensity of the defects depends on the types and numbers of contaminating microorganisms, and on the products associated with the organisms. Bassette *et al.* (1986) stated that milk produced in mammalian glands is considered sterile. However, as soon as milk travels through the udder, contamination occurs. The numbers and kinds of organisms associated with this contamination are influenced by (i) the environmental conditions of the dairy farm; (ii) the health of the animal; (iii) equipment; and (iv) personnel related to the collection, storage and transportation of the milk to processing plants. Likewise, the numbers and kinds of organisms associated with the finished dairy products are influenced by the time and temperature of processing, and by post-pasteurization contamination from equipment, containers, environment and personnel, as well as storage temperature and handling of the product by truck haulers, sales personnel, and consumers.

5.5.1 Acid flavor

Acid defect occurs in all dairy products, and terms such as 'high acid' and 'sour' are also used to describe various intensities of the same defect in different dairy products (Tables 5.1 and 5.2). The acid taints result from acid production by lactose-fermenting organisms, or from acids added in acidified products. The term 'high acid' is used more commonly in fermented dairy products (Table 5.2), whereas the term 'sour' is less frequently used today. Shipe (1980) stated that 'sour' was an appropriate term for fluid milk prior to the use of mechanical refrigeration and pasteurization because the predominant contaminating organism was the fast acid-producing *Streptococcus lactis*. The principal acid-producing bacteria in milk are *Streptococcus, Pediococcus, Leuconostoc, Lactobacillus* and members of the family Enterobacteriaceae (Gilmour and Rowe, 1981).

In milk, fermentation produces lactic acid and other by-products, which yield unpleasant, somewhat disagreeable odors. The sense of smell detects this odor easily at low levels of the acid. As the fermentation progresses and more acid is produced, the odor becomes less offensive and the acid taste more pronounced (Nelson and Trout, 1981). In fermented dairy products, excessive acid beyond what is considered desirable causes the high acid defect. This is one of the common defects found in aged Cheddar cheese.

5.5.2 Bitter flavor

Bitter taints can occur in almost all dairy products, but they can arise from a number of different causes. Bitterness may arise from the results of microbiological and enzymatic activities on proteins. Bitterness may also arise from accumulation of certain fatty acids (particularly capric and lauric acids) following milk fat lipolysis, and from certain alkaloids associated with weed intakes by the cow.

Bitter flavor in pasteurized milk is often associated with the growth of psychrotrophic bacteria and production of proteolytic enzymes, which can survive pasteurization and other heat treatments and remain active in the milk. Because the growth of psychrotrophs is rather slow at 4°C or lower, the resulting flavor defect usually become evident upon extended storage of milk (Shipe *et al.*, 1978).

Bitterness is a common defect in ripened hard cheese (Margalith, 1981). As a matter of fact, it is one of the most common defects in aged Cheddar cheese. Bitter flavor in hard cheese is generally attributed to a number of amino acids and peptides produced by contaminant bacteria acting on milk proteins, as well as by milk-coagulating agents added during manufacture. Bitter flavor in cultured dairy products can also be related to strains of the culture organisms used. A number of researchers have attempted to isolate and identify bitter components in hard cheese. Harwalkar and Elliott (1971) extracted an extremely bitter fraction from Cheddar cheese with chloroform and methanol. Champion and Stanley (1982) reported that valine and leucine occurred at higher levels in the bitter fraction than in the non-bitter fraction when separated by high-performance liquid chromatography, whereas lysine values were elevated in the bitter fraction separated by gel chromatography. Edwards and Kosikowski (1983) reported that bitter peptide fractions of Cheddar cheese rendered by milk-coagulating agents contained relatively large amounts of aliphatic, acidic and hydroxy amino acids, but small amounts of basic and aromatic acids. A comprehensive review on bitter components in dairy products can be found elsewhere (Schmidt, 1990).

5.5.3 Fruity flavor

A flavor defect that has been described as strawberry-like, ester-like or fruity occurs in pasteurized milk and other dairy products as a result of the growth of psychrotrophic bacteria. The term 'fruity' for this defect seems to have priority by virture of common usage (Bassette *et al.*, 1986). The ADSA scorecard lists fermented/fruity as the same category of defect. The fruity defect in pasteurized milk is produced primarily by *Pseudomonas fragi*, a gram-negative non-fermentative short rod. The organism is sensitive to heat; therefore, its presence in pasteurised milk products is most likely from post-pasteurization contamination. At 5 to 7°C, this organism outgrows most other species in milk (Shipe *et al.*, 1978). The organism is aerobic; therefore, proliferation and off-flavor development are enhanced by aeration of milk. In agitated milk inoculated with *P. fragi*, fruity aroma has been noted when the plate counts reached 5×10^8 CFU/ml (Reddy *et al.*, 1969). This flavor defect is usually evident in pasteurized milk stored for an extended period of time under refrigeration. The flavor defect in cottage cheese is also known to be associated with psychotrophic bacteria (Margalith, 1981). In Cheddar cheese, however, the flavor defect is more often associated with the starter culture used. The fruity flavor is one of the more common defects in cheese and has been described as pineapple-, apple- or pear-like in the literature (Bills *et al.*, 1965). Perry (1961) reported that among three strains of *Streptococcus lactis* and *Streptococcus cremoris* used as single-strain cultures for manufacturing Cheddar cheese, the *S. lactis* cultures consistently produced cheese with a flavor described as fruity or dirty, whereas the *S. cremoris* cultures produced cheese with normal flavor. Vedamuthu *et al.* (1964) also reported that the fruity defect of Cheddar cheese resulted from use of mixed-strain starter cultures that contained predominantly *S. lactis* or *Streptococcus diacetylactis*. However, the cultures of predominantly *S. cremoris* did not produce the defect.

The chemical compounds associated with the fruity aroma in milk and other dairy products are ethyl esters. Reddy *et al.* (1968) isolated ethyl acetate, ethyl butyrate and ethyl hexanoate from distillates of the cultures of *Pseudo-*

monas fragi that allowed a typical fruity aroma to develop. They reported that the most abundant esters with fruity odors were ethyl butyrate and ethyl hexanoate. Cultures that developed a strong fruity aroma were found to contain these compounds in concentrations of 0.35 and 0.50 ppm, respectively. Wellnitz-Ruen *et al.* (1982) reported that commercial milks judged to be fruity by sensory analysis were found to contain ethyl butyrate and ethyl hexanoate concentrations of 0.026–0.152 and 0.20–0.268 ppm, respectively.

Bills *et al.* (1965) reported that the most important constituents of fruity flavor in Cheddar cheese were also ethyl butyrate and ethyl hexanoate. They observed that these two esters were present at levels from two to ten times greater in the fruity samples than in normal cheese. Much higher levels of ethanol also were found in the fruity samples. Similar results were reported on the enhancement of a fruity aroma by addition of ethanol to the *Ps. fragi* cultures (Reddy *et al.*, 1968). It appears that excessive production of ethanol may be responsible for the enhanced esterification of some free fatty acids, which results in higher concentrations of ethyl esters. The formation of esters in milk or cheese is related to esterases produced by culture or psychrotrophic organisms.

5.5.4 Green flavor

Under certain ill-defined conditions, lactic cultures develop a flavor defect described as green or yogurt-like. In fermented milk products, the ratio of diacetyl to acetaldehyde is important for the flavor defect. Lindsay *et al.* (1965) reported that the diacetyl-to-acetaldehyde ratio was approximately 4 to 1 in desirably flavored mixed-strain butter cultures. When the ratio of the two compounds was lower than 3 to 1, a green flavour was detected.

5.5.5 Malty flavor

Malty flavor occasionally occurs in a number of different dairy products, including milk, butter, cottage cheese, Cheddar cheese and yogurt (Margalith, 1981; Nelson and Trout, 1981), although the defect is listed in the current ADSA scorecards only for cottage cheese. The flavor defect has been described as cooked, burnt, caramel, grape nuts and malty (Shipe *et al.*, 1978). It may develop in raw milk or in a cultured dairy product as a result of the metabolism of *Streptococcus lactis* subsp. *maltigenes* (Gordon *et al.*, 1963). The defect is due principally to the production of 3-methyl butanal from leucine (Jackson and Morgan, 1954; Morgan *et al.*, 1966). The other compounds involved are 2-methyl propanal, 2-methyl propanol and 3-methyl butanol.

5.5.6 Moldy or musty flavor

A moldy or musty flavor defect often resembles the odor of a damp, poorly ventilated cellar. In Cheddar cheese, the moldy defect is attributed to the growth of mold on cheese surfaces because of lost integrity of the cheese package and the admittance of air. In cottage cheese, the musty defect is attributed to microbial contaminants, particularly molds and psychrotrophic bacteria. In butter, the primary cause of the off-flavor is the growth of psychrotrophic bacteria, particularly *Pseudomonas taetrolens*. A musty taint may also be the result of storing cream in a damp, musty-smelling space or poorly ventilated room, or of using improperly cleaned cream separators or unwashed cream cans (Bodyfelt *et al.*, 1988).

5.5.7 Miscellaneous microbial flavors

Fermented, unclean and putrid flavors in milk may be related to the growth of psychrotrophic bacteria. Many of the organisms produce extracellular proteinases and lipases, which can survive pasteurization and act directly on micellar casein or on the fat globules in milk (Bassette *et al.*, 1986). The combinations of these enzymes give rise to mixtures of off-flavor taints, which are difficult to characterize. Different mixtures of lipid and protein degradation products are probably responsible for fermented, unclean and putrid flavors (Shipe, 1980). Protein degradation by psychrotrophs may also produce bitter flavor in milk products as discussed previously.

It was reported that Cheddar cheese (39% moisture) containing high levels of *Candida* species developed a 'fermented, yeasty' flavor defect after six months of storage (Horwood *et al.*, 1987). The cheese contained elevated levels of ethanol, ethyl acetate, and ethyl butyrate.

A strain of *Pseudomonas taetrolens* isolated from ripened munster cheese was responsible for a potato-like off-flavor (Gallois *et al.*, 1988).

5.6 Heat-induced flavors

The off-flavors associated with heat treatment can be potential flavor defects for almost all dairy products (Tables 5.1 and 5.2). Shipe *et al.* (1978) have classified the off-flavors for milk into four types: (i) cooked or sulfurous; (ii) heated or rich; (iii) caramelized; and (iv) scorched. The off-flavors in various dairy products are usually associated with the heat treatment of milk or cream that is used for manufacturing the products. Heating time and temperature of products, as well as heating methods used, govern the nature and intensity of the off-flavor. For example, heating milk to pasteurization time and temperature imparts a slight cooked or sulfurous taint, whereas heating above the normal pasteurization temperatures generates pronounced cooked flavors (Gould and Sommer, 1939; Bassette *et al.*, 1986). Cooked flavor is the most common terminology used to describe the heat-induced off-flavor in various dairy products (Tables 5.1 and 5.2). Ultra-high temperature processed (UHT) milk (135 to 150°C for several seconds) contains strong, unpleasant taste and smell, resembling hydrogen sulfide, carbon disulfide and boiled cabbage flavors, immediately after processing (Ashton, 1965). However, the unpleasant smell and taste dissipate rapidly and, after several days of refrigerated storage, a rich or heated flavor is perceptible in the UHT milk. It appears that the terms cooked and heated flavors are used indistinguishably for UHT milk on many occasions, however. The UHT milk normally contains stronger cooked flavor than pasteurized milk, although it may fluctuate from 'mild' to 'pronounced' (Hostettler, 1981). The intensity of cooked flavor in UHT milk may be dependent on the age of the milk, as well as on storage temperature. The cooked flavor usually declines sharply during the first month after processing and continuously declines through room temperature storage, although it still remains in the product after several months of storage (Rerkrai *et al.*, 1987). At refrigeration temperature, however, the cooked flavor dissipates more slowly and stays relatively high for a long period of time. The sensitivity to cooked flavor varies considerably from one group of people to another. Burton (1988) stated that some people prefer a strongly heated flavor in milk, perhaps because they have traditionally drunk boiled or in-container sterilized milk. To such people, the cooked flavor in UHT milk would not be considered undesirable. A caramelized flavor is often associated with retorted and autoclaved milk, whereas a scorched flavor can result from exceptionally large amounts of 'burn-on' in a heat exchanger. This flavor also occurs in dry milk powders subjected to abnormally high temperature processing (Shipe *et al.*, 1978).

The chemical compounds associated with heat-induced flavors are complex. In the literature, a variety of volatile flavor compounds are implicated for cooked, heated and caramelized flavors (Bassette *et al.*, 1986). Sulfur compounds have been considered as major contributors to heat-induced flavors in milks. It is well known that a number of sulfur compounds can be readily formed in milk from sulfur-containing amino acids, if a sufficient heat above the pasteurization temperature is applied. Sulfur compounds, such as hydrogen sulfide and methyl mercaptan, were implicated for cooked flavor in pasteurized milk as early as the 1930s (Gould and Sommer, 1939). Jaddou *et al.* (1978) reported that cabbage-like flavor in UHT milk was correlated with total volatile sulfur compounds and concluded that compounds such as hydrogen sulfide, carbonyl sulfide, methanethiol, carbon disulfide, and methyl disulfide could be responsible for the off-flavor. Behaviour of hydrogen sulfide in UHT milk during aseptic storage is illustrated in Figure 5.2 (Slinkard, 1976). The analysis was based on the headspace concentration of H_2S in 50 ml UHT milk (indirect-processed) that was exposed to a 55°C water-bath for 1 h in a 125 ml serum bottle. H_2S was detected by a GC flame photometric detector (FPD) at 394 nm. The column used was a 10.97 m FEP Teflon tubing (3 mm o.d. × 2 mm i.d.) containing Haloport-F Teflon powder (40/60 mesh, F&M Scientific). It shows that the initial concentration of H_2S was about 0.5 µg per 5 ml headspace. At 22°C storage, the concentration of H_2S decreased rapidly and was not detectable after 15 days. However, when the UHT milk was stored at

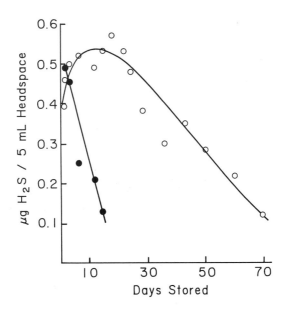

Figure 5.2   Hydrogen sulfide concentration in the headspace of UHT milk during storage at 4°C
(○) and 22°C (●). Redrawn and adapted from Slinkard (1976).

4°C, the concentration of H₂S increased slightly to a maximum at about 10
days and then showed a slow decline through 70 days. Considering that
cabbage-like flavor declines rapidly within a few days of processing (Thomas
et al., 1975) but the concentration of hydrogen sulfide can still be detectable
for up to 15 days at 22°C, hydrogen sulfide may be involved with cooked
flavor in UHT milk.

Cooked flavor in UHT milk may be caused by a combination of many
different types of flavor compounds. This may include some sulfur compounds, methyl ketones and lactones. As illustrated in Figure 5.3, some sulfur
compounds dissipate rapidly at room temperature storage. However, some
other sulfur compounds are relatively slow to disappear or show no sign of
decrease in the milk even after 29 days (Jeon, 1976). Methyl ketones with odd
numbers of carbon atoms are well known to be heat-induced products, originating from β-keto alkanoic acids in milk fat (Kinsella, 1969). However, their
significance in heated milk flavor is not established. Diacetyl in UHT milk is
in the range of 16 to 38 ppb (Scanlan *et al.*, 1968; Jeon *et al.*, 1978) and is
one of the most significant flavor compounds. Diacetyl imparts a nutty flavor,
has very low threshold (12 ppb), and has been implicated in giving 'rich'
flavor in UHT milk (Scanlan *et al.*, 1968). Lactones may also be important in
exerting characteristic heated flavor in UHT milk. A number of lactones were
found in UHT milk, including γ-decalactone, δ-decalactone and δ-dodecalac-
tone (Badings and Neeter, 1980; Coulibaly, 1990). These lactones have a

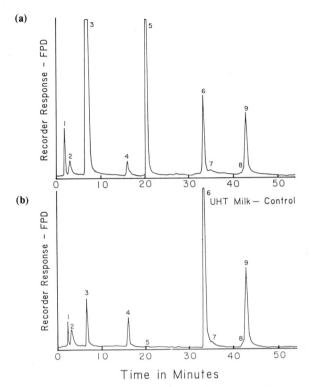

Figure 5.3 Effect of dissolved oxygen concentration on changes of sulfur compounds in UHT milk during storage. (a) UHT milk with ascorbic acid added, stored for 29 days at 22°C, dissolved oxygen 1.0 mg/l; (b) UHT milk control, stored for 29 days at 22°C, dissolved oxygen 3.4 mg/l. 1 = CH_3SCH_3; 2 = solvent; 3 = CH_3SSCH_3; 4 = unknown; 5 = isopropyldisulfide; 6 = unknown; 7 = unknown; 8 = benzothiazole; 9 = unknown. Redrawn and adapted from Jeon (1976).

peach or coconut-like flavor but could give different odor to milk when their flavor is mixed with others. More studies are probably needed to understand the role of lactones and other compounds in UHT milk flavor.

5.7 Oxidized flavor

Oxidized flavor is the general term applied to the flavor defect produced as a result of a reaction between molecular oxygen and milk lipids. The off-flavor is more common in milk and non-fermented dairy products, particularly those that can be stored for a long period of time. Several terms have been used to describe the off-flavor in dairy products, including cardboard, metallic, oily, stale, tallowy, painty and fishy. This wide difference in the off-flavor description is due mainly to variation in its character and intensity among many dairy products at the various stages of oxidation. For example, descriptive terms such as cardboard or papery are often used to characterize the off-flavor at the early stage of metal-induced oxidation of fluid milk. In ice cream, the off-flavor may be described as flat at the very early stage of oxidation. With a moderate progress of the reaction, the flavor may often be described as metallic, papery or cardboard. In the most intense stage of the oxidation, oily, tallowy, painty or fishy are common descriptors. Butter undergoes a continuous change in flavor defects during storage, and associative terms such as metallic, oily, tallowy, painty and fishy have been used to describe the various stage of the flavor defect development. The ADSA scorecard for butter currently contains three different entries for oxidized off-flavor: (i) metallic; (ii) oxidized; and (iii) tallowy. In concentrated and dry milk products, the oxidized flavor is often described as metallic, cardboard, oily or tallowy, depending on its intensity. In UHT milk, the flavor defect is often described as stale/oxidized. More details on the sensory perception of the oxidized flavors in dairy products can be found elsewhere (Bodyfelt et al., 1988).

5.7.1 Mechanisms and conditions

There are many excellent review articles and books on the mechanism of lipid oxidation (Korycka-Dahl and Richardson, 1980; Nawar, 1985; Weihrauch, 1988). In butter, which represents an aqueous concentration of phospholipid dispersed in fat, both components are susceptible, with the phospholipid being most readily oxidized (Patton, 1962). In dried dairy products and in anhydrous milk fat, the triglycerides may serve as a major reactant in oxidation (Day, 1966). In fluid milk, the oxidized flavor arises primarily from the oxidation of polyunsaturated fatty acids in the phospholipids of the fat globule membrane. However, milk varies considerably in its susceptiblity to the development of oxidized flavor. Principal catalysts for the oxidation of milk fat are copper, light and, of less importance, iron (Thomas, 1981). Several other substances are known to be involved in the autoxidation reactions. A low level of ascorbic acid in milk promotes copper-induced oxidation (Krukovsky, 1961). Some metalloproteins, notably milk peroxidase and xanthine oxidase, also promote autoxidation, but not by their enzymatic action. The cytochrome of the fat globule membrane may be involved in the initiation, perhaps by formation of singlet oxygen (Walstra and Jenness, 1984).

Although there has been a marked decrease in the incidence of copper-induced oxidized flavor in fluid milk because of the near elimination of copper alloys from dairy equipment, the role of the copper ion is still significant in oxidized flavor development. Naturally, milk contains a fair amount of copper (usually 20–25 µg/kg milk), part of which is in the fat globule membrane (10 µg/100 g fat globules). Copper can come from contamination, such as from the surface of the udder, from milking and processing equipment, and from water. This contaminant copper is mostly between 20 and 500 µg/kg milk and is much more active as a catalyst than natural copper ions in milk (Walstra and Jenness, 1984).

Oxidized flavor is also induced by light, particularly fluorescent light and sunlight, through photooxidation of lipid components. This off-flavor should be distinguished from 'sunlight' flavor, which is attributed to the degradation of serum proteins. The extent of deterioration of dairy products appears to be dependent on the intensity of the source and the length of exposure, as well as on the wavelength involved (Aurand et al., 1966; Dunkley et al., 1962). Although ultraviolet light is generally believed to be the cause, the literature is not clear about the relative importance of different wavelengths, particularly those emitted by fluorescent light (Emmons et al., 1986). Du and Armstrong (1970), using filters, found that most oxidation at 443 and 576 nm was from Warm-White fluorescent light, whereas most at 576 nm was from Cool-White light. Sattar et al. (1977), using sharp-cut filters on Cool-White fluorescent light, found large increases in peroxides in milk fat as the bands changed from 595 to 560 nm and went below 415 nm. Emmons et al. (1986) reported that yellow plastics, which transmitted very little light below 500 nm from Cool-White fluorescent light, did not appreciably reduce oxidation of butter.

Light also catalyzes oxidative deterioration of vitamin A in milk, which results in an off-flavor that has been described as haylike, strawlike, and raspberry (Coulter and Thomas, 1968). The reaction occurs frequently in lowfat milks and skim milk fortified with vitamin A. Recently, Fellman et al. (1991) reported that the off-flavor development was faster in skim and 2% milks fortified with an oil-based vitamin A than in an aqueous-based one, when each milk was exposed to the same intensity of fluorescent light. Although the mechanism for vitamin A destruction by light is not well understood, loss of vitamin A activity may result from an opening of the β-ionone ring portion of the vitamin A molecule (Wishner, 1965).

Dissolved oxygen plays an important role in the development of oxidized/stale flavor in UHT milk during storage. Zadow and Birtwistle (1973) reported that UHT milk with a high initial oxygen content (pO_2 142 mmHg) developed oxidized or rancid flavors after extended storage, although its flavor was good during the first few weeks of storage. Thomas et al. (1975) observed that UHT milk with a high initial oxygen content (8.9 ppm) developed a stale/oxidized flavor sooner than milk with a low initial oxygen content (1.0 ppm). However, it is questionable whether dissolved oxygen significantly affects development of stale/oxidized flavor in direct UHT milk during storage, because the concentration of dissolved oxygen is very low in such commercially processed milks (Rerkrai et al., 1987).

In packaged dry whole milk, elimination of oxygen is effective in preventing or retarding the onset of oxidation during extended periods of storage. This is usually done by vacuum packaging or replacement of available oxygen with an inert gas. Literature indicates that inert gas packaging to an oxygen level of 3–6% significantly increases the storage life of whole milk powders (Lea *et al.*, 1943; Greenbank *et al.*, 1946). Reports also indicate that inert gas packaging at a 0.5–1.0% oxygen level extensively retards development of recognizable tallowy flavor in milk powders (Lea *et al.*, 1943; Schaffer *et al.*, 1946).

Heat treatment and storage temperatures affect the development of oxidized flavor in various dairy products. Pasteurization of fluid milk increases susceptibility to the development of copper-induced and light-induced oxidized flavors (Smith and Dunkley, 1962). However, heating to higher temperatures (76°C or higher) reduces the susceptibility. The inhibitory effect of high heat treatment on oxidative deterioration of fluid milk and its products has been primarily attributed to the formation of sulfhydryl compounds by heat treatment of serum proteins (Gould and Sommer, 1939; Josephson and Doan, 1939). The sulfhydryl groups lower the oxidation-reduction potential of the dairy products. The effect of storage temperature on development of oxidized flavor is anomalous for fluid milks. Dunkley and Franke (1967) reported that more intense oxidized flavors and higher TBA (2-thiobarbituric acid) values were observed with fluid milks stored at 0°C than with milks stored at 4°C or 8°C. This may be related to a decrease in bacterial growth at the lower temperature. Lower bacterial growth could reduce the competition for oxygen and/or the production of reducing substances. The control of bacterial growth has led to prolonged storage of milk, which provides enough time for chemical and enzymatic reactions to occur and may contribute to the flavor problem (Bassette *et al.*, 1986). In other dairy products, however, low storage temperatures tend to decrease or inhibit oxidative deterioration. This is particularly true with dry whole milk and UHT milk products. Pyenson and Tracy (1946) reported that a storage temperature of 2°C significantly retarded the development of oxidative deterioration of dry whole milk as compared to samples stored at 38°C in an atmosphere of air. Downey (1969) reported that oxidative deterioration of UHT cream occurred two to three times faster at 18°C than at 10°C, whereas at 4°C little or no oxidation occurred. A number of researchers indicated that stale/oxidized flavor development in UHT milk is highly dependent on storage time and temperature (Renner and Schmidt, 1981; Rerkrai *et al.*, 1987). The rate of oxidative flavor deterioration of butter or butter oils also increases as storage temperature increases from subzero to room temperatures (Sattler-Dornbacher, 1963; Hamm *et al.*, 1968).

Many other conditions affect oxidized flavor development. These may include feed, stage of lactation, homogenization of milk, acidity (particularly in butter) and antioxidants.

5.7.2 Reaction products

Lipid oxidation leads to the production of various reaction end-products that impart oxidized flavor in milk and other dairy products. As shown in Table 5.6, most chemical compounds isolated from various off-flavored dairy products (oxidized, metallic, etc.) belong to four classes of carbonyls: (i) *n*-alkanals; (ii) *n*-alk-2-enals; (iii) *n*-alk-2,4-enals; and (iv) alk-2-ones. These carbonyls posses strong and characteristic odors and impart objectionable off-flavors to dairy products even at extremely low concentrations (Day *et al.*, 1963). Threshold values for most of these carbonyls in milk or oil are in parts per million to parts per billion (Kinsella, 1969). Consequently, the degree of lipid oxidation required to produce off-flavors is quite small. In addition, these aldehydes and ketones are known to have a synergistic or additive effect at sub-threshold concentrations in dairy products (Day, 1966). Considering that most of these carbonyls are at sub-threshold concentrations in oxidized dairy products, most off-flavors are believed to result from the combined effect of various carbonyls and other oxidative compounds. On the other hand, the particular oxidized off-flavor (metallic, tallowy, etc.) in various dairy products appears to be due to the dominance of certain carbonyls in that particular product.

Table 5.6 Some descriptive flavors and chemical compounds associated with oxidized dairy products

Flavor	Compounds	Reference
Oxidized		
Cardboard	Octanal, 2-heptenal, 2,4-heptadienal, oct-1-en-3-one	1,4,5,8
Spontaneous	2,4-Decadienal	7
Sunlight	Alk-2-enals (C6–C11)	9
Metallic	Oct-1-en-3-one	8
Oily	*n*-Alkanals (C6–C7), 2-hexenal, 2-heptanone	2
Fishy	Oct-1-en-3-one, *n*-alkanals (C5–C10), *n*-alk-2-enals (C3, C5, C6, C8, C9), 2,4-heptadienal	2
Painty	*n*-Pentanal, *n*-alk-2-enals (C5–C10)	2,3
Stale/oxidized	*n*-Alkanals (C3, C5–C9)	6
Tallowy	*n*-Alkanals (C7–C9), *n*-alk-2-enals (C7, C9) 2-heptanone	2,3

1 Forss *et al.* (1955a, b); 2 Forss *et al.* (1960a, b); 3 Forss *et al.* (1960c); 4 Hammond and Hill (1964); 5 Hammond and Seals (1972); 6 Jeon *et al.* (1978); 7 Parks *et al.* (1963); 8 Stark and Forss (1962); 9 Wishner and Keeney (1963)

Literature on the compounds isolated and identified from oxidized dairy products is abundant. Forss *et al.* (1955a,b) reported that unsaturated carbonyls, particularly 2-octenal, 2-nonenal, 2,4-heptadienal, and 2,4-nonadienal, contributed to a basic and characteristic cardboard flavor in skim milk in which oxidation was induced by copper. Stark and Forss (1962) isolated oct-1-en-3-one from oxidized dairy products and identified it as the compound responsible for metallic flavor. Day *et al.* (1963) agreed that oct-1-en-3-one gave a metallic flavor to milk, but stressed that this flavor was distinct from the oxidized flavor of milk. Hammond and Hill (1964) reported that a typical oxidized flavor was produced if an aldehyde such as 2-heptenal, octanal or

2,4-heptadienal was added to milk along with oct-1-en-3-one. Later, Hammond and Seals (1972) also reported that addition of 1–10 ppm of oct-1-en-3-one and octanal to homogenized milk produced a cardboard flavor. Stark and Forss (1966) isolated *n*-alkanols (C2–C9) in oxidized butter. Parks *et al.* (1963) implicated alk-2,4-enals, especially 2,4-decadienal, in the off-flavor associated with spontaneously oxidized fluid milk. Wishner and Keeney (1963) concluded that alk-2-enals (C6–C11) were important contributors to the oxidized flavor in milk exposed to sunlight. Bassette and Keeney (1960) ascribed the cereal-like flavor in dry skim milk to homologous saturated aldehydes resulting from lipid oxidation, in conjunction with products of the browning reaction. Forss *et al.* (1960a,b) reported that the fishy flavor in butterfat and washed cream is a mixture of an oily fraction and oct-1-en-3-one, the compound responsible for the metallic flavor. The oily fraction from washed cream contained *n*-hexanal, *n*-heptanal, and 2-hexenal, whereas the oily fraction from fishy butterfat contained these three aldehydes plus 2-heptanone. Forss *et al.* (1960c) also reported that the same type of carbonyl compounds were isolated from butterfat with tallowy and painty flavors as from fishy flavored butterfat or washed cream. They observed relative increases in *n*-heptanal, *n*-octanal, *n*-nonanal, 2-heptanone, hept-2-enal and non-2-enal in the tallowy butterfat, and relative increases in *n*-pentanal and alk-2-enals (C5–C10) in the painty butterfat. However, their sensory results indicated that *n*-octanal and *n*-nonanal provided elements in the tallowy flavor, whereas *n*-pentanal and pent-2-enal were important in the painty flavor. Jeon *et al.* (1978) observed significant increases in 2-methyl ketones (C3–C13), *n*-aldehydes (C3, C5–C9), and 1-butanol in UHT milk stored at various temperatures for 5 months. Although 2-methyl ketones were most abundant, they concluded that saturated aldedyes were the significant contributors to stale/oxidized off-flavor in stored UHT milk. Other researchers observed an increase in stale off-flavor paralleling increases in acetaldehyde, propanal, pentanal and hexanal (Mehta and Bassette, 1978; Bassette and Jeon, 1983). However, stale flavor development in UHT milk may also be related to other chemical reactions, such as non-enzymatic browning reactions and lipolysis of milk fat.

5.8 Miscellaneous off-flavors

This section describes off-flavors that are encountered occasionally in milk and dairy products, or those that cannot be either attributed to a specific cause or specifically defined in sensory terms.

Absorbed flavor.   This term is applicable to flavors that are absorbed directly from the environment. Some volatile substances may be absorbed directly from the air. For example, fat-soluble substances, such as turpentine and other volatile solvents, are readily absorbed into fats in dairy products from surrounding air (Anon, 1965).

Acetaldehyde flavor.   This term is applied to a flavor defect caused by an excessive production of acetaldehyde in milk cultures or yogurt and frequently referred to as coarse, green or green-apple defect. Within a normal range (5–40 ppm), acetaldehyde is primarily responsible for the characteristic plain yogurt flavor. However, it tends to give an undesirable flavor above the normal range (Bodyfelt *et al.*, 1988).

Astringent flavor.   This term has been used to describe a dry, puckery, oral sensation, which involves the sense of touch or feel rather than taste. Terms such as rough, chalky or powdery have also been used to describe this sensation. The defect is not very common in fluid milk. However, it may occur in milk products that have been processed at a high temperature. In this regard, astringency has been attributed to heat-altered whey proteins and, to some extent, to milk salts. Protein particles with low mineral contents are responsible for astringency in acidified milk products. Astringent flavors may also be produced when milk is fortified with iron salts, especially ferrous salts (Bassette *et al.*, 1986). Harwalkar *et al.* (1989) reported that the astringent off-flavor components isolated from UHT-sterilized milk were γ-casein-like breakdown products of casein.

Chalky flavor.   This term is used to describe a tactual effect, which is similar if not identical to astringent. This off-flavor is common in concentrated milk products, suggesting that the inclusion of fine, insoluble, chalk (powder) particles is the cause (Bodyfelt *et al.*, 1988). The chalky off-flavor is as much an objectionable mouthfeel sensation as it is an off-taste.

Cheesy flavor.   This term is used to describe a flavor defect in butter, milk cultures and cultured milk, which resembles Cheddar cheese flavor. Cheesy flavor in cultures is very uncommon and is more often associated with cultures that have been held for some time (Nelson and Trout, 1981).

Chemical flavors.   The off-flavors included in this group are caused by the contamination of milk and milk products with a variety of chemicals that are associated with cleaners, sanitizers and disinfectants (Shipe, 1980). The most frequent contaminants are chlorine and iodine compounds from sanitizers. Occasionally, milk may also be contaminated with traces of phenolic compounds from disinfectants or pesticides. A chlorophenol flavor in milk may be attributed to a reaction product formed as a result of chlorine treatment of water, and phenol in a water supply.

Diacetyl flavor.   This term is used to note an overall lack of flavor balance in cottage cheese because of domination of the distinct aroma of diacetyl, which masks other important flavor notes. It is often characterized by the presence of a harsh flavor and/or excess aroma, which indicates lack of balance for cottage cheese (Bodyfelt *et al.*, 1988).

Flat. This term is used to describe an absence of the characteristic flavor and aroma. The defect may occur in butter, cottage cheese, cultured milk, dry milk products and Cheddar cheese. In milk, addition of as little as 3–5% water can produce this defect. A flat taste also occurs in ice cream and creamed cottage cheese during an early or intermediate state of the development of an oxidized off-flavor (Bodyfelt *et al.*, 1988).

Foreign flavor. This term refers to a flavor defect that cannot be identified either by cause or chemical nature. Consequently, the defect usually refers to atypical or most of the unusual flavor taints occurring in dairy products. In milk, however, it is believed that foreign and medicinal flavors are caused by exposure of the milk or cows to medications, disinfecting materials, sanitizers, fly sprays, gasoline or many other compounds commonly used on the farm or in the processing plant. These materials may enter the milk supply directly, as with medications used on the cows' udders, or from improperly-rinsed sanitizing utensils, or disinfectant sprays drifting into open milk containers (Bassette *et al.*, 1986).

Gluey flavor. This term is uncommonly used but refers to the off-flavor related to stale or glue-like flavor in dry milk powders (Lea *et al.*, 1943).

Lacks freshness. This term is generally used to indicate a loss of the fine, pleasing taste qualities typically noted in excellent or high quality products. In milk, the term is often used to describe an 'old' taste or a flavor deterioration that cannot be positively identified. In ice cream, this defect is generally assumed to result from either a general flavor deterioration of the mix during storage or the use of one or more marginal quality dairy ingredients in mix formulations. In yogurt, the defect is related to the use of less than the freshest or highest quality dairy ingredients (Bodyfelt *et al.*, 1988).

Salty flavor. This term refers to an excessive saltiness in milk and other dairy products. In milk, it is most commonly found from cows in late lactation and occasionally from cows with mastitis (Anon, 1965). In some other dairy product, it is due to oversalting of products during manufacture.

Sunlight flavor. This term is applied to describe off-flavor associated with protein degradation in milk by exposure to sunlight. Terms such as activated, burnt, burnt feather, burnt protein, scorched, cabbage and mushroom have also been used in the literature. This off-flavor must be distinguished from oxidized flavor arising in milk by light-induced lipid oxidation. Sunlight flavor has been attributed to several compounds, including methional, methanethiol, methyl sulfides, hydrogen sulfide and some aldehydes (Bassette *et al.*, 1986).

Whey or whey taint. This term is used to describe the presence of whey flavor either because of its retention in butter and Cheddar cheese, or because of excessive use of whey products in ice cream.

Storage flavor. This term is used to describe off-flavor that is similar to a stale cream flavor or a lack of freshness, but is less pronounced and even less persistent than stale flavor (Nelson and Trout, 1981).

Yeasty flavor. This term refers to the off-flavor that resembles yeast fermentation in various dairy products. In Cheddar cheese this off-flavor may be identified by its sour, bread dough, yeasty or somewhat 'earthy' taste and aroma. In creamed cottage cheese, the defect may resemble 'vinegar-like', whereas in butter it may be perceived as fruity, vinegary, yeasty and even slightly fragrant (Bodyfelt *et al.*, 1988).

Acknowledgement

Contribution No. 92-204-B from the Kansas Agricultural Experiment Station, Kansas State University, Manhattan, KS 66506.

References

Al-Shabibi, M. M. A., Langner, E. H., Tobias, J. and Tuckey, S. L. (1964). Effect of added fatty acids on the flavor of milk. *J. Dairy Sci.* **47**, 295–296.

Anon (1965). *Grade 'A' Pasteurized Milk Ordinance: 1965 Recommendations of the US Public Health Service.* Public Health Service Publication 229, US Department of Health, Education, and Welfare, Washington, DC.

Ashton, T. R. (1965). Practical experience: The processing and aseptic packaging of sterile milk in the United Kingdom. *J. Soc. Dairy Technol.* **18**, 65–85.

Aurand, L. W., Singleton, J. A. and Noble, B. W. (1966). Photooxidation reactions in milk. *J. Dairy Sci.* **49**, 138–143.

Badings, H. T. and Neeter, R. (1980). Recent advances in the study of aroma compounds of milk and dairy products. *Neth. Milk Dairy J.* **34**, 9–30.

Bassette, R. and Jeon, I. J. (1983). Effect of process- and storage-times and temperatures on concentrations of volatile materials in ultra-high-temperature steam infusion processed milk. *J. Food Prot.* **46**, 950–953.

Bassette, R. and Keeney, M. (1960). Identification of some volatile carbonyl compounds from nonfat dry milk. *J. Dairy Sci.* **43**, 1744–1750.

Bassette. R., Fung, D. Y. C. and Mantha, V. R. (1986). Off-flavors in milk. *CRC Critical Reviews in Food Science and Nutrition* **24**, 1–52.

Bills, D. D. and Day, E. A. (1964). Determination of the major free fatty acids of Cheddar cheese. *J. Dairy Sci.* **47**, 733–738.

Bills, D. D., Morgan, M. E., Libbey, L. M. and Day, E. A. (1965). Identification of compounds responsible for fruity flavor defect of experimental Cheddar cheeses. *J. Dairy Sci.* **48**, 1168–1173.

Bills, D. D., Scanlan, R. A., Lindsay, R. C. and Sather, L. (1969). Free fatty acids and the flavor of dairy products. *J. Dairy Sci.* **52**, 1340–1345.

Bodyfelt, F. W., Tobias, J. and Trout, G. M. (1988). *The Sensory Evaluation of Dairy Products.* Van Nostrand Reinhold, New York, pp. 141–151; 204–209; 230–235; 271–277; 290–295; 339–416; 460–461.

Burton, H. (1988). *Ultra-High-Temperature Processing of Milk and Milk Products.* Elsevier Applied Science, London, p. 261.

Champion, H. M. and Stanley, D. W. (1982). HPLC separation of bitter peptides from Cheddar cheese. *Can. Inst. Food Sci. Technol. J.* **15**, 383–388.

Cole, D. D., Harper, W. J. and Hankinson, C. L. (1961). Observations on ammonia and volatile amines in milk. *J. Dairy Sci.* **44**, 171–173.

Coulibaly, K. (1990). Solid-phase extraction of volatile flavor compounds from foods. *M.S. Thesis,* Kansas State University, Manhattan, KS.

Coulter, S. T. and Thomas, E. L. (1968). Enrichment and fortification of dairy products and margarine. *J. Agric. Food Chem.* **16**, 158–162.

Day, E. A. (1966). Role of milk lipids in flavors of dairy products. In *Flavor Chemistry.* Ed. R. F. Gould. American Chemical Society, Washington, DC, pp. 94–120.

Day, E. A., Lillard, D. A. and Montgomery, M. W. (1963). Autoxidation of milk lipids. III. Effect on flavor of the additive interactions of carbonyl compounds at subthreshold concentrations. *J. Dairy Sci.* **46**, 291–294.

Dougherty, R. W., Shipe, W. F., Gudnason, G. V., Ledford, R. A., Peterson, R. D. and Scarpellino, R. (1962). Physiological mechanisms involved in transmitting flavors and odors to milk. I. Contribution of eructated gases to milk flavor. *J. Dairy Sci.* **45**, 472–476.

Downey, W. K. (1969). Lipid oxidation as a source of off-flavor development during the storage of dairy products. *J. Soc. Dairy Technol.* **22**, 154–162.

Du, C. T. and Armstrong, J. G. (1970). Light-induced oxidation in milkfat. *Can. Inst. Food Technol. J.* **3**, 167–170.

Duncan, S. E., Christen, G. L. and Penfield, M. P. (1991). Rancid flavor of milk: Relationship of acid degree value, free fatty acids, and sensory perception. *J. Food Sci.* **56**, 394–397.

Dunkley, W. L. and Franke, A. A. (1967). Evaluating susceptibility of milk to oxidized flavor. *J. Dairy Sci.* **50**, 1–9.

Dunkley, W. L. Franklin, J. D. and Pangborn, R. M. (1962). Effects of fluorescent light on flavor, ascorbic acid and riboflavin in milk. *Food Technol.* **16**, 112–118.

Edwards, J. and Kosikowski, F. V. (1983). Bitter compounds from Cheddar cheese. *J. Dairy Sci.* **66**, 727–734.

Emmons, D. B., Froehlich, D. A., Paquette, G. J., Buttler, G., Beckett, D. C., Modler, H. W., Brackenridge, P. and Daniels, G. (1986). Light transmission characteristics of wrapping materials and oxidation of butter by fluorescent light. *J. Dairy Sci.* **69**, 2248–2267.

Fellman, R. L., Dimick, P. S. and Hollender, R. (1991). Photo-oxidative stability of vitamin A fortified 2% lowfat milk and skim milk. *J. Food Prot.* **54**, 113–116.

Forss, D. A., Pont, E. G. and Stark, W. (1955a). The volatile compounds associated with oxidized flavor in milk. *J. Dairy Res.* **22**, 91–102.

Forss, D. A., Pont, E. G. and Stark, W. (1955b). Further observations on the volatile compounds associated with oxidized flavor in skim milk. *J. Dairy Res.* **22**, 345–348.

Forss, D. A., Dunstone, E. A. and Stark, W. (1960a). Fish flavor in dairy products. II. The volatile compounds associated with fishy flavour in butterfat. *J. Dairy Res.* **27**, 211–219.

Forss, D. A., Dunstone, E. A. and Stark, W. (1960b). Fishy flavour in dairy products. III. The volatile compounds associated with fishy flavour in washed cream. *J. Dairy Res.* **27**, 373–380.

Forss, D. A., Dunstone, E. A. and Stark, W. (1960c). The volatile compounds associated with tallowy and painty flavors in butter. *J. Dairy Res.* **27**, 381–387.

Gallois, A., Kergomard, A. and Adda, J. (1988). Study of the biosynthesis of 3-isopropyl-2-methoxypyrazine produced by *Pseudomonas taetrolens. Food Chem.* **28**, 299–309.

Gilmour, A. and Rowe, M. T. (1981) Micro-organisms associated with milk. In *Dairy Microbiology Vol. 1 The Microbiology of Milk.* Ed. R. K. Robinson. Applied Science Publishers, London, pp. 35–75.

Gordon, D. J., Morgan, M. E. and Tucker, J. S. (1963). Differentiation of *Streptococcus lactis* var. *maltigenes* from other lactic streptococci. *Appl. Microbiol.* **11**, 171–177.

Gould, I. A. Jr. and Sommer, H. H. (1939). Effect of heat on milk with especial reference to the cooked flavor. *Mich. Agr. Exp. Sta. Tech. Bull.* **164**, 1–48.

Greenbank, G. R., Wright, P. A., Deysher, E. F. and Holm, G. E. (1946). The keeping quality of samples of commercially dried milk packed in air and in inert gas. *J. Dairy Sci.* **29**, 55–61.

Hamm, D. L., Hammond, E. G. and Hotchkiss, D. K. (1968). Effect of temperature on rate of autoxidation milk fat. *J. Dairy Sci.* **51**, 483–491.

Hammond, E. G. and Hill, F. D. (1964). The oxidized metallic and grassy flavor components of autoxidized milk fat. *J. Amer. Oil Chem. Soc.* **41**, 180–184.

Hammond, E. G. and Seals, R. G. (1972). Oxidized flavor in milk and its simulation. *J. Dairy Sci.* **55**, 1567–1569.

Harwalkar, V. R. and Elliott, J. A. (1971). Isolation of bitter and astringent fraction from Cheddar cheese. *J. Dairy Sci.* **54**, 8–11.

Harwalkar, V. R., Boutin-Muma, B., Cholette, H., McKellar, R. C., Emmons, D. B. and Klassen, G. (1989). Isolation and partial purification of astringent compounds from ultra-high temperature sterilized milk. *J. Dairy Res.* **56**, 367–373.

Heath, H. B. and Reineccius, G. A. (1986). *Flavor Chemistry and Technology.* AVI Publishing Co., Inc., Westport, CT, pp. 245–249.

Horwood, J. F., Stark, W. and Hull, R. R. (1987). A 'fermented, yeasty' flavor defect in Cheddar cheese. *Aust. J. Dairy Technol.* **42**, 25–26.

Hostettler, H. (1981). Appearance, flavor and texture aspects: Developments until 1972. In *IDF New Monograph on UHT Milk.* International Dairy Federation, Brussels, Belgium, Document 133, pp. 11–24.

Jackson, H. W. and Morgan, M. E. (1954). Identity and origin of the malty aroma substance from milk cultures of *Streptococcus lactis* var. *maltigenes. J. Dairy Sci.* **37**, 1316–1324.

Jaddou, H. A., Pavey, J. A. and Manning, D. J. (1978). Chemical analysis of flavor volatiles in heat-treated milks. *J. Dairy Res.* **45**, 391–403.

Jeon, I. J. (1976). Identification of volatile flavor compounds and variables affecting the development of off-flavor in ultra-high-temperature processed sterile milk. *Ph.D. Dissertation*, University of Minnesota, Minneapolis, MN.

Jeon, I. J., Thomas, E. L. and Reineccius, G. A. (1978). Production of volatile flavor compounds in ultra-high-temperature processed milk during aseptic storage. *J. Agr. Food Chem.* **26**, 1183–1188.

Josephson, D. V. and Doan, F. J. (1939). Observations on cooked flavor in milk—its source and significance. *Milk Dealer* **29**(2), 35–54.

Kinsella, J. E. (1969). The flavor chemistry of milk lipids. *Chemistry and Industry* January 11, 36–42.

Korycka-Dahl, M. and Richardson, T. (1980). Oxidative changes in milk: Initiation of oxidative changes in foods. *J. Dairy Sci.* **63**, 1181–1198.

Kristofferson, T. and Gould, I. A. (1958). Characteristic Cheddar cheese flavor in relation to hydrogen sulfide and free fatty acids. *J. Dairy Sci.* **41**, 717. (Abstract).

Krukovsky, V. N. (1961). Review of biochemical properties of milk and the lipid deterioration in milk and milk products as influenced by natural varietal factors. *J. Agr. Food Chem.* **9**, 439–447.

Kwak, H. S., Jeon, I. J. and Perng, S. K. (1989). Statistical patterns of lipase activities on the release of short-chain fatty acids in Cheddar cheese slurries. *J. Food Sci.* **54**. 1559–1564.

Law, B. A., Sharp, M. E. and Chapman, H. R. (1976). The effect of lipolytic gram-negative psychrotrophs in stored milk on the development of rancidity in Cheddar cheese. *J. Dairy Res.* **43**, 459–468.

Lea, C. H., Moran, T. and Smith, J. A. B. (1943). The gas packaging and storage of milk powder. *London J. Dairy Res.* **13**, 162–215.

Lin, J. C. C. and Jeon, I. J. (1987). Effects of commercial food grade enzymes on free fatty acid profiles in granular Cheddar cheese. *J. Food Sci.* **52**, 78–83 and 87.

Lindsay, R. C., Day, E. A. and Sandine, W. E. (1965). Green flavor defect in lactic starter cultures. *J. Dairy Sci.* **48**, 863–869.

Margalith, P. Z. (1981). *Flavor Microbiology.* Charles C. Thomas Publisher, Springfield, IL, pp. 33–118.

Mehta, R. S. and Bassette, R. (1978). Organoleptic, chemical and microbiological changes in ultra-high-temperature sterilized milk stored at room temperature. *J. Food Prot.* **41**, 806–810.

Morgan, M. E. and Pereira, R. L. (1962). Volatile constituents of grass and corn silage. II. Gas-entrained aroma. *J. Dairy Sci.* **45**, 467–471.

Morgan, M. E. and Pereira, R. L. (1963). Identity of grassy aroma constituents of green forages. *J. Dairy Sci.* **46**, 1420–1422.

Morgan, M. E., Lindsay, R. C., Libbey, L. M. and Pereira, R. L. (1966). Identity of additional aroma constituents in milk cultures of *Streptococus lactis* var. *maltigenes. J. Dairy Sci.* **49**, 15–18.

Morr, C. V. and Ha, E. Y. W. (1991). Off-flavors of whey protein concentrates: a literature review. *Int. Dairy J.* **1**, 1–11.

Muir, D. D., Kelly, M. E. and Phillips, J. D. (1978). The effect of storage temperature on bacterial growth and lipolysis in raw milk. *J. Soc. Dairy Technol.* **31**, 203–208.

Nawar, W. W. (1985). Lipids. In *Food Chemistry.* Ed. O. R. Fennema, Marcel Dekker, Inc., New York, pp. 139–244.

Nelson, J. H. (1972). Enzymatically produced flavors for fatty systems. *J. Am. Oil Chem. Soc.* **49**, 559–562.

Nelson, J. A. and Trout, G. M. (1981). *Judging Dairy Products.* AVI Publishing Co., Inc., Westport, CT, pp. 100 and 209.

Ohren, J. A. and Tuckey, S. L. (1965). Relation of fat hydrolysis to flavor development in Cheddar cheese. *J. Dairy Sci.* **48**, 765. (Abstract).

Park, R. J. (1969). Weed taints in dairy produce. I. *Lepidium* taint. *J. Dairy Res.* **36**, 31–35.

Park, R. J., Armitt, J. D. and Stark, W. (1969). Weed taints in dairy produce. II. *Coronopus* or land cress taint in milk. *J. Dairy Res.* **36**, 37–46.

Parks, O. W., Keeney, M. and Schwartz, D. P. (1963). Carbonyl compounds associated with the off-flavor in spontaneously oxidized milk. *J. Dairy Sci.* **46**, 295–301.

Patton, S. (1962). Dairy products. In *Symposium on Foods: Lipids and Their Oxidation.* Eds. H. D. Schultz, E. A. Day and R. O. Sinnhuber. AVI Publishing Co., Inc., Westport, CT, pp. 190–201.

Patton, S. (1963). Volatile acids and the aroma of Cheddar cheese. *J. Dairy Sci.* **46**, 856–858.

Patton, S., Forss, D. A. and Day, E. A. (1956). Methyl sulfide and the flavor of milk. *J. Dairy Sci.* **39**, 1469–1470.

Perry, K. D. (1961). A comparison of the influence of *Streptococcus lactis* and *Str. cremoris* starters on the flavour of Cheddar cheese. *J. Dairy Res.* **28**, 221–229.

Pyenson, H. and Tracy, P. H. (1946). A spectrophotometric study of the changes in peroxide value of spray-dried whole milk powder during storage. *J. Dairy Sci.* **29**, 1–12.

Reddy, M. C., Bills, D. D., Lindsay, R. C. and Libbey, L. M. (1968). Ester production by *Pseudomonas fragi.* I. Identification and quantification of some esters produced in milk cultures. *J. Dairy Sci.* **51**, 656–659.

Reddy, M. C., Bills, D. D. and Lindsay, R. C. (1969). Ester production by *Pseudomonas fragi.* II. Factors influencing ester levels in milk cultures. *Appl. Microbiol.* **17**, 779–782.

Renner, E. and Schmidt, R. (1981). Chemical and physico-chemical aspects. In *New Monograph on UHT Milk 1981.* Document 133. International Dairy Federation, Brussels, Belgium, pp. 49–64.

Rerkrai, R., Jeon, I. J. and Bassette, R. (1987). Effect of various direct ultra-high-temperature heat treatments on flavor of commercially prepared milks. *J. Dairy Sci.* **70**, 2046–2054.

Roberts, W. M. (1959). Problems involved in flavor removal. *J. Dairy Sci.* **42**, 560–563.

Sattar, A., deMan, J. M. and Alexander, J. C. (1977). Wavelength effect on light-induced decomposition of vitamin A and β-carotene in solutions and milk fat. *Can. Inst. Food Sci. Technol.* **10**, 56–60.

Sattler-Dornbacher, S. (1963). Studien zum Redox-Potential in Butter. *Milchwiss. Ber.* **13**, 53–74.

Scanlan, R. A., Lindsay, R. C., Libbey, L. M. and Day, E. A. (1968). Heat-induced volatile compounds in milk. *J. Dairy Sci.* **51**, 1001–1007.

Scanlan, R. A., Sather, L. A. and Day, E. A. (1965). Contribution of free fatty acids to the flavor of rancid milk. *J. Dairy Sci.* **48**, 1582–1584.

Schaffer, P. S., Greenbank, G. R. and Holm, G. E. (1946). The rate of autoxidation of milk fat in atmospheres of different oxygen concentration. *J. Dairy Sci.* **29**, 145–150.

Schmidt, R. H. (1990). Bitter components in dairy products. In *Bitterness in Foods and Beverages.* Ed. R. L. Rouseff. Elsevier Science Publisher, Amsterdam, The Netherlands, pp. 183–204.

Shipe, W. F. (1980). Analysis and control of milk flavor. In *The Analysis and Control of Less Desirable Flavors in Foods and Beverages.* Ed. G. Charalambous. Academic Press, New York, pp. 202–203.

Shipe, W. F. and Senyk, G. F. (1981). Effects of processing conditions on lipolysis in milk. *J. Dairy Sci.* **64**, 2146–2149.

Shipe, W. F., Ledford, R. A., Peterson, R. D., Scanlan, R. A., Geerken, H. F., Dougherty, R. W. and Morgan, M. E. (1962). Physiological mechanisms in transmitting flavors and odors to milk. II. Transmission of some flavor components of silage. *J. Dairy Sci.* **45**, 477–480.

Shipe, W. F., Bassette, R., Deane, D. D., Dunkley, W.L., Hammond, E. G., Harper, W. J., Kleyn, D. H., Morgan, M. E., Nelson, J. H. and Scanlan, R. A. (1978). Off flavors of milk: nomenclature, standards, and bibliography. *J. Dairy Sci.* **61**, 855–869.

Slinkard, M. S. (1976). Changes in the headspace concentration of volatile sulfur compounds during aseptic storage of ultra-high-temperature processed milk. *M.S. Thesis*, University of Minnesota, Minneapolis, MN.

Smith, G. J. and Dunkley, W. L. (1962). Copper binding in relation to inhibition of oxidized flavour by heat treatment and homogenization. *Int. Dairy Congr.* **A**, 625–632.

Stark, W. and Forss, D. A. (1962). A compound responsible for metallic flavour in dairy products. I. Isolation and identification. *J. Dairy Res.* **29**, 173–180.

Stark, W. and Forss, D. A. (1966). *n*-Alkan-1-ols in oxidized butter. *J. Dairy Sci.* **33**, 31–36.

Thomas, E. L. (1981). Trends in milk flavors. *J. Dairy Sci.* **64**, 1023–1027.

Thomas, E. L., Burton, H., Ford, J. E. and Perkin, A. G. (1975). The effect of oxygen content on flavour and chemical changes during aseptic storage of whole milk after ultra-high-temperature processing. *J. Dairy Res.* **42**, 285–295.

Toan, T. T., Bassette, R. and Claydon, T. J. (1965). Methyl sulfide production of *Aerobacter aerogenes* in milk. *J. Dairy Sci.* **48**, 1174–1178.

Vedamuthu, E. R., Sandine, W. E. and Elliker, P. R. (1964). Influence of milk citrate concentration on associative growth of lactic streptococci. *J. Dairy Sci.* **47**, 110. (Abstract).

Walstra, P. and Jenness, R. (1984). *Dairy Chemistry and Physics*. John Wiley & Sons, New York, pp. 78–83.

Weihrauch, J. L. (1988). Lipids of milk: deterioration. In *Fundamentals of Dairy Chemistry*. Ed. N. P. Wong. Van Nostrand Reinhold Co., New York, pp. 215–278.

Wellnitz-Ruen, W., Reineccius, G. A. and Thomas, E. L. (1982). Analysis of the fruity off-flavor in milk using headspace concentration capillary column gas chromatography. *J. Agric. Food Chem.* **30**, 512–514.

Wishner, L. A. (1965). Light-induced oxidation in milk. *J. Dairy Sci.* **47**, 216–221.

Wishner, L. A. and Keeney, M. (1963). Carbonyl pattern of sunlight-exposed milk. *J. Dairy Sci.* **46**, 785–788.

Woo, A. H. and Lindsay, R. C. (1982). Rapid method for quantitative analysis of individual free fatty acids in Cheddar cheese. *J. Dairy Sci.* **65**, 1102–1109.

Zadow, J. G. and Birtwistle, R. (1973). The effect of dissolved O_2 on the changes occurring in the flavour of ultra-high-temperature milk during storage. *J. Dairy Res.* **40**, 169–177.

6 Oxidative pathways to the formation of off-flavours

S. P. KOCHHAR

6.1 Introduction

Food oils and fats comprise mainly triglycerides, which are triesters of fatty acids and glycerol. In addition to triglycerides (neutral or simple lipids), they also contain minor components such as free fatty acids, mono- and diglycerides, phospholipids, cerebrosides, sterols, terpenes, waxes, carotenoids, chlorophylls, naturally occurring antioxidants tocopherols (vitamin E), trace metals, etc. (Swern, 1979; Belitz and Grosch, 1987). The glycero-phospholipids (also called phosphatides or gums) are classed as any lipid containing one phosphate and one or two fatty acids attached to the glycerol moiety. Lipids are generally soluble in organic solvents and only sparingly soluble in water. Normal processing (i.e. different stages of refining) of oils and fats removes almost all undesirable minor components such as coloured compounds, free fatty acids, trace metals, etc. but retains the major neutral lipids and most of the natural antioxidants tocopherols present in the oils (Hoffmann, 1989). Freshly refined and deodorised oils have bland flavours, but develop unacceptable flavours on storage. In contrast to edible fats, low levels of a wide variety of lipid classes are present in milk and lean meats. Most fruits and vegetables contain only small amounts of simple and compound lipids.

With a few exceptions (e.g., milk, coconut and palm kernel oils, fish oils), the large proportion of fatty acids present (as esterified to glycerol) in oils and food lipids consists of palmitic (C16:0), stearic (C18:0), oleic (C18:1), linoleic (C18:2) and linolenic (C18:3) acids. Small amounts of branched-chain, cyclic, odd-numbered carbon atoms, hydroxy-and oxo-fatty acids are also present in certain edible fats. Fatty acids containing two or more double bonds are termed polyunsaturated fatty acids (PUFAs). Of the PUFAs, linoleic, linolenic, arachidonic, ecosapentaenoic, and docosahexaenoic containing respectively, two, three, four, five and six double bonds are important nutritionally. Linoleic and linolenic acids are essential because they cannot be synthesised by the body and must be provided in the diet. Most vegetable oils and nuts are the principal source of these essential fatty acids (EFAs). The PUFAS containing one or more methylene-interrupted double bonds in the *cis* configuration are very susceptible to oxidation with atmospheric oxygen.

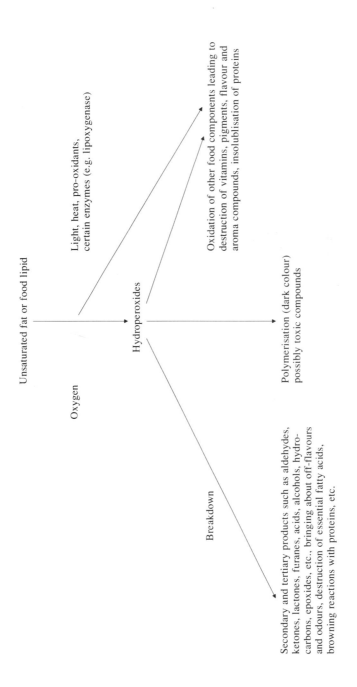

Figure 6.1 Overall picture of lipid oxidation in foodstuffs.

In foodstuffs, deterioration of unsaturated lipids can be caused both by enzymic and non-enzymic (oxidative) mechanisms. Oxidative deterioration, known as oxidative rancidity, of edible oils and food lipids is one of the major causes of food spoilage. This is of great concern to the food industry because it leads to the development of objectionable flavours and odours in oils and lipid-containing foods, which reduces their shelf-life or renders them unfit for consumption. In addition, oxidative deterioration brings about changes in the colour, texture and consistency, and losses in nutritional quality (e.g. reduction in oxidation products may be potentially toxic) (Billek, 1979; Sanders, 1990).

The overall picture of the effects of lipid oxidation in foodstuffs is illustrated in Figure 6.1. It should be pointed out that primary products of lipid oxidation, called hydroperoxides, are tasteless and odourless. It is the decomposition of hydroperoxides and secondary volatile oxidation products which gives rise to off-flavours in oils, fats and lipid-containing foods. Many terms such as beany, green, metallic, oily, fishy, bitter, fruity, soapy, painty, grassy, buttery, tallowy, oxidised or rancid, etc. have been used to describe undesirable flavours developing in food products.

This chapter deals with the important reaction pathways for the development of off-flavours, mainly via temperature, light and metal-catalysed oxidation of unsaturated fats or lipids. Since, from a practical viewpoint, it is not possible to remove all oxygen from foodstuffs, deterioration of oils and oil-containing foods cannot be avoided. However, this detrimental phenomenon can be restricted in order to maintain food quality and prolong the shelf-life of fats and fat-containing foodstuffs. For this purpose, understanding of these reaction pathways may provide a basis for preventing flavour deterioration and/or predicting the course of reactions that produce flavour problems in food products.

6.2 Unsaturated lipids as off-flavour precursors

It is generally agreed that the main reaction involved in oxidative deterioration of food lipids is between oxygen and unsaturated fatty acids, which may or may not be a part of the oil or phospholipid. This process of autoxidation leads to the development of various unpleasant odours and flavours, generally called rancid, in oils and lipid-containing foods, and thus renders them unpalatable. In some foods, however, (e.g. butter, cheese, cucumber, cooked chicken, etc.), a limited degree of lipid oxidation under certain conditions is desirable in order to produce typical and characteristic flavours or pleasant fried-food aromas. It is well established (Lundberg, 1961, 1962; Scott, 1965; Labuza, 1971; Kochhar and Meara 1975; Frankel, 1980) that the fundamental mechanisms of lipid oxidation involve free-radical chain reactions.

6.2.1 Free-radical autoxidation

The reaction of unsaturated lipids/fatty acids with oxygen to form hydro-peroxides (off-flavour precursors) is generally a free-radical process involving three basic stages.

6.2.1.1 Initiation

$$RH \xrightarrow[\text{metals}]{\text{heat, light}} R^{\bullet} + H^{\bullet} \atop \text{(free radicals)} \tag{6.1}$$

$$ROOH + M^{3+} \longrightarrow ROO^{\bullet} + H^{\bullet} + M^{2+} \tag{6.2}$$

$$ROOH + M^{2+} \longrightarrow RO^{\bullet} + {}^{\bullet}OH + M^{3+} \tag{6.3}$$

$$2ROOH \longrightarrow RO^{\bullet} + ROO^{\bullet} + H_2O \tag{6.4}$$

6.2.1.2 Propagation

$$R^{\bullet} + O_2 \longrightarrow ROO^{\bullet} \tag{6.5}$$

$$ROO^{\bullet} + RH \xrightarrow{\text{slow}} ROOH + R^{\bullet} \tag{6.6}$$

$$RO^{\bullet} + RH \longrightarrow ROH + R^{\bullet} \tag{6.7}$$

6.2.1.3 Termination

$$R^{\bullet} + R^{\bullet} \longrightarrow 2R \tag{6.8}$$

$$R^{\bullet} + ROO^{\bullet} \longrightarrow ROOR \tag{6.9}$$

$$ROO^{\bullet} + ROO^{\bullet} \longrightarrow ROOR + O_2 \tag{6.10}$$

where RH, $R^{\bullet}$, $RO^{\bullet}$, $ROO^{\bullet}$, ROOH and M represent unsaturated fatty acid or ester with H attached to allylic carbon atom, alkyl radical, alkoxy radical, peroxyl radical, hydroperoxide and transition metal, respectively. The formation of free radicals ($RO^{\bullet}$, $ROO^{\bullet}$) in the initiation stage can take place by thermal or photo-decomposition of hydroperoxides, by metal catalysis and/or by ultraviolet irradiation. The initiation process can also start, via reaction (6.1), at a carbon atom in the α-position with respect to the double bond, by a loss of hydrogen radical in the presence of trace metals, heat or light. The resulting lipid or alkyl radicals react with oxygen to form peroxy radicals in the propagation reaction (6.5), which is very fast in the presence of air. The peroxy radical $ROO^{\bullet}$ abstracts a hydrogen from the α–CH_2 group of another unsaturated lipid molecule by the reaction (6.6) to form hydroperoxide. At ambient temperature, reaction (6.6) is the slowest and rate-determining step of oxidation (Ragnarsson and Labuza, 1977). The termination stage involves the formation of non-radical, stable products by the interaction of $R^{\bullet}$ and $ROO^{\bullet}$ through reactions (6.8) to (6.10). In the presence of air, termination reaction (6.10) is the most important. Termination reactions (6.8) and (6.9) become more important when the oxygen concentration is low and away from the surface of the fat or food lipid.

6.2.2 Photo-sensitised oxidation

Since the reaction between unsaturated fatty acids (RH) and oxygen is thermo-dynamically difficult (activation energy of about 35 kcal/mol is needed), production of the first few radicals necessary for the onset of autoxidation can occur through photo-sensitised oxidation. This photoxidation process involves interaction between a double bond and an excited singlet oxygen produced from ordinary triplet oxygen ($^3O_2$) by light (e.g. visible, ultraviolet or X-ray) in the presence of a sensitiser such as chlorophyll, pheophytin, myoglobin and/or erythrosine (Labuza, 1971; Chan, 1977). The photo-sensitised oxida-tion process is illustrated as follows:

$$\text{Sens} \xrightarrow[h\nu]{\text{light}} {}^1\text{Sens}* \xrightarrow{\hspace{2cm}} {}^3\text{Sens}* \qquad (6.11)$$

$$^3\text{Sens}* + {}^3O_2 \xrightarrow{\hspace{2cm}} {}^1O_2^* + {}^1\text{Sens} \qquad (6.12)$$

$$^1O_2^* + RH \xrightarrow{\hspace{2cm}} ROOH \qquad (6.13)$$

where Sens, ^{1}Sens$*$ and ^{3}Sens$*$ represent sensitiser, excited singlet state and excited triplet state respectively. The oxygen molecule becomes activated to the singlet state by the transfer of energy from the excited triplet state photosensitiser. The excited singlet oxygen ($^1O_2^*$) thus produced is highly reactive, and reacts with methyl linoleate 10^3 to 10^4 times faster than the normal oxygen (Rawls and van Santen, 1970). Light has been found to be a very important factor affecting production of off-flavours in high linolenic acid oils such as soyabean and rapeseed oils (Tokarska et al., 1986). The metal-catalysed decomposition of these rapidly formed hydroperoxides has been suggested to initiate free radical autoxidation (reactions (6.2) and (6.3)).

Foote (1976) has postulated another mechanism, where the triplet sensitiser forms a Sens–oxygen complex, which reacts with a substrate to produce a hydroperoxide, and regenerates the sensitiser as:

$$^3\text{Sens} + {}^3O_2 \xrightarrow{\hspace{2cm}} {}^1[\text{Sens}-O_2] \qquad (6.14)$$

$$^1[\text{Sens}-O_2] + RH \xrightarrow{\hspace{2cm}} ROOH + {}^1\text{Sens} \qquad (6.15)$$

The riboflavin-sensitised photoxidation of unsaturated lipids involves this type of reaction mechanism. The reaction proceeds by hydrogen abstraction and produces the same conjugated diene hydroperoxides from linolenate as does free radical autoxidation (Chan, 1977).

Carotenoids, naturally present in many crude oils and some foods, quench singlet oxygen ($^1O_2$) and thus protect lipids against light-induced oxidation. However, these compounds would impart yellow colour to an oil, which is not acceptable to the consumer.

Refining and bleaching of oils effectively remove natural photosensitisers and thus reduce deterioration of oils by singlet oxygen. Obviously, storage of oils in the dark, or use of suitable packaging or containers that are absorbent to the light energy necessary for photosensitisation will protect the oils against singlet oxygen deterioration. It should be mentioned that chlorophyll and pheophytin show beneficial (antioxidant) effects on oils in the dark (Endo *et al.*, 1985a; Werman and Neeman, 1986). It has been suggested that these photosensitisers, in the dark, may act as hydrogen donors to break the free-radical chain reaction (Endo *et al.*, 1985b).

6.2.3 Factors affecting the rate of lipid oxidation

There are many factors that influence the rate of autoxidation of unsaturated fats or food lipids. Some of these factors, such as the degree of unsaturation of lipid, transition metals and biological catalysts, heat, etc., increase the autoxidation rate. Others, e.g. storage under an inert gas, antioxidants and synergists, decrease the autoxidation rate. In addition, foods also contain numerous non-lipid components that can affect the rate of lipid oxidation. Labuza (1971) has written an excellent review on the complicated kinetics of lipid oxidation in food systems. The effect of any given factor on autoxidation depends mainly on the nature of the reaction conditions. The involvement of various important factors is discussed in the following sections.

6.2.3.1 Fatty acid composition. As mentioned earlier, food oils and fats contain mixtures of saturated and unsaturated fatty acids esterified to glycerol. Because of the presence of double bonds, unsaturated fatty acids are chemically more reactive than the saturated fatty acids. This reactivity increases as the number of double bonds increases. For example, relative rates of oxidation of arachidonic, linolenic, linoleic and oleic are approximately 40:20:10:1, respectively. The position and geometry of double bonds also affect the chemical reactivity. *Cis* acids oxidise more readily than their *trans* isomers, and conjugated double bonds are more reactive than non-conjugated ones. Autoxidation of saturated fatty acids is extremely slow at low temperatures. At elevated temperatures (e.g. in frying operation), oxidation of saturated fatty acids can also take place.

6.2.3.2 Free fatty acids. Free fatty acids (FFAs) oxidise at a slightly higher rate than when they are esterified to the glycerol moiety. The pro-oxidant effect of FFAs on the autoxidation of many fats has been reported by some researchers (Olcott, 1958; Popov and Mizev, 1966). The time during which an oil's natural resistance to oxidation, due to the presence of naturally occurring antioxidants, inhibits the onset of rapid oxidation is generally known as the induction period. A decrease in the induction period of oils and fats was observed when as little as 0.1% FFA was added (Hartman et al., 1975). Moreover, the effectiveness of tocopherols is decreased by the addition of FFAs. This is probably due to FFA accelerating the decomposition of hydroperoxides to the extent that phenolic antioxidants do not function effectively. The presence of relatively high amounts of FFAs in oils can increase the uptake of catalytic trace metals (Fe and Cu) from storage equipment and thus enhance the rate of oil oxidation. Stahl and Sims (1986) have shown that peroxide value is an unreliable index of oxidative stability of oils containing high amounts of FFAs (greater than 3% as oleic).

6.2.3.3 Oxygen concentration.   At high oxygen pressure (i.e. when the supply of oxygen is unlimited), the rate of lipid oxidation is independent of oxygen pressure. At low oxygen pressure, the oxidation rate is approximately proportional to the oxygen pressure. However, other factors such as surface area and temperature also influence the effect of oxygen pressure on the autoxidation rate.

6.2.3.4 Surface area.   The rate of oxidation increases in proportion to the fat surface exposed to oxygen or air. As the ratio of available surface to fat volume increases, any reduction in the oxygen pressure becomes less effective in decreasing the rate of oxidation. In oil-in-water emulsions, for example low fat spreads, mayonnaise and related products, the rate of oxidation is controlled by the rate at which oxygen diffuses into the lipid phase.

6.2.3.5 Temperature.   Generally, the autoxidation rate increases as the temperature is increased. The solubility of oxygen in an oil decreases with an increase in temperature, therefore the effect of oxygen concentration on the rate of oxidation becomes less evident. The temperature dependence of lipid oxidation can be best described by the Arrhenius equation:

$$K = A\,e^{-E_a/RT} \tag{6.16}$$

where K is the reaction rate constant, A is an entropy constant, which does not depend on temperature, E_a is the activation energy, R is gas constant (1.986 kcal/mole/°C) and T is the absolute temperature in degrees Kelvin. The activation energy stays approximately constant provided that the mechanism pathway of the reaction does not change. In other words, every reaction has its specific activation energy, E_a. Hamm et al. (1968) investigated the influence of low temperatures (-27°C to 50°C) on the rate of oxidation of milk fat.

Their results showed that the same off-flavours (tallowy or oxidised-metallic) were produced during the autoxidation of the fat, suggesting the same reaction mechanism occurring at these temperatures. The rate constant plots were quite linear down to $-10°C$. In fact, the Arrhenius relationship does not always apply to the complex oxidative reactions of oils and fats. In multicomponent oil- and lipid-containing foods, the effect of temperature is not a simple one. This is due to the fact that the oxidation rates of the various reactions making up the chain reaction do not vary in the same proportion with the temperature. In other words, it may not be true that only one reaction is responsible for loss of flavour quality or development of off-odours, since other reactions causing quality changes may be more important at higher temperatures. Moreover, there may be a critical temperature above which the rate of one reaction becomes faster than that of the second reaction, both being responsible for oil deterioration (Labuza and Kamman, 1983). Therefore, the rate-limiting reaction at one temperature may not be the limiting reaction at a different temperature. When the oxidation phase of the reaction (after the induction phase) has begun, the lowering of temperature might be of little use for deterioration of fats. Also, at low temperatures, the mechanism of oxidation is a little different because *cis–trans* isomerisation during the oxidation process is limited. For example, at 25°C the linoleic acid mainly forms the *trans–trans* hydroperoxide, while at 0°C mainly *cis–trans* hydroperoxide is formed (Curda and Poulsen, 1978).

6.2.3.6 Moisture.   The rate of lipid oxidation in foods depends strongly on water content (Karel, 1980). The parameter water activity (a_w), defined as

$$a_w = p/p_o \qquad (6.17)$$

where p is partial pressure of water in food and p_o is vapour pressure of water, governs the distribution of water between different food components. In dried foods, e.g. evaporated whole milk, containing very low moisture (a_w less than 0.1) oxidation proceeds very rapidly. Increasing the water activity to about 0.3 retards lipid oxidation and often gives a minimum rate. This protective effect of moisture is thought to occur by decreasing the catalytic activity of trace metals, by quenching free radicals, and by promoting non-enzymic browning, which produces compounds with antioxidant activities. At higher water activities (0.55–0.85), the rate of oxidation is accelerated again, probably as a result of increased mobilisation of the catalysts present in foodstuffs.

6.2.3.7 Pro-oxidants.

Trace metals.   Trace metals are naturally occurring components of all food materials of both plant and animal origin. They are present in trace amounts (10^{-3} to 500 mg/kg), both in free and bound forms. Some transition metals, particularly those possessing two or more valency states with a suitable oxidation-reduction potential between them (such as Co, Cu, Fe, Mn and Ni), catalyse lipid oxidation. If present in active state, even at very low concentration (especially in the case of Cu at a level of 0.01 mg/kg), they can accelerate the rate of oxidation and thereby give rise to the off-flavour problem in soyabean oil, rapeseed oil or highly unsaturated fish oils. Several mechanisms for metal-catalysed lipid oxidation have been proposed. These include production of free radicals by direct reaction with lipid, RH, described as:

$$M^{2+} + RH \longrightarrow M^+ + R^\bullet + H^+ \qquad (6.18)$$

The metal ion in its lowest valency state can react readily with oxygen, forming an oxygen radical ion:

$$M^+ + O_2 \longrightarrow M^{2+} + \overline{O_2} \underset{+H^+}{\overset{-e}{\diagdown\diagup}} \begin{matrix} {}^1O_2 \\ HO_2^\bullet \end{matrix} \qquad (6.19)$$

The oxygen anion can either lose an electron to give singlet oxygen, or can react with a proton to form a peroxy radical, a good chain initiator (Uri, 1961).

Another mechanism of producing free radicals, necessary for the initiation step, is by decomposing lipid hydroperoxides as illustrated by reactions (6.2) and (6.3). Moreover, heavy metals not only catalyse decomposition of hydroperoxides (which results in development of off-flavours and off-odours) but also take part in further oxidation of secondary products (Pokorny, 1987).

Haematin compounds. Haem (Fe^{2+}) and haemin (Fe^{3+}) compounds, distributed widely in food tissues, are strong pro-oxidants. Catalase and peroxidase are the most important haem proteins (enzymes) present in many plant foods. Haemoglobin, myoglobin and cytochrome C present in animal tissue accelerate the rate of lipid autoxidation. These oxidative reactions are often considered to be responsible for the development of rancid or undesirable flavours during storage of fish, poultry and cooked meat. In fact, both the unsaturated lipids and ferrous-containing haem proteins present in meat and meat products may oxidise in the presence of oxygen. The former oxidation reaction leads to the formation of rancid odours and flavours whilst the latter leads to discoloration of the meat product. In fresh meat, oxidation of the bright red oxymyoglobin (MbO_2) pigments results in the formation of the undesirable, brown metmyoglobin (metMb). The problem of development of warmed-over flavour in cooked meat during refrigerated storage for a short time has been related to the oxidation of the phospholipids. On the other hand, oxidative deterioration of the triglycerides is considered to be responsible for the off-flavour developing in frozen raw meats with storage (Poste *et al.*, 1986; Mottram, 1987; Lyon, 1988).

Labuza (1971) reported the order of oxidative activity of haematin compounds to be cytochrome C > haemin > haemoglobin > catalase. Heat treatment of haem enzymes such as catalase and peroxidase leads to severe losses of enzymatic activity through denaturation of protein, but their pro-oxidative activity in lipid oxidation increases several times. This may be due to the increased exposure of the central metal ion, iron (Eriksson *et al.*, 1971). It is the spin state and not the valency state of iron in haem compounds that is responsible for catalysing lipid autoxidation. The mechanism of haematin-catalysed lipid autoxidation is given in Figure 6.2. Haematin compounds catalyse hydroperoxide decomposition by forming a complex with the hydroperoxide. The complex is then broken down into two free radicals, one of which is able to initiate a free-radical chain reaction. The haematin radical abstracts hydrogen from a lipid molecule, thus regenerating the haematin molecule and producing an alkyl radical for chain reaction. When the peroxide value of a lipid system is very low, other reaction mechanisms that involve changes in the valency state of iron become important in the autoxidation process (Pokorny, 1987). Moreover, the haematin compounds are strong pro-oxidants in aqueous food materials only.

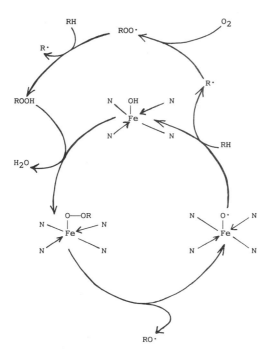

Figure 6.2   Reaction pathway of haematin-catalysed lipid oxidation.

6.2.3.8 Lipoxygenase (enzyme). The basic chemistry of enzyme-catalysed oxidation of food lipids such as in cereal products, and many fruits and vegetables is the same as for autoxidation, but the enzyme lipoxygenase (LPX) is very specific for the substrate, and for the method of substrate oxidation (Belitz and Grosch, 1987). LPX type I, from many natural sources e.g. soyabean, potato, tomato, wheat or maize germ, prefers free fatty acids containing a 1,4-*cis*-pentadiene group as the substrate. Therefore, the preferred substrates are linoleic and linolenic for the plant LPX, and arachidonic or ecosapentaenoic for the animal enzyme. Moreover, the reactions catalysed by LPX are also characterised by all the features of enzymic catalysis: peroxidation specificity, occurrence of a pH maximum, susceptibility to heat treatment and a high reaction rate in the range of 30–35°C.

Lipoxygenase is a metal-bound protein with a Fe atom in its active centre. Generally, LPX is activated by hydroperoxide and, during this activation, Fe^{2+} is oxidised to Fe^{3+}. The type I lipoxygenase-catalysed oxidation reaction is illustrated by the scheme given in Figure 6.3. The pentadienyl radical bound

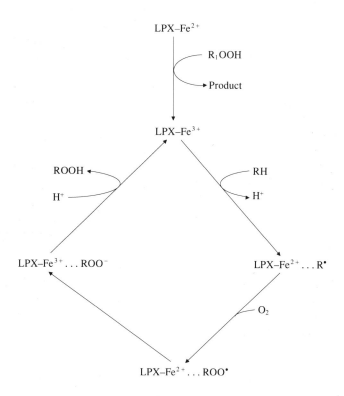

Figure 6.3 Schematic reaction pathway of type I lipoxygenase-catalysed lipid autoxidation. RH = linoleic acid.

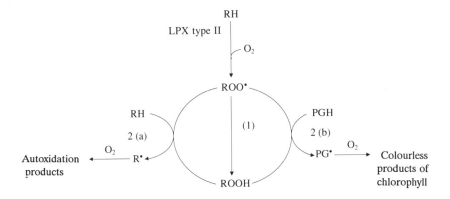

Figure 6.4 Schematic co-oxidation pathways of type II lipoxygenase-catalysed lipid autoxidation. RH = linoleic acid or its glyceride ester; PGH = chlorophyll; (1) = main catalysis pathway; 2(a) and 2(b) = co-oxidation pathways.

to enzyme is rearranged into a conjugated diene system, which follows an absorption of oxygen. The peroxy radical formed is then reduced by the enzyme and the hydroperoxide formed is released. The structures of these hydroperoxides (precursors of desirable or undesirable flavours) are the same as those obtained by the autoxidation process.

The type II lipoxygenase enzyme is present in gooseberry, soyabean and legumes, e.g. peas. The type II LPX acts more like a catalyst of autoxidation, with much less reaction specificity for linoleic acid. It can also react with unesterified substrate and thus does not require prior release of free fatty acids by lipolytic action in foods. Moreover, LPX type II can co-oxidise pigments such as carotenoids and chlorophyll present in foods and thus can decompose these pigments (PGH) into colourless products. The involvement of LPX type II, in the co-oxidation reaction mechanism is presented schematically in Figure 6.4. The free peroxyl radical, not converted to hydroperoxide, released by LPX type II can abstract a H atom from the pigment compound and thus involve these colour components in the oxidation process.

The type 2(b) co-oxidation pathway can explain the decrease in green pigments of ripe olives after harvesting, and of virgin olive oil on storage (Minguez-Mosquera *et al.*, 1990). The fatty acid hydroperoxides produced by LPX type II also have the same structures as those obtained by the non-catalysed autoxidation.

6.2.3.9 Antioxidants.   It is well known that antioxidants, when present either naturally or by addition, retard the onset or slow down the rate of oxidation of oils or food lipids. In addition, they protect oil-soluble vitamins, carotenoids and other nutritive ingredients. They also delay undesirable change brought about by oxidation of foods, for example discoloration in meat and meat products (Gregory, 1984; Miles *et al.*, 1986), and browning or 'scald' on fruits and vegetables (Winell, 1976; Olek *et al.*, 1983).

Primary food antioxidants are those compounds which terminate and inhibit the free-radical chain reactions of autoxidation. Natural and synthetic tocopherols, alkyl gallates, butylated hydroxytoluene (BHT), butylated hydroxyanisole (BHA), tertiary butyl hydroxyquinone (TBHQ), etc. belong to this group. Several review articles dealing with the antioxidant kinetics and mechanism of action have been published (Uri, 1961; Scott, 1965; Labuza, 1971; Kochhar, 1988; Gordon, 1990).

Phenolic (primary) antioxidants, whether naturally occurring (e.g. tocopherol or flavonoids) or permitted synthetic compounds (e.g. BHT, BHA, TBHQ or gallates), inhibit chain reactions by acting as hydrogen donors or free-radical acceptors. The reaction mechanisms of a primary antioxidant (AH) are described as:

$$AH + RO^\bullet \longrightarrow ROH + A^\bullet \tag{6.20}$$

$$AH + ROO^\bullet \rightleftharpoons ROOH + A^\bullet \tag{6.21}$$

$$RH + A^\bullet \longrightarrow AH + R^\bullet \tag{6.22}$$

$$AH + ROO^\bullet \rightleftharpoons [ROO^\bullet\ AH]\ \text{complex} \tag{6.23}$$

$$[ROO^\bullet\ AH] \longrightarrow \text{Non-radical products} \tag{6.24}$$

$$A^\bullet + A^\bullet \longrightarrow AA \tag{6.25}$$

$$A^\bullet + R^\bullet \longrightarrow RA \tag{6.26}$$

$$A^\bullet + ROO^\bullet \longrightarrow ROOA \tag{6.27}$$

The inhibitory reaction (6.21) is more important than reaction (6.20) or (6.23), and influences the overall inhibition rate constant. The stable resonance hybrid of the antioxidant free radical $A^\bullet$, and the non-radical reaction (6.24) to (6.27) products thus produced are incapable of initiation or propagation of the chain reactions.

Chelating agents or sequestrants, for example citric acid and isopropyl citrate, amino acids, phosphoric acid, tartaric acid, ascorbic acid and ascorbyl palmitate, ethylene diamine tetraacetic acid (EDTA), etc., which chelate metallic ions such as copper and iron, promote lipid oxidation through a catalytic action (Sherwin, 1976; Dziezak, 1986). The chelators are sometimes referred to as synergists since they greatly enhance the action of phenolic antioxidants. It is suggested that the synergist (SH) regenerates the primary antioxidant according to the reaction mechanism described as:

$$SH + A^\bullet \longrightarrow AH + S^\bullet \tag{6.28}$$

Phospholipids, ascorbic acid (or vitamin C) and ascorbyl palmitate are known as antioxidant synergists in oils and food systems (Cort, 1974; Pongracz, 1982; Dziedzic and Hudson, 1984; Kwon *et al.*, 1984). Ascorbyl palmitate (AP) and ascorbic acid (AA) also act as oxygen scavengers (Klaui and Pongrancz, 1982), and the reactions lead to dehydro-products, as described:

$$2AA + O_2 \longrightarrow 2 \text{ dehydro–AA} + 2H_2O \qquad (6.29)$$

$$2AP + O_2 \longrightarrow 2 \text{ dehydro–AP} + 2H_2O \qquad (6.30)$$

The chain-breaking reactions between AP and hydroperoxides, alkyl or peroxy radicals have also been described by Sedlacek (1975).

The use of an antioxidant, and the selection of which permitted antioxidant to use, depends on understanding the reaction mechanisms outlined in this section. For maximum efficiency, a combination of primary antioxidant, synergist and metal-chelating agent is often used. Moreover, the application of chelators and synergists prevents decomposition of hydroperoxides (off-flavour precursors), by their sequestering action on catalytic trace metals, thus retarding the development of rancid off-flavours and odours in oils, fats and lipid-containing foods.

6.2.4 Formation of hydroperoxides (off-flavour precursors)

Lipid hydroperoxides are the primary oxidation products of unsaturated fatty acids. The ease of formation of hydroperoxides depends, among other things, on the number of double bonds or allyl groups present. The relative rate of autoxidation of oleate:linoleate:linolenate was observed to be in the order of 1:40–50:100 on the basis of oxygen uptake, and in the order of 1:12:25 on the basis of peroxide formation. Gunstone (1984) has reported the relative rates of oxidation of fatty acids to be:

	C18:1	C18:2	C18:3
Autoxidation ($^3O_2$)	1	27	77
Photoxidation ($^1O_2$)	30×10^3	40×10^3	70×10^3

In the process of autoxidation, the CH_2-interrupted diene system is considerably more reactive than the isolated double bond present in oleate, but in photoxidation, relative rates of oxidation of oleate, linoleate and linolenate are not remarkably different. Photo-oxygenation is not a chain reaction. It proceeds by an 'ene' reaction mechanism in which the singlet oxygen ($^1O_2$) adds to an olefinic carbon atom, with consequent migration of the double bond and alteration from *cis* to *trans* configuration. The reaction is unaffected by antioxidants and there is no induction period.

As shown above, photoxidation is a quicker reaction than autoxidation. Photoxidation of methyl oleate can be 30 000 times quicker than its autoxidation and, for linoleate, photoxidation can be about 1600 times faster. Many researchers (Frankel, 1979, 1980; Paquette *et al.*, 1985; Chan and Coxon, 1987) have reviewed the hydroperoxidation of unsaturated fatty acids. The mechanisms for the formation of monohydroperoxides by free radical autoxidation and photo-sensitised oxidation of oleate, linoleate and linolenate are summarised in the following sections.

6.2.4.1 Monohydroperoxides produced by autoxidation ($^3O_2$). It is well recognised that the free-radical reaction mechanism of hydroperoxy formation involves the abstraction of hydrogen, by a peroxy radical, from the α-methylene group of a fat molecule. This is favoured because of the formation of a very stable allyl radical in which the electrons are delocalised over either three carbon atoms (oleate) or five carbon atoms (linoleate and linolenate). Due to resonance stabilisation, this results in the formation of isomeric hydroperoxides. The mechanisms for the formation of isomeric hydroperoxides by autoxidation are presented in Figures 6.5 to 6.7. During the last 15 years, owing to developments and availability of modern analytical tools such as gas chromatography–mass spectrometery (GC–MS), high-performance liquid chromatography (HPLC) and nuclear magnetic resonance (NMR), many workers (Lercker, 1980; Porter *et al.*, 1980, 1981; Frankel *et al.*, 1984) have been successful in separating and quantifying the relative distribution of isomeric hydroperoxides that form during oxidation of unsaturated fatty acids.

In the case of oleate, the hydrogen abstraction on C-8 and C-11 produces two allylic radicals. These intermediates react with oxygen at the end carbons to produce a mixture of 8-, 9-, 10- and 11-allylic hydroperoxides. Figure 6.5 presents the reaction scheme for the formation of these four evenly distributed isomeric hydroperoxides. However, the studies based on GC–MS (Frankel *et al.*, 1977a) and HPLC (Chan and Levett, 1977a) have shown that the mechanism of oleate autoxidation is more complicated than that presented in Figure 6.5. The amounts of 8- and 11-hydroperoxides produced are slightly greater

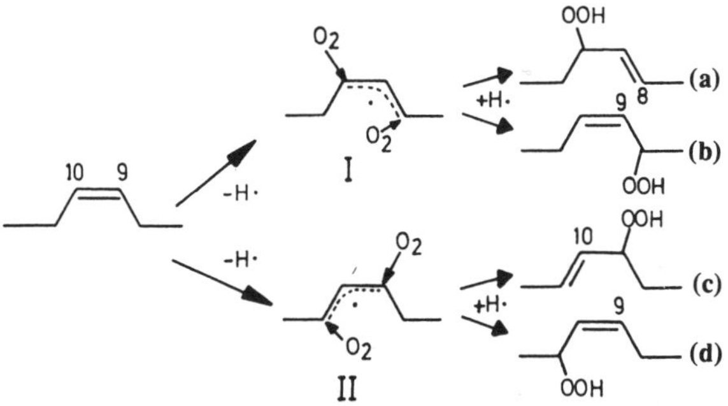

Figure 6.5 Reaction scheme for the formation of four isomeric hydroperoxides from oleate autoxidation: (a) 10-OOH; (b) 8-OOH; (c) 9-OOH; (d) 11-OOH.

than those of 9- and 10-isomers. This suggests a somewhat greater reactivity of C-8 and C-11 with $^3O_2$. In reality, free-radical autoxidation of oleate produces a mixture of all eight *cis* and *trans* isomers of 8-, 9-, 10- and 11-hydroperoxides. At 25°C, the amounts of *cis* and *trans* 8- and 11-hydroperoxides are similar, but the 9- and 10-isomers are mainly *trans*. Frankel (1979) has postulated a modified mechanism for oleate autoxidation, which explains the stereochemistry of oleate hydroperoxide isomers. In addition, his findings may also provide valuable information about the influence of autoxidation temperatures on the stereochemistry of volatile decomposition products responsible for off-flavours and odours.

Autoxidation of linoleate involves hydrogen abstraction on the doubly reactive allylic C-11, with the formation of a pentadienyl radical. This intermediate radical reacts at both ends with oxygen to produce a mixture of conjugated 9- and 13-diene hydroperoxides. The formation of these two major isomers is described in the reaction scheme presented in Figure 6.6. Chan and Levett (1977b) demonstrated from an autoxidised linoleate experiment, that four major geometric isomers—out of a possible eight isomers—were formed, with the hydroperoxide group at 9- and 13-positions. The *cis/cis* isomer and the *cis/trans* isomers with the *cis* double bond adjacent to the -OOH group were not produced to any appreciable extent. Belitz and Grosch (1987) have suggested that the monoallylic groups in linoleate (at 8- and 14-positions) can also react to a small extent during the autoxidation of linoleate. This will yield four additional hydroperoxides (8-, 10-, 12- and 14-OOH), each isomer having two isolated double bonds. The amount of these minor hydroperoxide isomers is about 4% of the total (Table 6.1).

Autoxidation of linolenate is more complex than that of linoleate since two separate 1,4-diene systems are present. Hydrogen abstraction on the two active

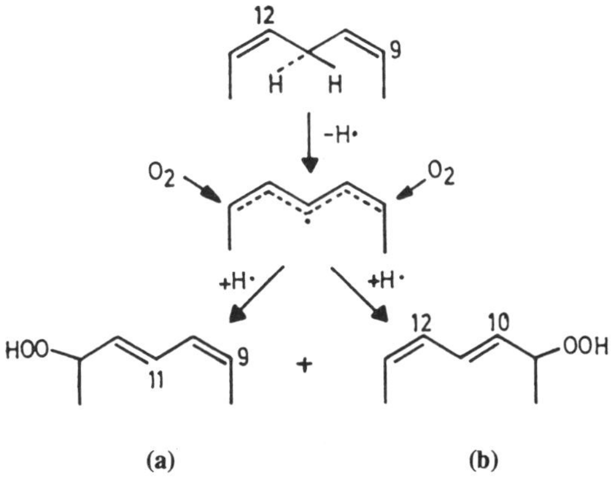

Figure 6.6 Reaction scheme for the formation of two major hydroperoxides from linoleate autoxidation: (a) 13-OOH; (b) 9-OOH.

Table 6.1 Proportions of monohydroperoxides formed by autoxidation and photoxidation of unsaturated fatty acids

Fatty acid	Monohydroperoxides			
	Position of		Proportion (%)	
	-OOH group	Double bond	Autoxidation	Photoxidation
Oleic	8	9	27	
	9	10	23	50
	10	8	23	50
	11	9	27	
Linoleate	8	9, 12	1.5	
	9	10, 12	46.5	31
	10	8, 12	0.5	18
	12	9, 13	0.5	18
	13	9, 11	49.5	33
	14	9, 12	1.5	
Linolenate	9	10, 12, 15	37	23
	10	8, 12, 15		13
	12	9, 13, 15	8	12
	13	9, 11, 15	10	14
	15	9, 12, 16		13
	16	9, 12, 14	45	25

Data compiled from Belitz and Grosch (1987); Frankel (1984); Gunstone (1984)

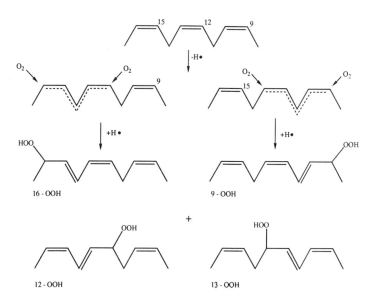

Figure 6.7 Reaction scheme for the formation of four isomeric hydroperoxides from linolenate autoxidation.

methylene groups (C-11 and C-14) produces two pentadienyl radicals. These intermediates react with oxygen at the end carbons to produce conjugated dienes with hydroperoxides on C-9 and C-13, or C-12 and C-16, with the third double bond remaining unaffected (Figure 6.7). Using GC-MS analysis, the presence of all eight geometric *trans/cis* and *trans/trans* diene isomers has been confirmed by Frankel *et al.* (1977b). These results showed that the amount of the 9- and 16-hydroperoxides was significantly higher than that of 12- and 13-hydroperoxides (Table 6.1). The formation of 9- and 16- hydroperoxide isomers is favoured because they cannot react with oxygen to form hydroperoxy peroxide, diperoxide or prostaglandin-like bicyclo endoperoxides (Frankel *et al.*, 1977b; Neff *et al.*, 1981; Chan *et al.*, 1982), while the 12- and 13-OOH isomers can produce such cyclic peroxides.

6.2.4.2 Monohydroperoxides produced by photoxidation ($^{1}O_2$). As mentioned earlier, photo-sensitised oxidation of unsaturated fats or fatty acids proceeds by a different mechanism from free-radical autoxidation. Singlet oxygen reacts with double bonds by concerted addition, and thus gets attached at either end carbon of a double bond, which is shifted to produce an allylic hydroperoxide in *trans* configuration. According to the above reaction mechanism, oleate yields a mixture of 9- and 10-hydroperoxides with allylic *trans* double bonds. Schematically, the formation of these two isomers is presented in Figure 6.8. Similarly, linoleate produces a mixture of 9-, 10-, 12- and 13-hydroperoxides (Figure 6.9), and linolenate furnishes a mixture of 9-, 10-, 12-, 13-, 15-, and 16-hydroperoxide isomers.

The proportions of hydroperoxide isomers formed by autoxidation and photo-oxygenation of oleic, linoleic and linolenic acids are given in Table 6.1. It is interesting to note that the internal isomers of autoxidised linolenate (12- and 13-OOH), and of photoxidised linoleate (10- and 12-OOH) and linolenate

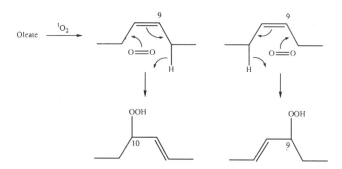

Figure 6.8   Reaction scheme for the formation of isomeric hydroperoxides of oleate by photoxidation.

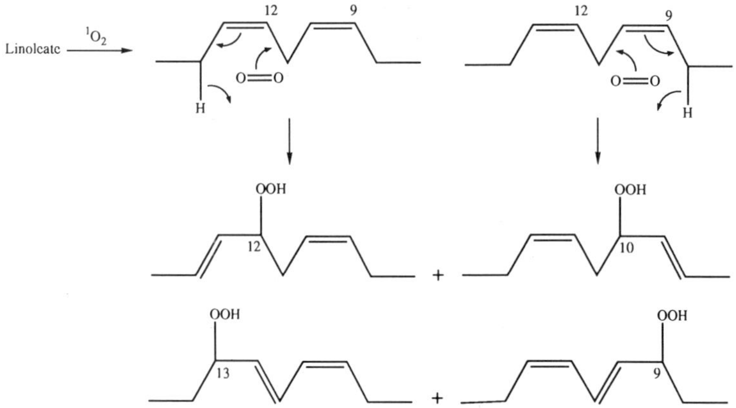

Figure 6.9 Reaction scheme for the formation of isomeric hydroperoxides of linoleate by photoxidation.

(10-, 12-, 13- and 15-OOH) are produced in significantly lower concentrations than the external isomers. This uneven formation of hydroperoxides is due to the fact that the internal isomeric hydroperoxides have a homoallylic structure that allows 1,3-cyclisation to form hydroperoxy cyclic peroxides (Chan *et al.*, 1982). The formation of isomeric hydroperoxides from autoxidation and photo-sensitised oxidation of highly unsaturated fatty acids, such as arachidonic and eicosapentaenoic acids, have also been reported in the literature (Frankel, 1984; Yamauchi *et al.*, 1985).

6.3 Decomposition of hydroperoxides and types of off-flavour compounds

Monohydroperoxides of unsaturated fatty acids formed by autoxidation, photoxidation or enzymic catalysis are very unstable and break down readily into a wide variety of volatile and non-volatile products. The decomposition products include aldehydes, ketones, alcohols, acids, hydrocarbons, lactones, furans and esters. For example, Smouse and Chang (1967) identified a variety of flavour compounds from reverted but not rancid soyabean oil with a peroxide value of 4.3 mEq/kg. The reversion flavours are described as beany, grassy, buttery, fishy, painty or haylike. Table 6.2 lists the major volatile compounds, out of 74, identified in a flavour-deteriorated soyabean oil. Snyder *et al.* (1985) studied the formation of volatiles in eight different vegetable oils on storage at 60°C. On the basis of gas chromatography–mass spectrometry, 34 volatile components degraded from unsaturated fatty acid monohydroper-

Table 6.2 Major volatile compounds identified in flavour-deteriorated soyabean oil (Smouse and Chang, 1967)

Aldehydes (18)	*Alcohols* (8)
Pentanal	Ethanol
Hexanal	Pentanol
Octanal	Pent-1-en-3-ol
Nonanal	Oct-1-en-3-ol
2-Heptenal	
2-Octenal	*Esters* (2)
2-Nonenal	Ethyl acetate
2-*trans*,4-*trans*-Heptadienal	
Ketones (8)	*Hydrocarbons* (6)
Heptan-2-one	Dec-1-yne
Octan-2-one	

Note: figures in parentheses give the number of compounds identified in that class.

oxides were identified. The relative concentrations of these volatile compounds (mainly aldehydes and hydrocarbons) were found to increase with the level of oxidative deterioration as measured by peroxide value. The off-flavours associated with most of the compounds identified from oxidised edible oils and from fat-containing food products are compiled in Table 6.3. It should be emphasised that the formation of oxidative off-flavour compounds in a particular food depends on a number of factors such as condition and duration of oxidation, oxygen tension, metals, heat, water content, and nature and surface of lipid. It is quite clear from Table 6.3 that the volatile components which impart undesirable flavours to oils, fats and lipid-containing foods are primarily carbonyl compounds. It should be pointed out here that the precursors of carbonyl compounds, monohydroperoxides, begin to decompose as soon as they are formed. During the initial stages of autoxidation, their rate of formation exceeds their decomposition. However, the presence of trace metals, heat or haematin compounds accelerates the rate of decomposition of hydroperoxides.

It is widely accepted (Bell *et al.*, 1951; Badings, 1970; Frankel, 1980, 1983; Grosch, 1987) that hydroperoxide decomposition involves homolytic cleavage of the –OOH group, giving rise to an alkoxy radical and a hydroxy radical. The alkoxy radical undergoes β-scission of C–C bond, with formation of an aldehyde and alkyl or vinyl radical. A general reaction pathway for the homolytic cleavage of monohydroperoxides is illustrated in Figure 6.10. β-Scission of an allylic alkoxy radical can occur via two ways: (1) cleavage of the C–C bond on the side of the oxygen-bearing carbon atom, i.e. away from the double bond–scission (A); and (2) cleavage of the C–C bond between the double bond and the oxygen bearing carbon atom–scission (B). Scission (A) will result in the formation of an unsaturated aldehyde and an alkyl radical, while scission (B) will yield a vinyl radical and a saturated aldehyde compound. By reacting with a hydroxyl radical, the alkyl radical produces an alcohol and the vinyl

Table 6.3 Characteristic flavours and associated compounds isolated and identified in oxidised fats of various foodstuffs

Flavour	Compounds
Cardboards, tallowy	Octanal; alkanals (C9–C11); alk-2-enal (C8, C9); 2,4-dienals (C7, C10); nona-2-*t*,6-*t*-dienal
Fatty/oily	Alkanals (C5–C7); hex-2-enal; 2,4-dienals (C5–C10); 2-*t*-pentenylfuran
Painty	Alkanals (C5–C10); alk-2-enals (C5–C10); hepta-2-*t*,4-*t*-dienal; 2-heptanone; pent-2-*t*-enal
Oxidised	Oct-1-ene-3-one; octanal; hept-2-enal; 2,4-heptadienal; alkanols (C2–C9)
Fishy	Alkanals (C5–C10); alk-2-enals (C5–C10); hepta-2-*t*,4-*t*-dienal; 2-alkanones (C3–C11); oct-1-en-3-one; deca-2-*t*,4*c*,7-*t*-trienal; pent-1-en-3-one
Grassy	2-*t*-Hexenal; nona-2,6-dienal; 2-*c*-pentenylfuran
Mild, pine-like	3-*t*-Hexenal
Rotten apple	2-*t*,4-*c*-Heptadienal
Rancid (hazelnut)	2-*t*,4-*t*-Heptadienal
Green-beany	3-*c*-Hexenal
Beany	Alkanals; non-2-enal
Deep-fried fat	2-*t*,4-*t*-Decadienal
Sweet aldehydic	2-*t*,4-*c*-Decadienal
Mushroom, mouldy	Oct-1-en-3-one; oct-1-en-3-ol
Cucumber-like	Nona-2-*t*,6-*c*-dienal; non-2-*t*-enal
Melon-like	Nona-3-*c*,6-*c*-dienals; non-6-*c*-enal
Potato-like	Penta-2,4-dienal
Lemon	Nonanal
Sharp	Octanal; pentanal
Brown beans	Oct-2-enal
Metallic	Pent-1-en-3-one; oct-1-en-3-one; 2-*t*-pentenylfuran; 1-*c*,5-octadien-3-one
Rancid	2-*t*-Nonenal; volatile fatty acids (C4–C10)
Nutty	2-*t*,4-*t*-Octadienal; 2-*t*,2-*c*-octenal
Creamy	4-*c*-Heptenal
Buttery	2-*c*-Pentenylfuran; diacetyl; 2, 3-pentanedione
Fruity	Alkanals (C5, C6, C8, C10); aliphatic esters; isobutyric acid
Green, putty	3-*c*-Hexenal; 4-*c*-heptenal
Hardened, hydrogenation	6-*t*-Nonenal
Liquorice	2-*t*-1-Pentenylfuran; 2-*c*-1-pentenylfuran; 5-pentenyl-2-furaldehyde
Soapy/fruity	Alkanals (C7–C9)
Bitter	Pentanal; hexanal; 2-*t*-heptenal

t = trans, *c* = cis
Data compiled from Smouse and Chang (1967); Badings (1970); Forss (1972); Kochhar and Meara (1975); Swoboda and Peers (1977a, b); Chang *et al.* (1983); Grosch (1987)

radical forms 1-enol, which tautomerises to the corresponding aldehyde (Figure 6.10). It can be noticed from Figure 6.10 that the chemical structures of the volatile compounds formed are affected by the structure of the residue R_1, whether scission (A) or (B) takes place, and by involvement of oxygen in the breakdown reactions of the alkoxy radical. Several workers (Keeny, 1962; Wilkinson and Stark 1967; Parsons, 1973) have given theoretical considerations to the reaction routes of various alkoxy radicals involved in the forma-

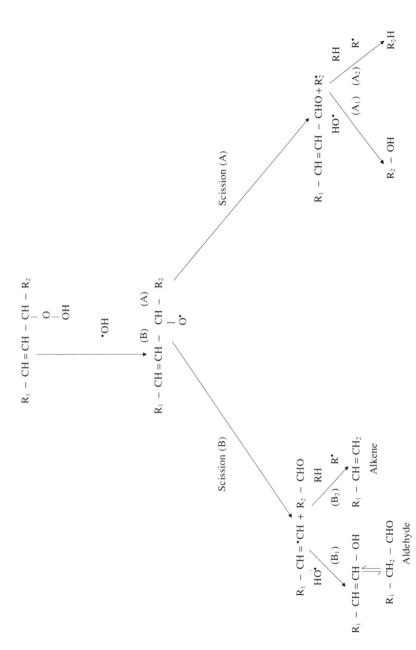

Figure 6.10 General reaction pathway for the homolytic cleavage of monohydroperoxides of unsaturated fats.

tion of volatile products. The domination of a particular reaction pathway during the decomposition of a particular monohydroperoxide depends on the oxidation state of the lipid, temperature, oxygen pressure, and the presence of catalysts, e.g. transition metals or haem proteins. In most cases, the application of the reaction sequences to the decomposition of monohydroperoxides of unsaturated fatty acids (given in Figure 6.10) shows agreement between the predicted compounds and the analytical results of the volatile products. The reaction routes discussed in the following sections explain mainly the degradation of monohydroperoxides of unsaturated fatty acids into volatile components, which are responsible for the development of off-flavours and odours in oils and lipid-containing foods.

6.3.1 Aldehydes

Aliphatic aldehydes are the most important volatile breakdown products because they are major contributors to unpleasant odours and flavours in food products. Some volatile aldehydes formed by autoxidation of unsaturated fatty acids, and the corresponding monohydroperoxides from which they originated—according to the reaction routes described in Figure 6.10—are listed in Table 6.4.

Table 6.4 Some volatile aldehydes obtained from various unsaturated fatty acid monohydroperoxides on the basis of β-scission reaction routes (Badings, 1970; Frankel, 1983)

Fatty acid	Monohydroperoxides	Scission route[a]	Aldehydes formed
Oleate	8-OOH	A; B, B₁	2-Undecenal; decanal
	9-OOH	A; B, B₁	2-Decenal; nonanal
	10-OOH	B	Nonanal
	11-OOH	B	Octanal
Linoleate	9-OOH	B, B₁	3-Nonenal
		A	2,4-Decadienal
	13-OOH	B	Hexanal
Linolenate	9-OOH	A	2,4,7-Decatrienal
		B, B₁	3,6-Nonadienal
	12-OOH	A	2,4-Heptadienal
		B, B₁	3-Hexenal
	13-OOH	A	3-Hexenal
	16-OOH	A	Propanal
Arachidonate	8-OOH	A	2,4,7-Tridecatrienal
		B, B₁	3,6-Dodecadienal
	9-OOH	A	3,6-Dodecadienal
	11-OOH	A	2,4-Decadienal
		B, B₁	3-Nonenal
	12-OOH	A	3-Nonenal
	15-OOH	A	Hexanal

[a] Breakdown of the alkoxyl radical in the reaction pathway given in Figure 6.10

Due to its large abundance in foodstuffs and its relatively high susceptibility to oxidation, linoleic acid and its glycerides are among the most important precursors of aldehyde compounds. Indeed, 25 volatile aldehydes identified by various workers from autoxidised linoleate at moderate temperatures are listed by Grosch (1987). The formation of two major aldehydes, hexanal and 2,4-decadienal can be explained by thermal decomposition of linoleate 13-OOH and 9-OOH hydroperoxides, respectively, through the scission routes (B) and (A). Chan *et al.* (1976) observed that decomposition of individual isomers (9- and 13-OOH) of methyl linoleate hydroperoxide gave rise to some four volatile compounds, such as hexanal, methyl octanoate, 2,4-decadienal isomers, and methyl 9-oxononanoate. However, the relative proportions of these cleavage products were different. These findings suggest that isomerisation of the -OOH group takes place during the homolytic cleavage of linoleate hydroperoxides. Some workers (Swoboda and Lea, 1965; Kimoto and Gaddis, 1969) showed that oxidation of fats containing linoleate favours the formation of hexanal under mild conditions and 2,4-decadienal at high temperatures. On the basis of gas chromatography and infrared studies, the amounts of *trans–cis* and *trans–trans* isomers of 2,4-decadienal were reported to be 28 and 72%, respectively (Hoffman, 1961). The 2-*trans*,4-*trans*-decadienal was considered to be derived from either the corresponding *trans*, *trans*-9-hydroperoxide or by isomerisation of 2-*cis*,4-*trans*-decadienal. As mentioned earlier, singlet oxidation of linoleate leads to the formation of four isomeric hydroperoxides, two of which (10-OOH Δ8,12 and 12-OOH Δ9,13) are not typical of the major conjugated isomers of autoxidation. The presence of 2-heptenal, probably arising from the 12-hydroperoxide isomer, has been detected in photoxidised linoleate, as well as in vegetable oils at very low levels of oxidation (Frankel and Neff, 1979; Frankel *et al.*, 1979). Figure 6.11 shows the reaction pathway for the formation of 2-heptenal through β-scission of linoleate 12-hydroperoxide Δ9,13.

Earlier workers (Badings, 1959; Gaddis *et al.*, 1961; Swoboda and Lea, 1965) put forward the reaction mechanism for the formation of 2-octenal through the decomposition of a non-conjugated 11-OOH linoleate hydroperoxide by the scission route (A). However, this monohydroperoxide has never been detected under a wide range of autoxidation conditions (Chan and Levett, 1977b; Frankel, 1979; Neff and Frankel, 1980). Another route for the formation of 2-octenal from the decomposition of 9-hydroperoxide of linoleate has been suggested by Schieberle and Grosch (1981). Figure 6.12 illustrates the reaction pathway for the production of 2-octenal and hexanal through B and B_1 cleavage of linoleate 9-hydroperoxide Δ10,12. The formation of 2-octenal occurs through an allylic radical, with formaldehyde and glyoxal as by-products. Cobb and Day (1965) isolated glyoxal from methyl linoleate oxidised at 45°C. The formation of 2-octenal has also been explained by further oxidation of 2,4-decadienal (Schieberle and Grosch, 1981). The peroxy radical

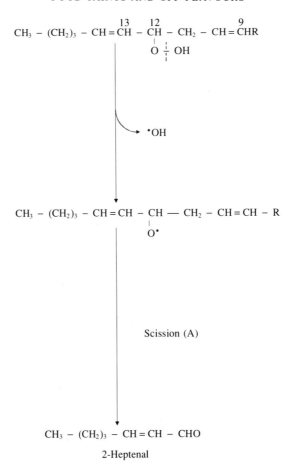

Figure 6.11 Reaction pathway for the formation of 2-heptenal by decomposition of linoleate 12-hydroperoxide.

attacks on the *n*-8 double bond, which produces 2-octenal via allylic peroxy intermediates:

$$CH_3(CH_2)_4 - CH = CH - CH = CH - CHO$$

ROO•

$$CH_3(CH_2)_4 - CH - CH - •CH - CH - $$

•OO•   OOR

$$CH_3(CH_2)_4 - CH = CH - CH - CH - $$

OO•   OOR

$$CH_3(CH_2)_4 - CH = CH - CHO \qquad (6.31)$$

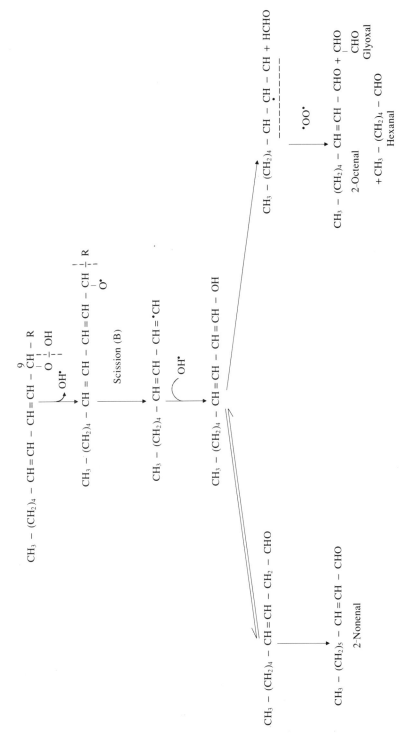

Figure 6.12 Reaction pathway for volatile aldehydes from decomposition of linoleate 9-hydroperoxide.

The formation of lower aldehydes can be explained by reaction of the alkyl radical (scission A, Figure 6.10) with oxygen to produce primary hydroperoxide:

$$R(CH_2)_x - CH_2^{\bullet} \xrightarrow{\quad O_2 \quad} R(CH_2)_x - CH_2OO^{\bullet}$$

$$\downarrow RH$$

$$R(CH_2)_x - CH_2OOH + R^{\bullet} \qquad (6.32)$$

The primary hydroperoxide produces an alkoxy radical by loss of hydroxyl radical. The alkoxy radical may form either an aldehyde by dehydration or lose a carbon and produce formaldehyde:

$$R(CH_2)_x - CH_2OOH \xrightarrow{\quad -OH^{\bullet} \quad} R(CH_2)_x - CH_2O^{\bullet}$$

$$R(CH_2)_x - CHO \qquad\qquad R(CH_2)_{x-1} - CH_2^{\bullet} + HCHO \qquad (6.33)$$

Lower aldehydes are then produced from the alkyl radicals of reaction (6.33) by repeating the reaction schemes.

Another reaction pathway for the formation of saturated aldehydes has been suggested by Parsons (1973). The reaction of the vinyl radical (scission B, Figure 6.10) with oxygen produces a vinyl hydroperoxide, which reacts with other lipid molecules to produce an aldehyde:

$$R_1 - CH = {}^{\bullet}CH \xrightarrow{\quad O_2 \quad} R_1 - CH = CH \xrightarrow{\quad RH \quad} R_1 - CH = CH$$
$$\qquad\qquad\qquad\qquad\qquad OO_{\bullet} \qquad\qquad\qquad\qquad OOH$$

$$R_1 - CH = CH \xrightarrow{\quad RH \quad} R_1 - CH_2 - CHO + OH^{\bullet} \qquad (6.34)$$
$$\qquad OOH$$

Frankel (1983) has reviewed other fragmentation schemes for cleavage of allylic hydroperoxides during the production of aldehydes. These include, for example, the thermal Hork-cleavage mechanism, Criegee's rearrangement, and Hawkins and Quin reactions. These reaction schemes involve the intermediate formation of alcohol ether:

$$R_1 - CH = CH - CH - R_2 \longrightarrow \left[R_1 - CH = CH - O - CH - R_2 \right]$$
$$\qquad\qquad\qquad O - OH \qquad\qquad\qquad\qquad\qquad\qquad OH$$

$$R_1 - CH_2 - CHO + R_2 - CHO \qquad (6.35)$$

Quantitative analysis of different volatile aldehydes (and ketones) produced from autoxidised oleic, linoleic, linolenic and arachidonic acids has been carried out by Badings (1970). The majority of aldehydes formed from autoxidation of these unsaturated fatty acids can be explained by the pathways of β-scission and other reaction schemes described above.

Processing of fats and oils can give rise to changes in fatty acid configuration. These changes can then lead to new precursors for the development of off-flavour compounds. For example, partial hydrogenation of linolenate in soyabean oil produces isomeric dienes, known as isolinoleic acids. Keppler and coworkers (1965, 1967) demonstrated that *cis,cis*-9,15 and 8,15-octadecadienoates are the precursors of an unpleasant 'hardening' flavour that develops during storage of hydrogenated oils. The aldehyde 6-*trans*-nonenal has been reported to be responsible for this off-flavour (Table 6.3). The pathway for the production of 6-*trans*-nonenal from the decomposition of 10-OOH derived from either the 9,15-diene (with allylic rearrangement) or the 8,15-diene (without allylic rearrangement) can be described (Figge, 1971) as:

$$CH_3 - CH_2 - \overset{15}{CH} = CH - (CH_2)_4 - \overset{10}{CH} - CH = \overset{8}{CH} - R$$
$$| $$
$$O - OH$$

$$\searrow \cdot OH$$

$$CH_3 - CH_2 - CH = CH - (CH_2)_4 - CH - CH = CH - R$$
$$|$$
$$O\cdot$$

Scission A, Figure 6.10

$$CH_3 - CH_2 - CH = CH - (CH_2)_4 - CHO \qquad (6.36)$$

Other compounds such as 2-*trans*,6-*trans*-octadienal, ketones, alcohols and lactones are also considered to contribute to the 'hydrogenation' flavour (Yasuda *et al.*, 1975). The 11-hydroperoxide of 12-*trans*,16-*trans*-octadecadienoate has been prosposed as the precursor of 2-*trans*,6-*trans*-octadienal. In addition, the decomposition of a wide range of isolinoleate hydroperoxides produces a mixture of aldehydes. Some of these aldehydes are similar to those formed from the linolenate hydroperoxides. The decomposition pathways for the production of several aldehydes are described in reactions (6.37) to (6.43).

$$CH_3 - \overset{17}{CH} - CH = CH - R$$
$$| $$
$$O \quad OH$$

$$CH_3 - CHO \qquad \text{Acetaldehyde} \qquad (6.37)$$

$$CH_3 - CH_2 - \overset{16}{CH} - CH = CH - R$$
$$|$$
$$O \quad OH$$

$$CH_3 - CH_2 - CHO \qquad \text{Propanal} \qquad (6.38)$$

$$CH_3 - CH = CH - \overset{15}{CH} - CH_2 - R$$
$$|$$
$$O \quad OH$$

$$CH_3 - CH = CH - CHO \qquad \text{2-Butenal} \qquad (6.39)$$

$$CH_3 - CH_2 - CH = CH - \overset{14}{CH} \vdots CH_2 - R$$
$$O \vdots OH$$

$$\downarrow$$

$$CH_3 - CH_2 - CH = CH - CHO \quad \text{2-Pentenal} \tag{6.40}$$

$$CH_3 - CH_2 - CH = CH - (CH_2)_4 - CH = CH - \overset{8}{CH} - CH_2 - R$$
$$O - OH$$

$$\downarrow$$

$$CH_3 - CH_2 - CH = CH - (CH_2)_4 - CH = CH - CHO$$
$$\text{2, 8-Undecadienal} \tag{6.41}$$

$$CH_3 - CH_2 - CH = CH - (CH_2)_3 - CH = CH - \overset{9}{CH} - CH_2 - R$$
$$O - OH$$

$$\downarrow$$

$$CH_3 - CH_2 - CH = CH - (CH_2)_3 - CH = CH - CHO$$
$$\text{2, 7-Decadienal} \tag{6.42}$$

$$CH_3 - CH_2 - CH = CH - (CH_2)_3 - \overset{11}{CH} \vdots CH = CH - R$$
$$O \vdots OH$$

$$\downarrow$$

$$CH_3 - CH_2 - CH = CH - (CH_2)_3 - CHO$$
$$\text{5-Octenal} \tag{6.43}$$

The unsaturated aldehydes are very potent flavour components, and have very low threshold values. For example, 6-*trans*-nonenal has one of the lowest threshold values, 5 µg/kg, for the melon or cucumber-like odour, and 0.3 µg/kg for taste. Many researchers have reported flavour threshold values of a variety of aldehydes in different media (Lea and Swoboda, 1958; Meijboom, 1964; Badings, 1970; Forss, 1972; Meijboom and Jongenotter, 1981). Table 6.5 lists flavour threshold values of selected aldehydes in various media. It can be noted that in paraffin oil, the taste threshold levels of these aldehydes are substantially lower than the odour levels. Generally, the lower members of saturated aldehydes give rise to sharp and irritating off-flavours, those in the intermediate range (C6–C9) contribute to fatty, tallowy, bitter, soapy-fruity, green flavours, and the higher members have citrus orange peel flavours. The flavours of 2-*trans*-alkenals have been described as green, painty or putty (C5–C7), nutty (C8) and tallowy/cucumbery (C9, C10). The flavour of 2-*trans*,4-*cis*-heptadecadienal has been described as being like rotten apples and that of its 2-*trans*,4-*trans* isomer like rancid hazelnuts (Keppler, 1977).

Table 6.5 Flavour threshold (mg/kg) of some selected aldehydes in various media

Aldehyde	Paraffin oil			Vegetable oil	Water
	Odour[a]	Taste[a]	Taste		
Propanal	3.6	1.6	1.0[b]		
Pentanal	0.24	0.15			
Hexanal	0.32	0.15	0.6[b]	0.3[b]	0.03[b]
Heptanal	3.2	0.042			
Octanal	0.32	0.068	0.6[b]	0.9[b]	0.005[b]
Nonanal	13.5	0.32			
Decanal	6.7	1.0	0.7[b]	0.6[b]	0.007[b]
2-t-Pentenal	2.3	0.32	1.0[c]		
2-t-Hexenal	10.0	2.5	0.6[c]		
2-t-Nonenal	3.2	0.1	0.4[c]	0.08[b]	0.006[b]
3-c-Hexenal	0.11	0.11	0.09[c]		
3-t-Hexenal	1.2	1.2	0.95[c]		
2-t,4-t-Hexadienal	0.27	0.036	0.04[c]		
2-t,4-t-Heptadienal	10.0	0.46	0.1[c]		
2-t,4-t-Octadienal	1.0	0.15			
2-t,4-c-Heptadienal	3.6	0.055	0.04[c]		
2-t,6-t-Nonadienal	0.21	0.018	0.02[c]		0.001[c]
2-t,6-c-Nonadienal	0.01	0.002	0.0015[c]		0.0001[c]
2-t,4-t-Decadienal	2.15	0.28	0.1[c]		0.0005[c]
2-t,4-c,7-c-Decatrienal			0.024[c]		

t = trans, c = cis
[a] Meijboom (1964); [b] Lea and Swoboda (1958); [c] Forss (1972)

The 2,4-decadienals have oily/fatty, deep-fat frying odours. Some unsaturated lactones, e.g. gamma-lactone of 4-hydroxy-2-nonenoic acid, also contribute to the pleasant flavour of deep-fried chicken. It should be pointed out that extremely low levels of many aldehydes contribute to desirable flavours of many foods. For example, 4-cis-heptenal in trace amounts (1 µg/kg) contributes a creamy flavour to butter. McGill et al. (1977) have reported odour threshold level of this aldehyde to be 0.04 µg/kg. Badings (1984) has mentioned octadecadienoic acid C18:2 Δ11,15 cis to be the precursor of this aroma compound. The pathway for the formation of 4-cis-heptenal by decomposition of linoleate 11-hydroperoxide Δ12,15 is shown in Figure 6.13. If the concentration of such aldehydes goes above a certain level, then oxidative flavour defects can occur. For example, the cold-storage flavour of white fish has been related to the presence of 4-cis-heptenal (McGill et al., 1974, 1977).

With a few exceptions, there is no generalised straightforward relationship between molecular structure of aldehydes and flavour intensity (Meijboom, 1964; Keppler, 1977). For instance, 3-cis-hexenal (0.09–0.11 mg/kg) and 4-cis-heptenal (0.0005–0.0016 mg/kg) are much more intense than the corresponding 3-trans-hexenal (0.6–2.5 mg/kg) and 4-trans-heptenal (0.1–0.32 mg/kg). On the other hand, 6-cis-nonenal has a higher flavour threshold value (0.002 mg/kg) than the trans isomer (0.0003 mg/kg), as reported by Keppler et al. (1965, 1967). It is worth stressing that the off-flavour aldehyde

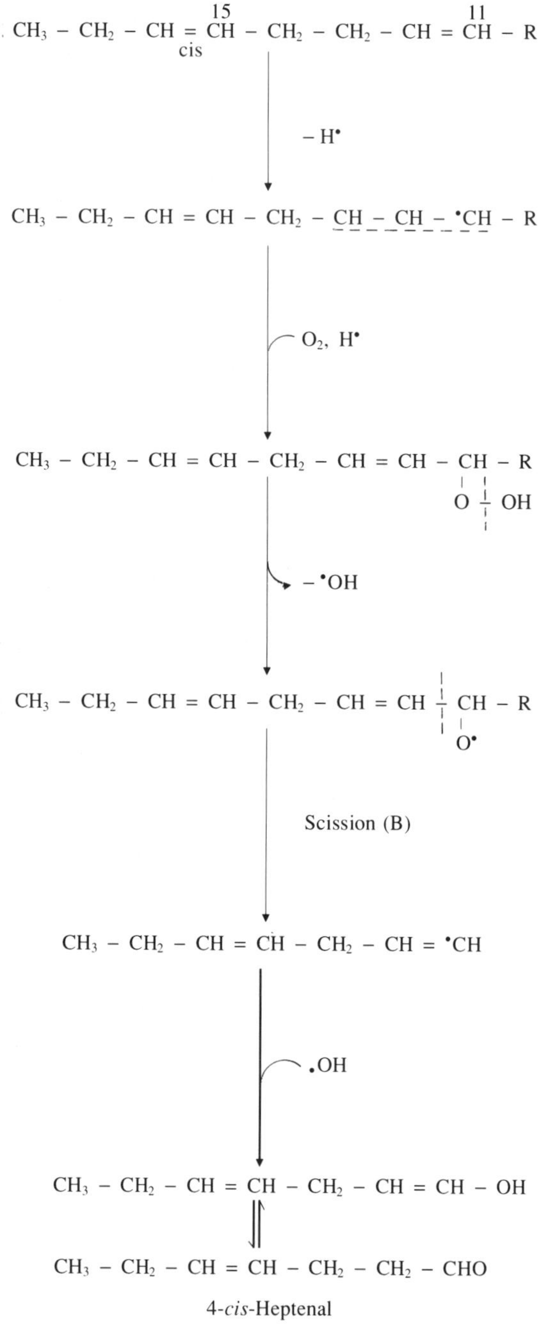

$$\overset{15}{CH_3 - CH_2 - CH = CH} - CH_2 - CH_2 - CH = \overset{11}{CH} - R$$
$$\underset{cis}{}$$

$$\downarrow \quad - H^{\bullet}$$

$$CH_3 - CH_2 - CH = CH - CH_2 - \underline{CH} - \underline{CH} - {}^{\bullet}\underline{CH} - R$$

$$\downarrow \quad O_2, \, H^{\bullet}$$

$$CH_3 - CH_2 - CH = CH - CH_2 - CH = CH - CH - R$$
$$O \mid OH$$

$$\downarrow \quad - {}^{\bullet}OH$$

$$CH_3 - CH_2 - CH = CH - CH_2 - CH = CH \mid CH - R$$
$$O^{\bullet}$$

$$\downarrow \quad \text{Scission (B)}$$

$$CH_3 - CH_2 - CH = CH - CH_2 - CH = {}^{\bullet}CH$$

$$\downarrow \quad {}^{\bullet}OH$$

$$CH_3 - CH_2 - CH = CH - CH_2 - CH = CH - OH$$
$$\Updownarrow$$
$$CH_3 - CH_2 - CH = CH - CH_2 - CH_2 - CHO$$

4-*cis*-Heptenal

Figure 6.13 Reaction scheme for the formation of 4-*cis*-heptenal by decomposition of linoleate 11-hydroperoxide.

compounds resulting from linoleate-containing fats require a much higher level of oxidation than those formed by deterioration of linolenate-containing oils or food lipids containing n-3 fatty acids. This is simply due to the low threshold values of the volatile aldehydes produced from oxidation of linolenate and fatty acids with an n-3 double bond (Keppler, 1977). The various unsaturated aldehydes originating from n-3 fatty acids, which have low taste values, are 3-*cis*-hexenal (0.09–0.11 mg/kg), 4-*cis*-heptenal (0.0005–0.0016 mg/kg), 2-*trans*,4-*cis*-heptadienal (0.04–0.06 mg/kg), 2-*trans*,6-*cis*-nonadienal (0.002 mg/kg) and 2-*trans*,4-*cis*,7-*cis*-decatrienal (0.024 mg/kg). Further oxidation of unsaturated aldehydes complicates the understanding of the origin of shorter chain aldehydes and other volatile products. For example, from autoxidation of pure 2,4-decadienal at room temperature, a complex mixture of volatile products such as butenal, hexanal, 2-heptenal, 2-octenal, benzaldehyde, glyoxal, *trans*-2-buten-1,4-dial, furan, ethanol, acrolein, pentane, benzene, and acetic, hexanoic, 2-octenoic and 2,4-decadienoic acids have been identified by Matthews *et al.* (1971). The oxidation of 2-nonenal at 45°C produces C2, C3, C7, C8 alkanals, glyoxal, and a mixture of C7, C8 and C9 α-keto aldehydes (Lillard and Day, 1964).

The alkanals (C2–C4), glyoxal, α-keto aldehydes (C5–C9), *cis*-2-buten-1,4-dial and malonaldehyde were the products of oxidation of 2,4-heptadienal. Several mechanisms to explain the formation of these compounds from oxidative degradation of aldehydes have been postulated (Loury, 1972; Loury and Forney, 1968; Forney, 1974; Schieberle and Grosch, 1981). The reaction route for the formation of α-keto aldehyde has been suggested to be via α-hydroperoxidation of 2,4-heptadecadienal.

$$CH_3 - CH_2 - CH=CH - CH - CHO$$
$$\downarrow O_2$$
$$CH_3 - CH_2 - CH=CH - {}^\bullet CH - CH - CHO$$
$$\downarrow RH \qquad\qquad O - O^\bullet$$
$$\downarrow R^\bullet$$
$$CH_3 - CH_2 - CH=CH - {}^\bullet CH - CH - CHO$$
$$\downarrow R^\bullet \qquad\qquad O \;\;\; OH$$
$$CH_3 - CH_2 - CH=CH - CH_2 - \underset{\parallel}{C} - CHO + ROH \tag{6.44}$$
$$O$$

The reaction pathway for the formation of short-chain aldehydes from oxidation of 2,4-heptadienal can be described as:

$$CH_3 - CH_2 - CH=CH - CH=CH - CHO$$
$$\downarrow O_2, RH$$
$$\downarrow R^\bullet$$
$$CH_3 - CH_2 - CH=CH \;|\; CH - {}^\bullet CH - CHO$$
$$\downarrow RH \quad O \;|\; OH$$
$$\downarrow R^\bullet$$
$$CH_3 - CH_2 - CH_2 - CHO + OHC - CH_2 - CHO \tag{6.45}$$
$$\text{Butanal} \qquad\qquad \text{Malonaldehyde}$$

Malonaldehyde is almost odourless and flavourless but its production forms the basis for the well-known thiobarbituric acid (TBA) test used for measuring fat oxidation. The TBA test is particularly used in assessing oxidative rancidity of meats and meat products. Also, this dialdehyde is preferentially formed by autoxidation of fatty acids with three or more double bonds. The various reaction pathways for the formation of malonaldehyde from autoxidation of linolenic acid via decomposition of bicyclic compounds have been proposed by some workers (Frankel, 1984; Belitz and Grosch, 1987). As mentioned earlier, during the autoxidation of unsaturated lipids many secondary products such as hydroperoxy epoxide, hydroperoxy cyclic peroxides and dihydroperoxide are also produced (Frankel *et al.*, 1982). The same cleavage pathways as discussed for monohydroperoxides have been used to explain their formation. For example, the production of hexanal from the 13-hydroperoxy,10, 12-cyclic peroxide of linoleate can be described as:

$$
\begin{array}{c}
\text{HO} - \text{O} \qquad \text{O} + \text{O} \\
\text{CH}_3 - (\text{CH}_2)_3 - \text{CH}_2 - \text{CH} - \text{CH} \quad \text{CH}_2 - \overset{9}{\text{CH}} = \text{CH} - \text{R} \\
\text{CH}_2
\end{array}
$$

$$\downarrow \; ^{\bullet}\text{H}$$

$$\text{CH}_3 - (\text{CH}_2)_3 - \text{CH}_3$$

Pentane

$$\text{CH}_3 - (\text{CH}_2)_4 - \text{CHO}$$

Hexanal $\qquad\qquad -\text{H}_2, \; -\text{O}_2$

$$\text{CH}_3 - (\text{CH}_2)_3 - \text{CH} = \text{CH} - \text{CHO}$$

2–Heptenal $\qquad\qquad\qquad\qquad\qquad$ (6.46)

The cleavage of the peroxy ring and carbon–carbon bond β to the double bond leads to the production of 2-heptenal. The formation of volatile ketones and furan cleavage products from the cyclic peroxides of autoxidised linoleate and linolenate are discussed in the following sections.

6.3.2 Ketones

Aliphatic ketones formed by autoxidation of unsaturated fatty acids also contribute to undesirable flavours of oils and food products. For example, Stark and Forss (1962) characterised oct-1-en-3-one as being responsible for metallic off-flavour in oxidised dairy products. Later, Cronin and Ward (1971) observed that the strong metallic flavour changes to an uncooked mushroom flavour when the concentration of oct-1-en-3-one is lowered from 10 to 1 mg/kg. Table 6.6 lists flavour properties of several saturated and unsaturated ketones. It can be seen that the odour thresholds of vinyl ketones are in the range of most potent aldehydes. The thresholds of methyl ketones (alkan-2-

Table 6.6 Flavour thresholds and descriptive flavours of some ketones

Compound	Flavour description	Threshold values (mg/kg)	
		Oil	Water
2-Butanone	Ethereal, unpleasant[a]		60[b]
2-Pentanone	Fruity, banana-like[a], pear drops[c]		2.3
2-Hexanone	Ethereal[c]		0.93
2-Heptanone	Spicy[a], rancid almonds[c]		0.65
2-Octanone	Green, fruity[a], ethereal[c]		0.15
2-Nonanone	Fruity, fatty[a], turpentine[c]		0.19
3-Buten-2-one	Sharp, irritating[a]	0.25[d]	
1-Penten-3-one	Sharp, fishy, oily, painty[a]	0.003	0.001[a]
1-Octen-3-one	Mouldy, mushroom, metallic[a, e]	0.0001	0.0001
1-c-5-Octadien-3-one	Metallic, musty, fungal[e]	4.4×10^{-4e}	1.2×10^{-6e}
3-t-,5-t-Octadien-2-one	Fatty, fruity[d]	0.3[d]	
3-t,5-c-Octadien-2-one	Fatty, fruity[d]	0.2	
3,5-Undecadien-2-one	Fatty, fried[d]	1.6	

c = cis, t = $trans$
[a] Forss (1972); [b] Siek et $al.$ (1971); [c] Kellard et $al.$ (1985); [d] Badings (1970); [e] Swoboda and Peers (1977b)

ones) are significantly higher than those of their isomeric aldehydes (Table 6.5) and vary with carbon chain length.

Long ago, methyl ketones were known to contribute to off-flavours in dairy products, e.g. the stale flavour of dried milk (Parks and Patton, 1961) and butter oil (Wyatt and Day, 1965). The formation of methyl ketones in dairy products has been reviewed by some workers (Day, 1966; Forss, 1969, 1971). Methyl ketones are also responsible for the development of off-flavours in vegetable oils and animal fats. Chang et $al.$ (1961) identified 2-heptanone in flavour-deteriorated soyabean oil. Mookherjee et $al.$ (1965) isolated the C3–C9 methyl ketones from stale crisps and showed that this group of compounds was present in larger quantity than aldehydes. These workers suggested that the autoxidation of aldehydes via the acyl radical is involved in the formation of ketones present in stale crisps. The acyl radical reacts with 1-alkene or with an alkyl radical:

$$R - C^{\bullet} + R_1 - CH = CH_2 \longrightarrow R - C - CH_2 - CH^{\bullet} - R_1$$
$$\overset{\|}{O} \qquad\qquad\qquad\qquad \overset{\|}{O}$$

$$\xrightarrow{\text{RCHO}} \quad R - C - CH_2 - CH_2 - R_1 + R - C^{\bullet}$$
$$\overset{\|}{O} \qquad\qquad\qquad\qquad \overset{\|}{O} \qquad (6.47)$$

The presence of 2-octanone in roasted peanuts (Walradt et $al.$, 1971) and cooked chicken (Nonaka et $al.$, 1967) has been detected.

Methyl ketones and secondary alcohols have been shown to be responsible for the development of rancid off-flavour in desiccated coconut (Kellard et $al.$, 1985). This type of flavour deterioration, called ketonic rancidity, is caused by partial oxidation of short-chain fatty acids resulting from the action of moulds. The reaction pathway for the conversion of octanoic acid to 2-heptanone and 2-heptanol can be described as:

$$CH_3 - (CH_2)_4 - CH_2 - CH_2 - COOH$$

$$\underset{ROOH}{\overset{ROO^\bullet}{\Big\downarrow}}$$

$$CH_3 - (CH_2)_4 - {}^\bullet CH - CH_2 - COOH$$

$$\underset{R^\bullet}{\overset{O_2,\ RH}{\Big\downarrow}}$$

$$CH_3 - (CH_2)_4 - \underset{\underset{OOH}{|}}{CH} - CH_2 - COOH$$

$$\Big\downarrow\ -H_2O;\ -CO_2$$

$$CH_3 - (CH_2)_4 - \underset{\underset{O}{\|}}{C} - CH_3 \qquad \text{2–Heptanone}$$

$$\Big\downarrow\ +H^+$$

$$CH_3 - (CH_2)_4 - \underset{\underset{OH}{|}}{CH} - CH_3$$

2–Heptanol

$$(6.48)$$

Another mechanism for this conversion has been suggested (Forney and Markovetz, 1971) as an aborted β-oxidation sequence:

$$\text{Fatty acid} \longrightarrow \alpha,\ \beta\text{-unsaturated acid}$$
$$\downarrow$$
$$\beta\text{-keto acid} \longleftarrow \beta\text{-hydroxy acid}$$
$$\downarrow$$
$$\text{methyl ketone} + CO_2 \qquad\qquad (6.49)$$

The flavour notes of 2-heptanone and 2-heptanol have been described to be rancid almonds and rancid coconut, respectively (Kellard *et al.*, 1985). These workers have reported the flavour of 2-nonanol to be stale and musty. Earlier workers (Langler and Day, 1964; Kinsella, 1969) identified a mixture of methyl ketones and lactones as being responsible for the stale off-flavour developing in vacuum-packed whole dried milk and stored butter. The even-numbered β-keto alkanoic acids esterified in the glycerides are reported to be the precursors of methyl ketones. Such glycerides are present at a level of about 0.4% in butterfat. The methyl ketones are produced from these precursors by hydrolysis and decarboxylation. The reaction is catalysed by heat, but can also occur in stored dairy products because of its low energy of activation. Moreover, Langler and Day (1964) showed that methyl ketones at concentrations far below their flavour threshold values interacted positively to produce unpleasant off-flavours. It should be pointed out that methyl ketones are formed in mould-ripened cheeses, which contributes to the formation of desirable flavours (Badings, 1984). The typical flavours of certain mould-ripened cheeses are due to the high concentration of methyl ketones, particularly 2-heptanone and 3-nonanone (Kinsella and Hwang, 1976). Moreover,

these compounds, along with short-chain fatty acids and lactones, are important flavour constituents of butter aroma. Any imbalance among these substances may cause flavour defects.

Other alkanones apart from the 2-alkanones have also been identified by
many researchers. For example, 4-heptanone was identified in flavour-
deteriorated hydrogenated soyabean oil (Schepartz and Daubert, 1950).
Nonaka et al. (1967) reported the presence of 3-octanone and 5-undecanone
in cooked chicken. The reaction mechanisms for the formation of these
ketones are not yet proposed.

The flavours of vinyl ketones are usually more unpleasant than those of
other saturated ketones. As mentioned earlier, 1-octen-3-one is responsible for
metallic flavour in oxidised butter oil, washed cream and safflower oil (Stark
and Forss, 1962). Many reaction mechanisms have been suggested for the
formation of this vinyl ketone from polyunsaturated fatty acids (Hammond
and Hill, 1964; Wilkinson and Stark, 1967). The reaction pathway for the
formation of 1-octen-3-one and 1-octen-3-ol from arachidonate 12-hydroper-
oxide via the β-scission route is illustrated in Figure 6.14. A similar reaction
pathway for the formation of 1-penten-3-one—responsible for the sharp and
fishy flavour in oxidised fats—from the decomposition of linoleic hydroperoxide has been proposed by Stark et al. (1967). Badings (1970) identified
1-octen-3-one and 1-penten-3-one from autoxidised arachidonic acid and linolenic acid, respectively.

Swoboda and Peers (1977a, b, 1978) have identified 1-cis-5-octadien-3-one
in the volatile compounds produced from butterfat after autoxidation in the
presence of copper and α-tocopherol. The development of off-flavours such
as metallic, musty/fungal, etc. in oxidised butterfat has been attributed to this
unsaturated ketone with an extremely low flavour threshold value (Table 6.6).
Linolenic acid and its n-3 homologues, for example 5,8,11,14,17-eicosapen-
taenoic acid and 7,10,13,16,19-docosapentaenoic acids (traces of which occur
in butterfat), have also been proposed as precursors for the formation of such
potent off-flavour carbonyl compounds. No reaction pathway for the formation of 1-cis-5-octadien-3-one was suggested. The presence of 3,5-octadiene-
2-one with a flavour note of fatty/fruity (Table 6.6) was identified from the
volatiles of autoxidised linolenic acid (Badings, 1970). The formation of this
conjugated unsaturated ketone has been explained by Frankel et al. (1982)
from their study on the thermal decomposition of hydroperoxy cyclic peroxides of autoxidised linoleate. The production of 3,5-octadien-2-one occurs
through cleavage of the peroxy ring of the 9-hydroperoxy-10,12- cyclic peroxide of linoleate:

$$CH_3 - CH_2 - CH = CH - CH = CH - \underset{\searrow}{CH} \overset{O - O}{\diagdown} \underset{CH_2}{\diagup} CH - \overset{9}{CH} - R$$
$$OOH$$

$$CH_3 - CH_2 - CH = CH - CH = CH - \underset{\underset{O}{\parallel}}{C} - CH_3$$

(6.50)

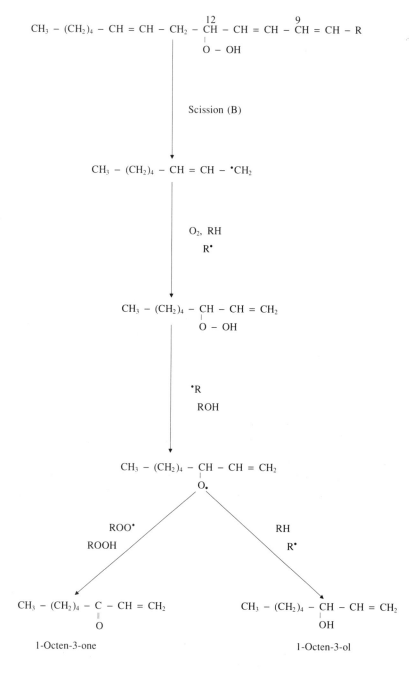

Figure 6.14 Reaction pathway for the formation of 1-octen-3-one and 1-octen-3-ol from arachidonate 12-hydroperoxide.

The identification of 3,5-octadien-2-one among other volatile decomposition products (including aldehydes) of methyl docosahexenoate after oxidation at 50°C for 28 h was also reported by Noble and Nawar (1975). The reaction route proposed for the formation of the unsaturated ketone via scission of the peroxy ring is described as:

$$
\begin{array}{c}
\overset{16}{\underset{\underset{O-O^\bullet}{|}}{CH_3 - CH_2 - (CH=CH)_2 - CH - CH_2 - CH=CH - R}} \\
\downarrow \\
CH_3 - CH_2 - (CH=CH)_2 - \underset{\underset{O \rule{2cm}{0.4pt} O}{|}}{CH} - CH_2 - {}^\bullet CH - CH - R \\
\downarrow \quad O_2,\ RH \\
CH_3 - CH_2 - (CH=CH)_2 - \underset{\underset{O \rule{2cm}{0.4pt} O}{|}}{CH} - CH_2 \overset{OOH}{\underset{}{\;CH}} - CH - R \\
\downarrow \\
CH_3 - CH_2 - CH=CH - CH=CH - \underset{\underset{O}{\|}}{C} - CH_3
\end{array}
$$

$$(6.51)$$

These findings show that hydroperoxy cyclic peroxides are also important sources of volatile off-flavour components formed by degradation of photoxidised/autoxidised fats and oils containing polyunsaturated fatty acids.

6.3.3 Furans

Several furan derivatives are reported to be responsible for flavour defects in reverted soyabean oil (Smouse and Chang, 1967; Krishnamurthy *et al.*, 1967; Ho *et al.*, 1978). These include 2-pentylfuran, 2-*cis*-pentenylfuran, 2-*trans*-pentenylfuran and 5-pentenyl-2-furaldehyde. At concentration levels of 1 to 10 mg/kg these compounds impart characteristic beany, grassy, metallic, liquorice and buttery flavours to various oils (Table 6.3). Chang *et al.* (1966) suggested the formation of 2-pentylfuran from 4-ketononanal, possibly derived from oxidised linoleic acid:

$$
\begin{array}{c}
CH_3 - (CH_2)_4 - \underset{\underset{O}{\|}}{C} - CH_2 - CH_2 - CHO \\
\downarrow \\
CH_3 - (CH_2)_4 - \overset{OH}{\underset{}{C}}=CH - CH=\overset{OH}{\underset{}{CH}} \\
\downarrow\ -H_2O \\
CH_3 - (CH_2)_4 - \underset{O}{\overset{\displaystyle CH \rule{1cm}{0.4pt} CH}{\diagdown \diagup}} \overset{\|}{} \overset{\|}{}
\end{array}
$$

$$(6.52)$$

A complete reaction scheme for the formation of 2-pentylfuran by the decomposition of 10-hydroperoxy linoleate Δ8,12 has been proposed by Ho *et al.* (1978):

$$CH_3 - (CH_2)_4 - CH = CH - CH_2 - \overset{10}{CH} \mid CH = \overset{8}{CH} - CH_2 - R$$
$$O \mid OH$$

Scission (A)

$$CH_3 - (CH_2)_4 - CH = CH - CH_2 - CHO$$

$$O_2, \ RH$$

$$CH_3 - (CH_2)_4 - CH - CH_2 - CH_2 - CHO$$
$$O \mid OH$$

$$R^\bullet$$

$$CH_3 - (CH_2)_4 - \underset{O}{\overset{\|}{C}} - CH_2 - CH_2 - CHO$$

$$CH_3 - (CH_2)_4 - \underset{OH}{C} = CH - CH = CH$$
$$OH \qquad\qquad OH$$

$$- H_2O$$

$$CH_3 - (CH_2)_4 - \underset{O}{C}\overset{CH - CH}{\underset{\diagdown\diagup}{\|\quad\quad\|}}CH \qquad\qquad (6.53)$$

The hydroperoxide involved here is the 10-OOH, which is not typical of linoleate autoxidation. However, it can arise from singlet oxidation of linoleic acid. An alternative reaction pathway involving singlet oxygen and the formation of a hydroperoxy cyclic peroxide has been suggested (Frankel, 1983) (see equation (6.54)). The cyclisation of the conjugated diene system in the 9-OOH of linoleate by singlet oxygen produces a cyclic peroxide, which decomposes into pentylfuraldehyde and then to pentylfuran by the liberation of formaldehyde. The pentenylfuran derivatives can be derived from the corresponding 9-hydroperoxide of linolenate. Consequently, cyclic peroxides formed by singlet oxygen may be important precursors of substituted furans and furaldehydes. Alternative mechanisms for the formation of *cis* and *trans* isomers of 2-(1-pentenyl)furan and 2-(2-pentenyl)furan by the decomposition of the corresponding 4-ketononenals of autoxidised linolenate have also been suggested by Ho *et al.* (1978). Later, Chang and coworkers (1983) positively identified the presence of *cis*- and *trans*-2-(1-pentenyl)furans and a mixture of *cis*- and *trans*-2-(2-pentenyl) in reverted soyabean oil.

6.3.4 Alcohols

Aliphatic alcohols make a small contribution to the off-flavours produced by the oxidative deterioration of food lipids. This is because their flavour thresh-

$$(6.54)$$

olds are significantly higher than those of the corresponding aldehydes. Moreover, many of the saturated and unsaturated alcohols (Forss, 1972) contribute to the desirable notes of fruits and vegetables, and some are commercially important flavour and fragrant compounds. Table 6.7 lists flavour properties

Table 6.7 Flavour thresholds and descriptive flavours of selected alcohols

Compound	Flavour description	Threshold values (mg/kg)	
		Oil	Water
1-Butanol	Oxidised		7.5[a]
1-Pentanol	Oxidised		4.5
1-Hexanol	Oxidised, green bean[b]		2.5
1-Heptanol	Oxidised, green bean[b]		0.52
2-Pentanol	Ethereal[c]		8.5
2-Hexanol	Turpentine[c]		6.7
2-Heptanol	Rancid coconut[c]		0.41
2-Nonanol	Musty, stale[c]		0.28
2-*t*-Hexen-1-ol	Sweet wine[b]		6.7
2-*t*-Octen-1-ol	Fatty		0.84
1-Penten-3-ol	Oxidised	4.2[d]	3[b]
1-Hexen-3-ol	Rubbery, rancid[e]	0.5[e]	
1-Octen-3-ol	Musty, foreign	0.9, 0.0075[d]	0.001

t = *trans*

[a] Siek *et al.* (1971); [b] Forss (1972); [c] Kellard *et al.* (1985); [d] Badings (1970);
[e] Evans *et al.* (1971)

of selected aliphatic alcohols. It can be noticed that as the chain length increases, the flavour becomes stronger. For example, 3-*cis*-hexen-1-ol has a fresh green-leaf odour and its *trans* isomer has a fatty green smell. 2-*trans*-Hexen-1-ol has a sweet wine odour and the higher C8–C12 2-alken-1-ols have fatty odours (Forss, 1972). The low flavour threshold vinyl alcohol, 1-octen-3-ol, has been reported to be responsible for the mushroom note in deteriorated milk fat and dairy products (Downey, 1969). Many workers (Hoffmann, 1962; Badings, 1970; Stark and Forss, 1964; Frankel, 1983) have identified the presence of 1-octen-3-ol from oxidised methyl linoleate, deteriorated butter and soyabean oil. The reaction scheme for the formation of this vinyl alcohol from arachidonate 12-hydroperoxide has already been described in Figure 6.14. Evans *et al.* (1971) have reported the odour threshold of 1-octen-3-ol in cottonseed oil to be 0.9 mg/kg, and that of 1-hexen-3-ol to be 0.5 mg/kg. When present at concentration of 2 mg/kg, 1-octen-3-ol produced musty/foreign odour while the latter gave rubbery/rancid flavour. The formation of 1-octen-3-ol in oxidised linoleate oils may be explained by the decomposition of non-conjugated linoleate 10-hydroperoxide, possibly produced by singlet oxygen (see equation (6.55)).

Stark *et al.* (1967) isolated 1-penten-3-ol from oxidised butter, while Smouse and Chang (1967) identified this from the volatiles of soyabean oil. Tressl *et al.* (1981) identified this compound from the volatiles formed by light-induced oxidation of methyl linoleate. The formation of 1-penten-3-ol from linoleate 13-hydroperoxide via the scission (B) route can be explained by a reaction pathway similar to that for the formation of 1-octen-3-ol from arachidonate (Figure 6.14).

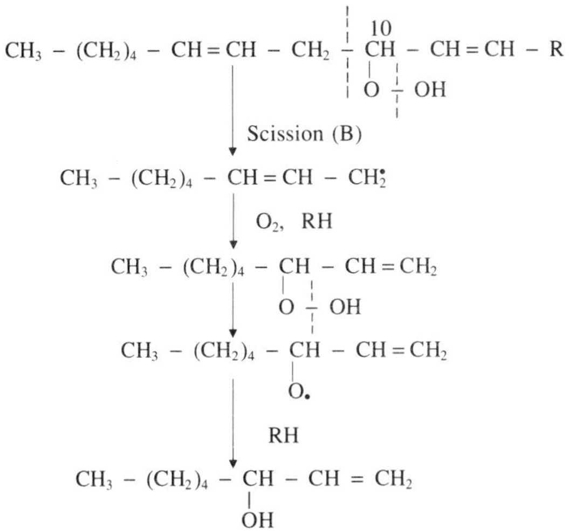

$$CH_3 - (CH_2)_4 - CH = CH - CH_2 \mid \overset{10}{CH} - CH = CH - R$$
$$\mid O \mid OH$$

Scission (B)

$$CH_3 - (CH_2)_4 - CH = CH - \overset{\bullet}{CH_2}$$

$$O_2, \; RH$$

$$CH_3 - (CH_2)_4 - CH - CH = CH_2$$
$$\mid O \mid OH$$

$$CH_3 - (CH_2)_4 - CH - CH = CH_2$$
$$\mid \overset{\bullet}{O}$$

RH

$$CH_3 - (CH_2)_4 - CH - CH = CH_2 \tag{6.55}$$
$$\mid OH$$

The production of 2-alkanols in deteriorated butter and other products (Kellard *et al.*, 1985) has already been described by reaction pathway (6.48). The formation of 1-alkanols and 1-alkenols can occur by the decomposition of the primary hydroperoxides of fatty acids via the alkoxy radical, which abstracts hydrogen from the substrate. For example, the formation of 1-pentanol, hexanal and pentanal from 13-OOH of the linoleate $\Delta 9,11$ can be described as:

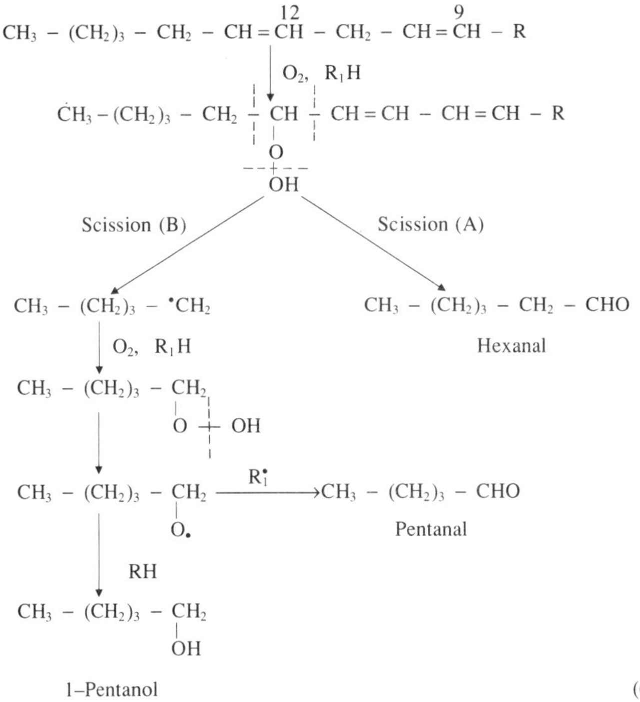

$$CH_3 - (CH_2)_3 - CH_2 - \overset{12}{CH} = CH - CH_2 - CH = \overset{9}{CH} - R$$

$$O_2, \; R_1H$$

$$\overset{\bullet}{CH_3} - (CH_2)_3 - CH_2 \mid CH \mid CH = CH - CH = CH - R$$
$$\mid O$$
$$--\mid-- $$
$$OH$$

Scission (B)   Scission (A)

$$CH_3 - (CH_2)_3 - {}^{\bullet}CH_2 \qquad\qquad CH_3 - (CH_2)_3 - CH_2 - CHO$$

$$O_2, \; R_1H \qquad\qquad\qquad\qquad\qquad Hexanal$$

$$CH_3 - (CH_2)_3 - CH_2$$
$$\mid O + OH$$

$$CH_3 - (CH_2)_3 - CH_2 \xrightarrow{R_1^{\bullet}} CH_3 - (CH_2)_3 - CHO$$
$$\mid \overset{\bullet}{O} \qquad\qquad\qquad\qquad Pentanal$$

RH

$$CH_3 - (CH_2)_3 - CH_2$$
$$\mid OH$$

1–Pentanol   (6.56)

As mentioned, flavour thresholds of alcohols are significantly higher than those of the corresponding aldehydes (Tables 6.5 and 6.7). Consequently, the transformation of even a small amount of potent off-flavour aldehyde above the detection level into the alcohol will bring about considerable odour reduction. Hence, the quenching of off-flavours arising from lipid oxidation by the addition of alcohol dehydrogenase and NADH to certain foods has been suggested by some workers (Eriksson, 1975; Eriksson et al., 1977).

6.3.5 Acids

Aliphatic acids C1–C12 play significant role in the desirable aroma as well as in the off-flavours of many foodstuffs. Their role is particularly important in dairy products and lauric-rich oil-containing foods. The contribution of free fatty acids formed by hydrolytic/lipolytic rancidity to the off-flavours developing in such food products is discussed by Badings (1984). Because of the higher flavour thresholds of fatty acids (Table 6.8), they may not always be present in sufficient quantity to contribute to off-flavours arising from oxidative deterioration of unsaturated lipids. They are, however, produced during the autoxidation of higher fatty acids and their glyceride esters. For instance, Horvat et al. (1969) identified 12 fatty acids, including C1 and C5–C9, from autoxidation of methyl linoleate at room temperature. Several reaction pathways have been proposed to explain the formation of short-chain fatty acids. These include oxidation of fatty acids to shorter chain lengths, thermal degradation of esters, hydrolysis of triglyceride esters, thermal oxidation of an α-carbon atom of acyl groups of triglycerides, and autoxidation of aldehydes and ketones. Nawar (1969) discussed the formation of fatty acids from the non-oxidative thermal degradation of food lipids. During the autoxidation, the fatty acids are readily formed by further oxidation of aldehydes. These fatty acids are produced by the consecutive shortening of acyl chains derived from higher acids (e.g. oleic):

$$CH_3 - (CH_2)_6 - {}^\bullet CH_2 \xrightarrow{\ O_2,\ RH\ } CH_3 - (CH_2)_6 - CH_2OOH \qquad (6.57)$$

Table 6.8 Flavour descriptions and threshold values of some saturated fatty acids

Fatty acid	Flavour description	Threshold value (mg/kg)		
		Coconut oil[a]		Vegetable oil[b]
		Odour	Taste	Taste
Butyric (C4)	Buttery, cheesy, rancid[a]	35	160	0.6
Caproic (C6)	Fatty, rancid, goat-like	25	50	2.5
Caprylic (C8)	Soapy, rancid, musty	> 1000	25	350
Capric (C10)	Sour, cheesy, soapy	> 1000	15	200
Lauric (C12)	Fatty, soapy	> 1000	35	700

[a] Matheis (1990); [b] Feron and Govignon (1961)

The primary hydroperoxide formed is then converted into a short-chain fatty acid and hydrogen by the interaction with aldehyde through a peroxy-hydroxy intermediate (Frankel, 1983):

$$CH_3 - (CH_2)_6 - CH_2OOH + R - CHO \rightarrow CH_3 - (CH_2)_6 - CH_2OO - \overset{\overset{\displaystyle OH}{|}}{CH} - R$$

$$CH_3 - (CH_2)_6 - COOH + R - CHO + H_2 \qquad (6.58)$$

The primary hydroperoxide can also lose a hydroxy radical, to produce formaldehyde and a shorter alkyl radical:

$$CH_3 - (CH_2)_6 - CH_2OOH \longrightarrow CH_3 - (CH_2)_6 - CH_2O^\bullet$$

$$CH_3 - (CH_2)_6 - CH_2O^\bullet \longrightarrow CH_3 - (CH_2)_5 - CH_2^\bullet + HCHO \quad (6.59)$$

The shorter alkyl radical takes part in the consecutive reactions, which result in the formation of lower fatty acids. Unsaturated acids, e.g. 2- and 3-alkenoic, are also formed when oils and fats are heated in the presence of oxygen (Kawada et al., 1967). It is interesting to note that the presence of double bond decreases the fattiness of acids. Other types of acids such as keto-, hydroxy- and dicarboxylic acids have also been isolated from heated/frying oils (Kawada et al., 1967). Keto- and hydroxyacids are important precursors of desirable flavour compounds, such as methyl ketones and lactones, occurring in many natural foods. Moreover, a wide variety of fatty acids are also used for flavouring foods.

6.3.6 Hydrocarbons

Saturated hydrocarbons are known to possess higher threshold values and weak flavours, while C5–C10 alk-l-enes, alk-1-ynes and some alkadienes are observed to have moderately potent flavours (Evans et al., 1971). Table 6.9 lists flavour notes of selected hydrocarbons in cottonseed oil. For example, 1-hexyne gives buttery/rubbery flavour at the concentration of 0.5 mg/kg. Smouse et al. (1965) identified 1-decyne from autoxidised oils and reported its flavour threshold to be 0.1 mg/kg in cottonseed oil. The formation of 1-decyne from autoxidised oleate through Scission (B), followed by dispro-portionation, has been proposed as shown in equation (6.60).

The formation of 1-alkene occurs when the alkenyl radical abstracts a H atom from the substrate. According to Merritt et al. (1967), the hydrocarbons are mainly responsible for the off-flavours developing in irradiated fats. The decomposition of fats by autoxidation and gamma-irradiation results in completely different amounts of hydrocarbons and carbonyl compounds. Gamma-irradiation produces free radicals of random chain length followed by hydrogen abstraction, and by recombination and possible exchange, forming

Table 6.9 Flavour descriptions of selected hydrocarbons in cottonseed oil

Hydrocarbon	Concentration (mg/kg)	Flavour description[a]
Nonane	1000	Buttery, creamy
Nonane	2000	Grassy
1-Nonene	4	Buttery, nutty, rancid
1-Nonene	16	Rancid
1-Hexyne	0.5	Buttery, rubbery
1-Decyne	5	Buttery, beany
1-Decyne	10	Buttery, grassy, melon-like
1,3-Nonadiene	8	Buttery, beany, rancid
1,3-Nonadiene	16	Buttery, rancid

[a] Evans et al. (1971)

$$CH_3 - (CH_2)_7 - \overset{9}{CH} = CH - CH_2 - R$$
$$\downarrow O_2$$
$$CH_3 - (CH_2)_7 - CH = CH - CH - R$$
$$| \quad O - OH$$
$$\downarrow \text{Scission (B)}$$
$$CH_3 - (CH_2)_7 - CH = CH^\bullet$$
$$\downarrow -H^\bullet$$
$$CH_3 - (CH_2)_7 - C \equiv CH \qquad (6.60)$$

major amounts of hydrocarbons. The main hydrocarbons arising from specific radiolytic cleavage of fatty acids can be used as indicators of irradiated foods (Nawar and Balboni, 1970).

Ethane and pentane are the major hydrocarbons produced by thermal or metal-catalysed decomposition of linolenic and linoleic hydroperoxides (Dumelin and Tappel, 1977). Ethane is formed by homolytic cleavage of the 16-hydroperoxide of linolenate. Pentane is produced by the similar reaction pathway from the 13-hydroperoxide of linoleate. Some workers (Evans et al., 1969; Fioriti, 1977; Kochhar et al., 1978; Morrison et al., 1981) have obtained good correlations between pentane production, and the flavour score or peroxide value of oxidising oils containing linoleate. The decrease in flavour quality of mayonnaise with storage has been related to the increase in the amounts of pentane and isomers of 2,4-decadienal (Min and Tickner, 1982). The measurements of other volatile compounds, namely pentanal, hexanal, 2-hexenal, 2-heptenal, 2,4-heptadienal, etc. have also been used for evaluating flavour deterioration of oils and lipid-containing foodstuffs (Snyder et al., 1985; Selke and Frankel, 1987).

A careful selection of the method and sampling temperature is however needed to distinguish between off-flavour volatiles reflecting the actual flavour of the food lipid at time of testing, and the breakdown of flavour precursors indicating the formation of undesirable flavours on storage. The applications of various pathways described in this chapter, and the proper use of objective techniques for flavour evaluation can then offer a solution to flavour and odour problems arising from oxidative deterioration of oils and lipid-containing foods.

6.4 Conclusions

The oxidative reaction pathways of the breakdown of lipids into off-flavour compounds involve autoxidation and/or photo-sensitised oxidation of unsaturated fatty acids. The primary oxidation products are monohydroperoxides—flavourless and odourless—which are precursors of unpleasant odours and flavours developing in oils, fats and foods containing them. Decomposition of hydroperoxides into volatile components occurs through homolytic cleavage of the hydroperoxide group. Two reaction routes of β-scission of unsaturated alkoxy radical have been suggested to explain the formation of several volatile products associated with off-flavours. The domination of a particular pathway depends on the oxidation state of the oil or food lipid, temperature, oxygen pressure, the presence of pro- and antioxidative catalysts, and other factors.

The volatile aldehydes and vinyl ketones are mainly responsible for the potent off-flavours developing in food lipids because their threshold levels for taste and odour are very low. Other volatile oxidation products such as furan derivatives, vinyl alcohols, ketones, alcohols, alkynes, short-chain fatty acids, etc. also contribute to undesirable flavours to varying extents. The flavour response of these volatile compounds is affected markedly by the geometry and position of double bonds present. It should be pointed out that unsaturated aldehydes and ketones are susceptible to further oxidation, which gives rise to additional off-flavour compounds. Also, many non-volatile secondary products such as hydroperoxy epoxides, hydroperoxy cyclic peroxides and di-peroxides have been identified in the oxidised lipids. Decomposition of these secondary products would contribute further to the complex volatile products influencing the flavours and odours of lipid-containing foods. Speculative reaction pathways for decomposition of these non-volatile products have been proposed in many cases.

Despite the progress made in understanding the reaction pathways of several volatile oxidation products, we still do not know how to prevent or eliminate the formation of undesirable flavours in foods containing polyunsaturated fatty acids. The addition of antioxidants or the lowering of oxygen concentration, i.e. vacuum packaging, and good manufacturing practice can retard the development of off-flavours in food lipids. Further research is still needed to resolve the course of formation of many unexplored volatiles of an appallingly complex mixture of products of hydroperoxide decomposition, which can affect flavour quality of lipid-containing foods in extremely small quantities.

Acknowledgement

I would like to thank my wife, Dr Neena Kochhar, for her patience and help during the preparation and typing of the manuscript for this chapter.

References

Badings, H. T. (1959). Isolation and identification of carbonyl compounds formed by autoxidation of ammonium linoleate. *J. Am. Oil Chem. Soc.* **36**, 648–650.

Badings, H. T. (1970). Cold storage defects in butter and their relation to the autoxidation of unsaturated fatty acids. *Neth. Milk Dairy J.* **24**, 147–256.

Badings, H. T. (1984). Flavours and off-flavours. In *Dairy Chemistry and Physics*. Eds P. Walstra and R. Jenness. Wiley Interscience, London and New York, pp. 336–357.

Belitz, H. -D. and Grosch, W. (eds) (1987). *Food Chemistry*. Springer-Verlag, Berlin and Heidelberg, Chapter 3, pp. 128–200.

Bell, E. R., Raley, J. H., Rust, F. F., Seubold, F. H. and Vaughan, W. E. (1951). Reactions of free radicals associated with low temperature oxidation of paraffins. *Discuss. Faraday Soc.* **10**, 242–249.

Billek, G. (1979). Heated oils—Chemistry and nutritional aspects. *Nutr. Metab.* **24**, 200–210.

Chan, H. W.-S. (1977). Photo-sensitised oxidation of unsaturated fatty acid methyl esters. The identification of different pathways. *J. Am. Oil Chem. Soc.* **54**, 100–104.

Chan, H. W.-S. and Levett, G. (1977a). Oxidation of methyl oleate: Separation of isomeric methyl hydroperoxyoctadecenoates and methyl hydroxystearate by high performance liquid chromatography. *Chemistry and Industry (London)*, 692–693.

Chan, H. W.-S. and Levett, G. (1977b). Autoxidation of methyl linoleate. Separation and analysis of isomeric mixtures of methyl linoleate hydroperoxides and methyl hydroxy-linoleates. *Lipids* **12**, 99–104.

Chan, H. W.-S. and Coxon, D. T. (1987). Lipid hydroperoxides. In *Autoxidation of Unsaturated Lipids*. Ed. H. W.-S. Chan. Academic Press, London.

Chan, H. W.-S., Prescott, F. A. A. and Swoboda, P. A. T. (1976). Thermal decomposition of individual positional isomers of methyl linoleate hydroperoxide: Evidence of carbon oxygen bond scission. *J. Am. Oil Chem. Soc.* **53**, 572–576.

Chan, H. W.-S., Coxon, D. T., Peers, K. E. and Price, K. R. (1982). Oxidative reactions of unsaturated lipids. *Food Chem.* **9**, 21–34.

Chang, S. S., Brobst, K. M., Tai, H. and Ireland, C. E. (1961). Characterization of the reversion flavour of soybean oil. *J. Am. Oil Chem. Soc.* **38**, 671–674.

Chang, S. S., Smouse, T. H., Krishnamurthy, R. G., Mookherjee, B. D. and Reddy, B. R. (1966). Isolation and identification of 2-pentyl furan as contributing to the reversion flavour of soybean oil. *Chemistry and Industry* No. 46, 1926–1927.

Chang, S. S., Shen, G.-H., Tang, J., Jin, Q. Z., Shi, H., Carlin, J. T. and Ho, C.-T. (1983). Isolation and identification of 2-pentenylfurans in the reversion flavour of soybean oil. *J. Am. Oil Chem. Soc.* **60** (3), 553–557.

Cobb, W. Y. and Day, E. A. (1965). Further observations of the dicarbonyl compounds formed via autoxidation of methyl linoleate. *J. Am. Oil Chem. Soc.* **42**, 1110–1112.

Cort, W. M. (1974). Antioxidant activity of tocopherols, ascorbyl palmitate, and ascorbic acid and their mode of action. *J. Am. Oil Chem. Soc.* **51**, 321–325.

Cronin, D. A. and Ward, M. K. (1971). The characterisation of some mushroom volatiles. *J. Sci. Food Agric.* **22**, 477–479.

Curda, D. and Poulsen, K. P. (1978). Oxidation of lard at low temperatures. *Nahrung* **22** (1), 25–34.

Day, E. A. (1966). Role of milk lipids in flavours of dairy products. *Advan. Chem. Ser.* **56**, 94–120.

Downey, W. K. (1969). Lipid oxidation as a source of off-flavour development during the storage of dairy products. *J. Soc. Dairy Technol.* **22**, 154–161.

Dumelin, E. E. and Tappel, A. L. (1977). Hydrocarbon gases produced during *in vitro* peroxidation of polyunsaturated fatty acids and decomposition of performed hydroperoxides. *Lipids* **12**, 894–900.

Dziedzic, S. Z. and Hudson, B. J. F. (1984). Phosphatidyl ethanolamine as a synergist for primary antioxidants in edible oils. *J. Am. Oil Chem. Soc.* **61** (6), 1042–1045.

Dziezak, J. D. (1986). Antioxidants. *Food Technol.* **40** (9), 94–97; 100–102.

Endo, Y., Usuki, R. and Kaneda, T. (1985a). Antioxidant effects of chlorophyll and pheophytin on the autoxidation of oils in the dark. I. Comparison of the inhibitory effects. *J. Am. Oil Chem. Soc.* **62** (9), 1375–1378.

Endo, Y., Usuki, R. and Kaneda, T. (1985b). Antioxdant effects of chlorophyll and pheophytin on the autoxidation of oils in the dark. II. The mechanism of antioxidative action of chlorophyll. *J. Am. Oil Chem. Soc.* **62** (9), 1387–1390.

Eriksson, C. E. (1975). Aroma compounds derived from oxidised lipids. Some biochemical and analytical aspects. *J. Agric. Food Chem.* **23**, 126–128.

Eriksson, C. E., Olsson, P. A. and Svensson, S. G. (1971). Denatured haemoproteins as catalysts in lipid oxidation. *J. Am. Oil Chem. Soc.* **48**, 442–447.

Eriksson, C. E., Qvist, J. A. and Vallentin, K. (1977). Conversion of aldehydes to alcohols in liquid foods by alcohol dehydrogenase. In *Enzymes in Food and Beverage Processing*. Eds R. L. Ory and A. J. St. Angelo. ACS Symposium Series No. 47, pp. 132–142.

Evans, C. D., List, G. R., Hoffmann, R. L. and Moser, H. A. (1969). Edible oil quality as measured by thermal release of pentane. *J. Am. Oil Chem. Soc.* **46**, 501–504.

Evans, C. D., Moser, H. A. and List, G. R. (1971). Odour and flavour responses to additives in edible oils. *J. Am. Oil Chem. Soc.* **48**, 495–498.

Feron, R. and Govignon, M. (1961). The relation between free acidity and taste of edible oils. *Ann. Fals. Expert. Chim.* **54**, 308–314.

Figge, K. (1971). Dimeric fatty acid $[1-^{14}C]$ methyl esters. I. Mechanisms and products of thermal and oxidative-thermal reactions of unsaturated fatty acid esters. Literature review. *Chem. Phys. Lipids* **6**, 159–177.

Fioriti, J. A. (1977). Measuring flavour deterioration of fats, oils, dried emulsions and foods. *J. Am. Oil Chem. Soc.* **54**, 450–453.

Foote, C. S. (1976). In *Free Radicals in Biology*. Vol. 11. Ed. W. A. Pryor. Academic Press, New York, pp. 85–133.

Forney, F. W. and Markovetz, A. J. (1971). The biology of methyl ketones. *J. Lipid Res.* **12**, 383–395.

Forney, M. (1974). Formation of hydrocarbons and other volatile products by oxidation of *n*-nonanal and *n*-non-2-enal. *Rev. Fr. Corps Gras.* **21**, 429–436.

Forss, D. A. (1969). Flavours of dairy products: A review of recent advances. *J. Dairy Sci.* **52**, 832–840.

Forss, D. A. (1971). The flavour of dairy fats—A review. *J. Am. Oil Chem. Soc.* **48**, 702–710.

Forss, D. A. (1972). Odour and flavour compounds from lipids. In *Progress in Chemistry of Fats and other Lipids*. Vol. XIII, part 4. Ed. R. T. Holman. Pergamon Press, Oxford.

Frankel, E. N. (1979). Autoxidation. In *Fatty Acids*. Ed. E. H. Pryde. AOCS Monograph, American Oil Chemists Society, Champaign, Illinois, pp. 353–378.

Frankel, E. N. (1980). Lipid Oxidation. *Prog. Lipid Res.* **19**, 1–22.

Frankel, E. N. (1983). Volatile lipid oxidation products. *Prog. Lipid Res.* **22**, 1–33.

Frankel, E. N. (1984). Lipid oxidation: Mechanisms, products and biological significance. *J. Am. Oil Chem. Soc.* **61** (12), 1908–1917.

Frankel, E. N., Neff, W. E. and Bessler, T. R. (1979). Analysis of autoxidised fats by GC-MS. IV. Soya bean oil methyl esters. *Lipids* **14**, 39–46.

Frankel, E. N., Neff, W. E., Rohwedder, W. K., Khambay, B. P. S., Garwood, R. F. and Weedon, B. C. L. (1977a). Analysis of autoxidised fats by GC-MS. I. Methyl oleate. *Lipids* **12**, 901–907.

Frankel, E. N., Neff, W. E., Rohwedder, W. K., Khambay, B. P. S., Garwood, R. F. and Weedon, B. C. L. (1977b). Analysis of autoxidised fats by GC–MS. III. Methyl linolenate. *Lipids* **12**, 1055–1061.

Frankel, E. N. and Neff, W. E. (1979). Analysis of autoxidised fats by GC–MS. V. Photosensitised oxidation. *Lipids* **14**, 961–967.

Frankel, E. N., Neff, W. E., Selke, E. and Weisleder, D. (1982). Photosensitized oxidation of methyl linoleate: Secondary and volatile thermal decomposition products. *Lipids* **17**, 11–18.

Frankel, E. N., Garwood, R. F., Khambay, B. P. S., Moss, G. P. and Weedon, B. C. L. (1984). Stereochemistry of olefin and fatty acid oxidation. Part 3. The allylic hydroperoxides from the autoxidation of methyl oleate. *J. Chem. Soc. Perkin Trans. I*, 2233–2240.

Gaddis, A. M., Ellis, R. and Currie, G. T. (1961). Carbonyls in oxidising fat. V. The composition of neutral volatile monocarbonyl compounds from autoxidised oleate, linoleate, linolenate esters and fats. *J. Am. Oil Chem. Soc.* **38**, 371–375.

Gordon, M. H. (1990). The mechanism of antioxidant action *in vitro*. In *Food Antioxidants*. Ed. B. J. F. Hudson. Elsevier Applied Science, London and New York, pp. 1–18.

Gregory, D. (1984). Antioxidants: The radical answer. *Food, Flavourings, Ingredients, Packaging and Processing* **6**, 18–19, 21–22.

Grosch, W. (1987). Reactions of hydroperoxides—products of low molecular weight. In *Autoxidation of Unsaturated Lipids*. Ed. H. W.-S. Chan. Academic Press, London, pp. 95–139.

Gunstone, F. D. (1984). Reaction of oxygen and unsaturated fatty acids. *J. Am. Oil Chem. Soc.* **61** (2), 441–447.

Hamm, D. L., Hammond, E. G. and Hotchkiss, D. K. (1968). Effect of temperature on rate of autoxidation of milk fat. *J. Dairy Sci.* **51** (4), 483–491.

Hammond, E. G. and Hill, F. D. (1964). The oxidised-metallic and grassy flavour components of autoxidised milk fat. *J. Am. Oil Chem. Soc.* **41**, 180–184.

Hartman, L., Antunes, A. J., dos Santos Garruti, R. and Chaib, M. A. (1975). The effect of free fatty acids on the taste, induction periods and smoke points of edible oils and fats. *Lebensmittel. Wiss.-Technol.* **8**, 114–118.

Ho, C. T., Smagula, M. S. and Chang, S. S. (1978). The synthesis of 2-(1-pentenyl) furan and its relationship to the reversion flavour of soybean oil. *J. Am. Oil Chem. Soc.* **55**, 233–237.

Hoffmann, G. (1961). Isolation of two pairs of isomeric 2,4-alkadienals from soybean oil reversion flavour concentrate. *J. Am. Oil Chem. Soc.* **38**, 31–32.

Hoffmann, G. (1962). 1-Octen-3-ol and its relation to other oxidative cleavage products from esters of linoleic acid. *J. Am. Oil Chem. Soc.* **39**, 439–444.

Hoffmann, G. (ed.) (1989). *The Chemistry and Technology of Edible Oils and Fats and their High Fat Products*. Academic Press, London and New York, p. 187.

Horvat, R. J., McFadden, W. H., Ng, H., Lane, W. G., Lee, A., Lundin, R. F., Scherer, J. R. and Shepherd, A. D. (1969). Identification of some acids from autoxidation of methyl linoleate. *J. Am. Oil Chem. Soc.* **46**, 94–96.

Karel, M. (1980). Lipid oxidation, secondary reactions, and water activity of foods. In *Autoxidation in Food and Biological Systems*. Eds M. G. Simic and M. Karel. Plenum Press, New York and London, pp. 191–206.

Kawada, T., Krishnamurthy, R. G., Mookherjee, B. D. and Chang, S. S. (1967). Chemical reactions involved in the deep fat frying of foods. II. Identification of acidic volatile decomposition products of corn oil. *J. Am. Oil Chem. Soc.* **44**, 131–135.

Keeny, M. (1962). Secondary degradation products. In *Lipids and their Oxidation*. Eds. H. W. Scultz, E. A. Day and R. O. Sinnhuber. AVI Publishing Co. Inc., Westport, pp. 79–89.

Kellard, B., Busfield, D. M. and Kinderlerer, J. L. (1985). Volatile off-flavour compounds in desiccated coconut. *J. Sci. Food Agric.* **36**, 415–420.

Keppler, J. G. (1977). Twenty-five years of flavour research in a food industry. *J. Am. Oil Chem. Soc.* **54**, 474–477.

Keppler, J. G. and Horikx, M. M. (1967). Iso-linoleic acids responsible for the formation of the hardening flavour. *J. Am. Oil Chem. Soc.* **44**, 543–544.

Keppler, J. G., Schols, J. A., Feenstra, W. H. and Meijboom, P. W. (1965). Components of the hardening flavour present in hardened linseed oil and soybean oil. *J. Am. Oil Chem. Soc.* **42**, 246–249.

Kimoto, W. I. and Gaddis, A. M. (1969). Precursors of alk-2,4-dienals in autoxidised lard. *J. Am. Oil Chem. Soc.* **46**, 403–408.

Kinsella, J. E. (1969). The flavour chemistry of milk lipids. *Chemistry and Industry*, 36–42.

Kinsella, J. E. and Hwang, D. H. (1976). Enzymes of *Penicillium roqueforti* involved in the biosynthesis of cheese flavour. *Critical Reviews in Food Science and Nutrition*, **8**, 191–228.

Klaui, H. and Pongracz, von G. (1982). Ascorbic acid and derivatives as antioxidants in oils and fats. In *Vitamin C*. Eds J. N. Counsell and D. H. Hornig. Applied Science Publishers, London, Chapter 9, pp. 139–166.

Kochhar, S. P. (1988). Natural antioxidants — A literature survey. *Leatherhead Food RA Scientific and Technical Survey No. 165*.

Kochhar, S. P. and Meara, M. L. (1975). A survey of the literature on oxidative reactions in edible oils as it applies to the problem of off-flavours in foodstuffs. *Leatherhead Food RA Scientific and Technical Survey No. 87*.

Kochhar, S. P., Meara, M. L. and Weir, G. S. D. (1978). Effect of non-glyceride components on the flavour stability of edible oils. Presented at *14th World Congress International Society for Fat Research*, Brighton, England, 17–22 September.

Krishnamurthy, R. G., Smouse, T. H., Mookerjee, B. D., Reddy, B. R. and Chang S. S. (1967). Identification of 2-pentylfuran in fats and oils in relationship to the reversion flavour of soybean oil. *J. Am. Oil Chem. Soc.* **32**, 372–374.

Kwon, T. W., Snyder, H. E. and Brown, H. G. (1984). Oxidative stability of soybean oil at different stages of refining. *J. Am. Oil Chem. Soc.* **61** (12), 1843–1846.

Labuza, T. P. (1971). Kinetics of lipid oxidation in foods. *CRC Critical Reviews in Food Technology*. **2**, 355–405.

Labuza, T. P. and Kamman, J. F. (1983). Reaction kinetics and accelerated tests simulation as a function of temperature. In *Computer Aided Techniques in Food Technology*. Ed. I. Saguy. Marcel Dekker, Inc., New York, pp. 71–117.

Langler, J. E. and Day, E. A. (1964). Development and flavour properties of methyl ketones in milk fat. *J. Dairy Sci.* **47**, 1291–1296.

Lea, C. H. and Swoboda, P. A. T. (1958). The flavour of aliphatic aldehydes. *Chemistry and Industry*, 1289–1290.

Lercker, G. (1980). Auto-oxidation process, In *Autoxidation in Food and Biological Systems*. Eds. M. G. Simic and M. Karel. Plenum Press, New York and London, pp. 269–290.

Lillard, D. A. and Day, E. A. (1964). Degradation of monocarbonyls from autoxidising lipids. *J. Am. Oil Chem. Soc.* **41**, 549–552.

Loury, M. (1972). Possible mechanism of autoxidative rancidity. *Lipids* **7**, 671–675.

Loury, M. and Forney, M. (1968). Autoxidation of linoleic acid at temperature of 20° and 40°C. II. Theoretical considerations. *Rev. Fr. Corps Gras* **15**, 663–673.

Lundberg, W. O. (ed.) (1961). *Autoxidation and Antioxidants*. Vol. I. Interscience, New York.

Lundberg, W. O. (ed.) (1962). *Autoxidation and Antioxidants*. Vol. II. Interscience, New York.

Lyon, B. G. (1988). Descriptive profile analysis of cooked, stored and reheated chicken patties. *J. Food Sci.* **54** (4), 1086–1090.

Matheis, G. (1990). *Dragoco Review No. 4.*, pp. 131–149.

Matthews, R. F., Scanlan, R. A. and Libbey, L. M. (1971). Autoxidation products of 2,4-decadienal. *J. Am. Oil Chem. Soc.* **48**, 745–747.

McGill, A. S., Hardy, R., Burt, J. R. and Gunstone, F. D. (1974). Hept-*cis*-4-enal and its contribution to the off-flavour in cold stored cod. *J. Sci. Food Agric.* **25**, 1477–1489.

McGill, A. S., Hardy, R. and Gunstone, F. D. (1977). Further analysis of the volatile components of frozen cold stored cod and the influence of these on flavour. *J. Sci. Food Agric.* **28**, 200–205.

Meijboom, P. W. (1964). Relationship between molecular structure and flavour perceptibility of aliphatic aldehydes. *J. Am. Oil Chem. Soc.* **41**, 326–328.

Meijboom, P. W. and Jongenotter, G. A. (1981). Flavour perceptibility of straight chain, unsaturated aldehydes as a function of double-bond position and geometry. *J. Am. Oil Chem. Soc.* **58**, 680–682.

Merritt, C. Jr., Forss, D. A., Angelini, P. and Bazinet, M.L. (1967). Volatile compounds produced by irradiation of butterfat. *J. Am. Oil Chem. Soc.* **44**, 144–146.

Miles, R. S., Mackeith, F. K., Bechtel, P. J. and Novakofski, J. (1986). Effect of processing, packaging and various antioxidants on lipid oxidation of restructured pork. *J. Food Protect.* **49**, 222–225.

Min, D. B. and Tickner, D. B. (1982). Preliminary gas chromatographic analysis of flavour compounds in mayonnaise. *J. Am. Oil Chem. Soc.* **59** (5), 226–228.

Minguez-Mosquera, M. I., Gandul-Rojas, B., Garrido-Fernandez, J. and Gallardo-Guerrero, L. (1990). Pigments present in virgin olive oil. *J. Am. Oil Chem. Soc.* **67** (3), 192–196.

Mookherjee, B. D., Deck, R. E. and Chang, S. S. (1965). Relationship between monocarbonyl compounds and flavour of potato chips. *J. Agric. Food Chem.* **13**, 131–134.

Morrison, W. H., III, Lyon, B. G. and Robertson, J. A. (1981). Correlation of gas liquid chromatographic volatiles with flavour intensity scores of stored sunflower oils. *J. Am. Oil Chem. Soc.* **58**, 23–27.

Mottram, D. S. (1987). Lipid oxidation and flavour in meat and meat products. *Food Sci. Technol. Today* **1** (3), 159–162.

Nawar, W. W. (1969). Thermal degradation of lipids. A review. *J. Agr. Food Chem.* **17**, 18–21.

Nawar, W. W. and Balboni, J. J. (1970). Detection of irradiation treatment in foods. *J. Ass. Offic. Anal. Chem.* **53**, 726–729.

Neff, W. E. and Frankel, E. N. (1980). Quantitative analyses of hydroxystearate isomers from hydroperoxides by high pressure liquid chromatography of autoxidised and photosensitized-oxidized fatty esters. *Lipids* **15**, 587–590.

Neff, W. E., Frankel, E. N. and Weisleder, D. (1981). High pressure liquid chromatography of autoxidised lipids. II. Hydroperoxy-cyclic peroxides and other secondary products from methyl linolenate. *Lipids* **16**, 439–448.

Noble, A. C. and Nawar, W. W. (1975). Identification of decomposition products from autoxidation of methyl 4,7,10,13,16,19-docosahexaenoate. *J. Am. Oil Chem. Soc.* **52**, 92–95.

Nonaka, M., Black, D. R. and Pippen, E. L. (1967). Gas chromatographic and mass spectral analysis of cooked chicken meat volatiles. *J. Agric. Food Chem.* **15**, 713–717.

Olcott, H. S. (1958). The role of free fatty acids on antioxidant effectiveness in unsaturated oils. *J. Am. Oil Chem. Soc.* **35**, 597–599.

Olek, M., Declercq, B., Caboche, M., Blanchard, F. and Sudrand, G. (1983). Application of electrochemical detection to the determination of ethoxyquin residues by high performance liquid chromatography. *J. Chromatogr.* **281**, 309–313.

Paquette, G., Kupranycz, D. B. and van de Voort, F. R. (1985). The mechanism of lipid autoxidation. II. Non-volatile secondary oxidation products. *Can. Inst. Food Sci. Technol. J.* **18** (3), 197–206.

Parks, O. W. and Patton, S. (1961). Volatile carbonyl compounds in stored dry whole milk. *J. Dairy Sci.* **44**, 1–9.

Parsons, A. M. (1973). Influence of lipid oxidation on quality of lipid-containing food products. In *3rd International Symposium on Metal Catalysed Lipid Oxidation*. Institut des Corps Gras, Paris, 27–30 September 1973, pp. 148–170.

Pokorny, J. (1987). Major factors affecting the autoxidation of lipids. In *Autoxidation of Unsaturated Lipids*. Ed. H. W.-S. Chan. Academic Press, London, pp. 141–206.

Pongracz, von G. (1982). Stabilisation of cocoa butter substitute fats. *Fette Seifen AnstrMittel.* **84** (7), 269–272.

Popov, A. D. and Mizev, I. D. (1966). Pro-oxidative action of free fatty acids in the autoxidation of fats. *Rev. Fr. Corps Gras* **13** (10), 621–626.

Porter, N. A., Weber, B. A., Weenen, H. and Khan, J. A. (1980). Autoxidation of polyunsaturated lipids. Factors controlling the stereochemistry of product hydroperoxides. *J. Am. Chem. Soc.* **102**, 5597–5601.

Porter, N. A., Lehman, L. S., Weber, B. A. and Smith, K. J. (1981). Unified mechanism for polyunsaturated fatty acid autoxidation. Competition of peroxy radical hydrogen atom abstraction, β-scission, and cyclization. *J. Am. Chem. Soc.* **103**, 6447–6455.

Poste, L. M., Willemot, C., Butler, G. and Patterson, C. (1986). Sensory aroma scores and TBA values as indices of warmed-over flavour in pork. *J. Food Sci.* **51**, 886–888.

Ragnarsson, J. O. and Labuza, T. P. (1977). Accelerated shelf-life testing for oxidative rancidity in foods—A review. *Food Chem.* **2**, 291–308.

Rawls, H. R. and van Santen, P. J. (1970). A possible role for singlet oxygen in the initiation of fatty acid autoxidation. *J. Am. Oil Chem. Soc.* **47**, 121–125.

Sanders, T. A. B. (1990). Nutritional significance of rancidity. In *Symposium Proceedings, Rancidity in Foods*. No. 45, held at Leatherhead Food RA, 10 March 1988.

Schepartz, A. I. and Daubert, B. F. (1950). Flavour reversion in soybean oil. V. Isolation and identification of reversion compounds in hydrogenated soybean oil. *J. Am. Oil Chem. Soc.* **27**, 367–373.

Schieberle, P. and Grosch, W. (1981). Model experiments about the formation of volatile carbonyl compounds. *J. Am. Oil Chem. Soc.* **58**, 602–607.

Scott, G. (1965). *Atmospheric Oxidation and Antioxidants*. Elsevier Publishing Co., New York.

Sedlacek, B. A. J. (1975). The mechanism of the action of ascorbyl palmitate and other antioxidants on the autoxidation of fats. *Nahrung* **19** (3), 219–229.

Selke, E. and Frankel, E. N. (1987). Dynamic headspace capillary gas chromatographic analysis of soybean oil volatiles. *J. Am. Oil Chem. Soc.* **64** (5), 749–753.

Sherwin, E. R. (1976). Antioxidants for vegetable oils. *J. Am. Oil Chem. Soc.* **53**, 430–436.

Siek, T. J., Albin, I. A., Sather, L. A. and Lindsay, R. C. (1971). Comparison of flavour threshold of aliphatic lactones with those of fatty acids, esters, aldehydes, alcohols and ketones. *J. Dairy Sci.* **54**, 1–4.

Smouse, T. H. and Chang, S. S. (1967). A systematic characterisation of the reversion flavour of soya bean oil. *J. Am. Oil Chem. Soc.* **44**, 509–514.

Smouse, T. H., Mookherjee, B. D. and Chang, S. S. (1965). Identification of dec-l-yne in the initial autoxidation products of some vegetable oils. *Chemistry and Industry*. (London) 1301–1303.

Snyder, J. M., Frankel, E. N. and Selke, E. (1985). Capillary gas chromatographic analyses of headspace volatiles from vegetable oils. *J. Am. Oil Chem. Soc.* **62**, 1675–1679.

Stahl, H. D. and Sims, R. J. (1986). Tempeh oil-antioxidant (?). *J. Am. Oil Chem. Soc.* **63** (4), 555–556.

Stark, W. and Forss, D. A. (1962). A compound responsible for metallic flavour in dairy products. *J. Dairy Res.* **29**, 173–180.

Stark, W. and Forss, D. A. (1964). A compound responsible for mushroom flavour in dairy products. *J. Dairy Res.* **31**, 253–259.

Stark, W., Smith, J. F. and Forss, D. A. (1967). *n*-Pent-1-en-3-ol and *n*-pent-1-en-3-one in oxidised dairy products. *J. Dairy Res.* **34**, 123–129.

Swern, D. (ed.) (1979). *Bailey's Industrial Oil and Fat Products*. Vol. 1, 4th edn. John Wiley & Sons, New York.

Swoboda, P. A. T. and Lea, C. H. (1965). The flavour volatiles of fats and fat-containing foods. II. A gas chromatographic investigation of volatile autoxidation products from sunflower oil. *J. Sci. Food Agric.* **16**, 680–689.

Swoboda, P. A. T. and Peers, K. E. (1977a). Volatile odorous compounds responsible for metallic, fishy taint formed in butterfat by selective oxidation. *J. Sci. Food Agric.* **28**, 1010–1018.

Swoboda, P. A. T. and Peers, K. E. (1977b). Metallic odour caused by vinyl ketones formed in the oxidation of butterfat. The identification of octa-1,*cis*-5-dien-3-one. *J. Sci. Food Agric.* **28**, 1019–1024.

Swoboda, P. A. T. and Peers, K. E. (1978). *trans*-4,5-Epoxyhept-*trans*-2-enal. The major volatile compound formed by the copper and α-tocopherol induced oxidation of butterfat. *J. Sci. Food Agric.* **29**, 803–807.

Tokarska, B., Hawrysh, Z. J. and Clandinin, M. T. (1986). Study of the effect of antioxidants on storage stability of Canola oil using gas liquid chromatography. *Can. Inst. Food Sci. Technol. J.* **19** (3), 130–133.

Tressl, R., Bahri, D. and Engel, K.-H (1981). Lipid oxidation in fruits and vegetables. *ACS Symposium Series*. **170**, 213–232.

Uri, N. (1961). Mechanism of antioxidation. In *Autoxidation and Antioxidants*. Vol. 1. Ed. W. O. Lundberg. Interscience, New York and London. Chapter 4, pp. 55–106.

Walradt, J. P., Pittet, A. O., Kinlin, T. E., Muralidhara, R. and Sanderson, A. (1971). Volatile components of roasted peanuts. *J. Agric. Food Chem.* **19**, 972–979.

Werman, M. J. and Neeman, I. (1986). Oxidative stability of avocado oil. *J. Am. Oil Chem. Soc.* **63** (3), 355–360.

Wilkinson, R. A. and Stark, W. (1967). A compound responsible for metallic flavour in dairy products. II. Theoretical consideration of the mechanism of formation of oct-1-en-3-one. *J. Dairy Res.* **34**, 89–102.

Winell, B. (1976). Quantitative determination of ethoxyquin in apples by gas chromatography. *Analyst (London)* **101**, 883–886.

Wyatt, C. J. and Day, E. A. (1965). Evaluation of antioxidants in deodorized and non-deodorized butter oil stored at 30°C. *J. Dairy Sci.* **48**, 682–686.

Yamauchi, R., Yamada, T., Kato, K. and Ueno, Y. (1985). Autoxidation and photosensitised oxidation of methyl eicosapentaenoate: Secondary oxidation products. *Agric. Biol. Chem.* **49** (7), 2077–2082.

Yasuda, K., Peterson, R. J. and Chang, S. S. (1975). Identification of volatile compounds developed during storage of a deodorized hydrogenated soybean oil. *J. Am. Oil Chem. Soc.* **52**, 307–311.

7 Packaging material as a source of taints

P. TICE

7.1 Introduction

Modern consumer packaging in its various forms plays a number of important roles, such as protecting and preserving the quality of the product, easing handling, enhancing sales appeal and presenting information. For foods, the most important roles are providing protection from external contamination and preserving the quality.

In the past, many foods were supplied in bulk to the grocer, and the principal packaging material available to wrap and package these foods for the customer was paper. The paper wrap or bag, usually unsealed, provided only rudimentary protection for the food against external contaminants. Many of these contaminants could cause the food to become tainted. Cheese, butter and other dairy products were known to be particularly susceptible to tainting from soaps and soap powders, and all foods were kept well away from paraffin oil, which was in frequent use in rural areas for cooking and lighting.

Food packaging in use today is manufactured from a variety of materials, which include a wide range of plastics as well as the more traditional paper, board, metal and glass. In recent years, technical developments, particularly with plastics, have resulted in sophisticated packaging materials that perform efficiently the important functions of protection and quality preservation.

Although packaging now provides effective protection for packaged foods against external contamination, it has been well known for many years that the packaging itself can be a source of odorous and tainting substances. Any odours that arise from packaging materials are usually of very low intensity, but varied in nature. Although some are not unpleasant, they can be unacceptable if they are present at high levels and are not normally associated with the food product. Some packaging material odours, even when present at high levels, do not result in tainting of the food, but an unpleasant smell released when the food package is opened would, in most cases, deter the consumer from eating the contents. Usually, odorous substances will taint the packaged food and, in some cases, the tainting can be more severe than would be expected from the level of odour. This occurs when the food readily absorbs the substance as it is released from the packaging, displacing the normal distribution equilibrium for the substance between the packaging material and the sealed air space inside the pack. The degree of tainting occurring by this mechanism will depend on the storage time and temperature.

Tainting of packed foods can occur, as indicated previously, by vapour transfer of substances from the packaging to the food. However, when a food product is in immediate contact with the packaging material, tainting can also occur by direct migration of substances from the packaging into the food. Such substances can be odorous or non-odorous and the extent of migration will depend particularly on the nature of the food and the substances migrating. Generally, direct migration of substances from packaging materials occurs more readily with fatty foods. Fats and oils can penetrate the surfaces of some packaging materials, such as plastics, and promote the migration into the food product. With dry foods, particularly the finely powdered, which have a high surface area to volume ratio, migration of even non-volatile substances can still occur.

In 1960, it was stated that the actual number of cases reported of food contaminated by odour from packaging was extremely small (The Institute of Packaging, 1960). Of the complaints received by a major confectionery manufacturer, the majority were concerned with printed paper and board packaging, while film, both printed and unprinted, represented only a minor proportion. This is not surprising since, at that time, paper and board packaging was more extensively used than films and other plastics. Concern about the need to establish a better understanding of potential sources of odorous and tainting substances from packaging resulted in an investigation survey, the results of which were published in a monograph entitled *Survey of Odour in Packaging of Foods* (Harvey, 1963). This monograph lists the main types of materials used for food packaging and many of the individual material components that are known to have odorous and tainting properties. Printing inks and adhesives are particularly mentioned, as both contain solvents and certain resins with well-defined odours. Although in recent years there have been significant developments in packaging materials and, in many cases, a major change to plastics from the more traditional paper, board, metals and glass, the principal sources of odours and tainting substances remain the same. These will be covered in more detail in later sections of this chapter.

The monograph also gives advice on how to reduce odour hazards, including setting clear specifications for the packaging, good storage and the use of appropriate printing. Some of the recommendations were taken from an earlier publication *Odour from Paper and Board—Notes for Guidance* (British Carton Association, 1956). The basic precautionary criteria listed in the monograph still apply in the 1990s.

One of the principal problems of carrying out odour sensory tests on packaging materials using a panel of individual testers (described in detail in section 7.2), is that of obtaining comparable descriptions from the individual panel members on the nature of any odours detected. This difficulty has been recognised and, for paper and board packaging materials, the Finnish Pulp and

Paper Research Institute have established a set of odour descriptor terms based on odour descriptors published by Dravnieks and Bock (1978). The descriptor set consists of 140 terms arranged in a sequence, which facilitates independent consideration of each descriptor. The intensity of the odour associated with each descriptor is scored on a scale of 0 to 5. It was recognised, however, that the set of 140 descriptors was too large and that a much smaller set could be more effective (Soderhjelm and Parssinen, 1985).

To carry out an assessment on a paper or board material, portions are placed in bottles with a small amount of water and stored for 24 h at 23°C. A panel of ten is then asked to assess the odours and assign an intensity rating. The results are evaluated by computer and the odour descriptors are arranged into 24 groups with similar terms together. The groups of descriptors are further arranged to form a hedonic sequence, with very pleasant odours to the far left and the most unpleasant odours to the far right. This gives an odour 'fingerprint' for the material and a means of assessing the similarity or dissimilarity between two odours.

The classification of odours has also been discussed by Koszinowski and Piringer (1986). They pointed out that difficulties arise in classifying or describing an odour when it is due to several substances, as small changes in the relative amounts of the individual substances can significantly change the character of the odour. It was further pointed out that odours can arise from: (i) principal components of the packaging material; (ii) impurities; (iii) reaction products formed during manufacture; and (iv) environmental contamination. For substances of known chemical nature and origin, quantitative measurements can be carried out to confirm the cause of the odour. Where the odour is due to impurities, reaction products, or contaminants, the task of identification can be difficult. The wide concentration range covered is illustrated by the list compiled by Koszinowski and Piringer of the odour threshold values for a number of organic compounds (Table 7.1). There are also other publications that list odour threshold values (Fazzalari, 1978; van Gemert, 1984).

Table 7.1 Odour threshold values (OT) in air

OT (mg/m^2)	Compound
10^3	Octane, nonane
10^2	Ethanol, acetone
10^1	Toluene, ethyl acetate, methyl ethyl ketone
10^0	Vinyl acetate, acetic acid
10^{-1}	Styrene, mesityl oxide, methyl methacrylate
10^{-2}	Chlorophenol, eugenol, butanoic acid
10^{-3}	Butyl acrylate, 2-nonenal, ethyl mercaptan
10^{-4}	1-Octen-3-one, amyl mercaptan, ethyl acrylate
10^{-5}	1-Nonen-3-one
10^{-6}	Vanillin

Source: Koszinowski and Piringer, 1986

It is not easy to obtain a positive 'identification' for an odour, even with a trained panel of assessors. Instrumental techniques, usually based on gas chromatography/mass spectrometry, have been used for many years to separate individual volatile substances released from various packaging materials, and to identify their chemical composition (these techniques are discussed further in section 7.7). Such analyses do not, however, directly characterise the odour of the volatile substance or group of substances.

In research work at Warwick University to develop an 'electronic nose,' an instrument was constructed consisting of a 12-element array with SnO_2 Taguchi-type sensors placed inside a 5 l flask and connected to a microcomputer via a 12-bit interface. Each sensor processes different individual odour characteristics whilst at the same time being sensitive to a broad spectrum of gases. Interfering effects arising from moisture and temperature fluctuations are reduced by the inclusion of ceramic humidity and temperature sensors. It was considered that the instrument could find applications in the food and drinks industry monitoring product aromas. Likewise it could be used for quality control of food packaging (Shurmer, 1987; Anon, 1989).

7.2 Standard odour and taint assessment methods

7.2.1 Legislation

Packaging manufacturers, food packers and food retailers are well aware that food packaging should be free both of unacceptable odours, and of any tendency to taint the food, to ensure that the packaged food is bought and accepted by the consumer. In the UK these requirements are incorporated into the Food Safety Act, with the regulations contained in UK Statutory Instrument No. 1523 *The Materials and Articles in Contact with Food Regulations 1987* (UK Statutory Instrument, 1987). Statutory Instrument No. 1523 originated from EC Directive 76/893/EEC, which was the first 'framework' Directive on materials and articles intended to come into contact with foodstuffs (EC, 1976). Article 2 states that:

Materials and articles must be manufactured in compliance with good manufacturing practice so that, under their normal or foreseeable conditions of use, they do not transfer their constituents to foodstuffs in quantities which could:

- endanger human health
- bring about an unacceptable change in the composition of the foodstuffs or a deterioration in the organoleptic characteristics thereof.

A new EC framework Directive 89/109/EEC was passed in 1988 by the Council of The European Communities, but the aforementioned basic requirements with respect to protecting human health and preserving the organoleptic properties of the food remain the same (EC, 1989). The other EC Member States have also incorporated the regulations of EC Directive 76/893/EEC into their national laws. Since the definition of organoleptic is 'capable of being perceived by one or more sense organs', the regulation covers any deterioration in the taste of the food due to tainting by contaminating substances originating from the packaging, and any deterioration in the aroma of the food due to odours originating from the packaging.

In the United States there are also regulations that prohibit packaging from adversely affecting the organoleptic properties of the packaged food, similar to those in Europe. The Federal Food, Drug and Cosmetics Act, Section 402(a)(3) specifies that any 'indirect food additive' originating from packaging should not impart to the food 'odor or off-taste rendering it unfit for consumption'.

7.2.2 Test methods

In order to ensure that the packaging meets the requirements of the food packer and in turn the food consumer, and that it complies with the legal requirements, tests should be carried out on the packaging prior to use. These tests must assess any odours that are emitted from the packaging and determine their intensities, and also establish any propensity to cause tainting of the foods for which the packaging is intended.

In the following sections the procedures for British Standard odour and taint tests will be described in some detail, as they are similar to those used in most other standard odour and taint test methods (e.g. ASTM Standards E-460 (ASTM, 1988), E-462 (ASTM, 1989a) and E-619 (ASTM, 1989b); German DIN Standard 10 955 (DIN, 1983)). It is not suggested, however, that the BSI standard was the first of the taint and odour tests.

7.2.2.1 UK standard sensory test methods. In 1960, the British Standards Institution (BSI) was requested to produce an agreed standard for odour assessment of food packaging materials. The response was BS 3755, which was published in 1964 and entitled *Methods of Test for The Assessment of Odour from Packaging Materials used for Foodstuffs* (BSI, 1971). The committee responsible for drawing up the standard consisted of representatives from raw material suppliers, packaging converters and food manufacturers, as well as printing inks manufacturers and the relevant packaging associations.

The standard contained tests for odours from packaging, and also taint tests. The test procedures described were designed to be used where the packaging material was believed to be odorous. It was pointed out that with some packaging materials (such as paper and board), the intensity of odours varies according to the moisture content of the material and the surrounding atmosphere. To cater for this possibility, the standard included tests for the packaging material in the condition received, and tests in which the moisture content was changed by the addition of a quantity of water.

It was appreciated that the level of odorous substances that could be tolerated by a particular food depended on a number of factors, such as storage temperature and the period the food would be in the package. It was pointed out, therefore, that the conditions of test specified in the standard could be varied to suit the particular packaging and use.

Recommendations were given for the test room and for selecting the test panel, which it was said may consist of 2, 3, 4, 6 or 12 persons. As both the odour and taint tests require care and concentration by members of the panel to arrive at the best assessment result, the environment within the room used for the testing should be free of odours and any other distractions. Individual booths are recommended, and forms should be provided to record the results. Persons selected for the panel should have been tested to ensure that they have good sensitivity to odours, particularly to the odours likely to be present with the types of packaging normally tested. Likewise, panel members should also be able to readily detect the types of taints that could be expected from the packaging tested. It is often found, however, that an individual panel member is quite sensitive to particular odours or taints, but relatively insensitive to other odours and taints. With a different panel member, the reverse can occur.

For the taint testing, the standard stated that the actual food for which the package was intended should be used to estimate whether or not any taint would be produced during the shelf-life of the package. It was, however, realised that it would be practically difficult to carry out the test with some foods and it was recommended that in such cases 'sensitive' foods, such as milk chocolate, butter or icing sugar, be used. When using these alternative test foods, care has to be taken to ensure that the test is not over-sensitive, resulting in packaging being rejected unnecessarily.

The basic procedure for the odour test involves placing portions of the packaging material in jars, which are sealed and stored at room temperature for 24 h. For the 'wet' tests, a small quantity of water is also added to the jars. Members of the test panel are asked to open jars presented to them and make an assessment on any odours detected by placing them in one of three categories: 'A'—odourless; 'B'—slight odour; and 'C'—odour present at a level strong enough to render the sample unacceptable. As it is appreciated that there can be tainting without odour, samples with results in categories A and B are then subjected to a taint test. Samples falling into category C are normally rejected, although such samples can also, if desired, be subjected to the taint test.

Although packaging materials with strong odours might not cause tainting of the packaged foods, it is very likely that if a consumer detected a strong odour when the package was opened, the food would be rejected without tasting it.

For the taint tests, the basic procedure prescribed involves placing the packaging material in a container, with the test food suspended above the packaging by a suitable means. Where the food in the pack is in direct contact with the packaging, the procedure can be modified by also placing the test food on the surface of the packaging test specimen.

The standard specifies that a second container be prepared with the packaging material and the test food, but with the addition of a quantity of water. In a third container a further specimen of the test food is placed, to be used as the reference food for the taint testing by the panel. The containers are then stored at ambient temperature, usually for a period of 24 h.

The test foods that have been exposed to the packaging, both dry and wet, are then removed from the containers and each is divided into 12 portions. The reference food is also removed from its container, and two lots of 12 portions prepared. Pairs are prepared from two sets of 12 portions of test and reference food for the assessment by the panel. The pairs consist of either two 'exposed' portions, two 'reference' portions, or one of each. The panel members are presented with the pairs and asked to say whether they are the same or different. No information is required on whether the taste is pleasant or unpleasant.

For the purpose of interpretation of the results, it is assumed that the treatment of exposing the foodstuff to the packaging material has made a difference, and the hypothesis is tested assuming that a 'correct' answer will take this difference into account. The number of correct answers is compared with the possibility of obtaining such a result by chance. Tainting is considered to have occurred if there are 10 to 12 correct answers. With 6 or less correct answers the conclusion is 'no detectable taint'. If an intermediate number of correct answers is obtained—7, 8 or 9—the difference produced is considered not statistically significant and the test should be repeated. It is preferable to use the same panel and the two sets of results combined with the following conclusions: greater than 17 correct answers, taint has occurred; 17 or less correct answers, no detectable taint.

The BSI standard was re-issued in 1971 with some minor amendments and it is again to be reviewed. Users of this standard now often carry out the assessments with the triangular test instead of the paired test described above. The triangular test is contained in ISO Standard 4120–1983 *Sensory Analysis—Methodology—Triangular test* (ISO, 1983). As the name of this test implies, three portions of food are used for each individual test, with two being identical and one different. Usually the two are reference food and the one is a food portion that has been exposed to the packaging. The results are subjected to a similar statistical assessment.

7.2.2.2 ASTM standard test methods. The ASTM Standard E 462–84 (reapproved 1989) *Standard Test Method for Odor and Taste Transfer from Packaging Film* uses large test samples 1 yd^2 (0.9 m^2) with test conditions of at least 20 h at room temperature. Tests are carried out for odour directly from a test specimen confined in glass jars, and also for odours absorbed in mineral oil, water and butter. Water, butter and chocolate are used for testing of any off-flavours transferred from the packaging film. Assessments are made using any suitable category scale, but scales with either four or seven levels are suggested.

The ASTM Standard E 619–84 (reapproved 1989) *Standard Practice for Evaluating Foreign Odors in Paper Packaging* details the methods for the examination of odours from paper packaging, but mentions in the scope that the techniques described in the standard are also applicable to paper packaging that is printed and formed with adhesives, and to paper/plastic laminates.

The individual odour detection methods include direct examination without confinement of single sheets, sheet stacks and bundles, and examination with confinement at room temperature and at an elevated temperature of 52°C, either dry or after dampening with water. There are also procedures to examine the transfer of odours to mineral oil (as a simulant for fatty foods), test foods and odour-sensitive commercial products. Taste tests with butter, cream and milk chocolate are described for assessment of off-tastes produced by adsorption of odours from the paper packaging.

7.2.2.3 Other standard test methods. For taint testing of paper and board packaging, both printed and unprinted, the test method now most often used is a version of the Robinson Test, established by Lena Robinson in the early 1960s (Robinson, 1964). It is currently being considered for an EN (European Committee for Standardization) standard.

A Finnish version (Finnish Pulp and Paper Research Institute) follows broadly the procedure of the BSI method, discussed in section 7.2.2.1. However, the recommended exposure time is 48 h, the storage temperature is 22 ± 1°C, and the atmosphere within the vessels containing the packaging material and test food is maintained at 75% relative humidity by means of a saturated sodium chloride solution. Milk chocolate is the standard test food, but other test foods can be used as appropriate. For the assessment panel, a team of at least eight is recommended, with ten being the ideal number. The triangular test is the principal method of assessment, with two reference portions of food and one portion of food that has been exposed to the packaging.

The statistical significance of the difference between the sample and the reference test food is obtained from the significance table in the ISO Triangular Test standard. From the number of correct answers recorded and the number of tasters in the panel, it is determined whether any off-taste/tainting recorded by panel members is: (i) statistically significant; (ii) significant; (iii) almost significant; or (iv) not significant. The panel members are also asked to record the intensity of any off-flavours/tainting compared with the reference food on a 0 to 3 numerical scale—'0' for 'no off-flavour' and '3' for 'strong off-flavour'. The arithmetic mean of the individual numerical results is calculated. If two or more tasters give the same description of any off-taste/tainting, this can also be included in the report.

In the German DIN Standard 10 955, there are separate details for testing for off-flavours or taints transferred to foods by direct contact and through the enclosed air space of packaging. The test is applicable to packaging made from a variety of materials. The standard gives a recommended list of test foods for various food products for which the packaging containers are intended. These include: (i) unsalted butter for greasy food products containing water; (ii) icing sugar and butter biscuits for dry food products; and (iii) 10% v/v ethanol for alcoholic drinks. The recommended test conditions are 23°C for 20 ± 2 h but, to match longer storage times, the test time can be extended.

Three procedures are listed for the odour and off-taste sensory assessments. These are: (i) ranking test; (ii) triangular test and (iii) paired comparison test. The triangular test and the paired comparison test have already been described. The ranking test is used where a large number of samples are presented simultaneously and there are relatively big sensory differences in intensity. A scale of 0 to 4 is used for the off-odour assessments, with '0' for 'no perceptible off-odour' and '4' for 'strong off-odour'. A similar 0 to 4 scale is used for the off-flavour assessments.

Other countries, such as Switzerland, Sweden, The Netherlands and Australia have variations of the odour and taint tests already described. In addition, some individual companies, particularly companies that manufacture and market 'sensitive' food products such as chocolate confectionery, have developed their own odour and taint test methods for packaging materials. An example is the odour method of Cadbury Schweppes, a well-known confectionery company (Cadbury, 1991). The test method states that the odour panel should consist of at least six members. The test is carried out on sheet or film packaging taken from stacks or reels, with each panel member being given at least two layers to enable the nose to be put between the layers. Assessments are required on both sides of the sheet or film. Bags are assessed: (i) directly after opening; or (ii) after opening, air expelled, re-filled with odour-free air and stored for at least 24 h. A six-point scale is used, with '1' – 'very strong odour' and '6' – 'no odour'. Guidance is given for the panel members on the interpretation of the six odour ratings. A mean value from the panel assessment of 3.8 or greater is generally considered as acceptable.

7.2.3 Precautions in selecting, transporting and handling samples for sensory testing

As odours and food tainting can be caused by very low levels of substances in the packaging, it is essential that care is taken when selecting, transporting and handling samples of packaging for sensory testing. The test sample should be in a state representative of the majority of the packaging. Consequently, a sample that has been freely exposed to the air for a long period of time should not be used since any odorous and tainting substances might well have been lost due to volatilisation.

When taking test samples of a film, the outer turns of the reel should first be removed. Similarly, with stacked sheets and containers the top five to ten sheets or containers should be discarded. When taking samples from a production line, time should be allowed for the production process to stabilise.

Some types of packaging, particularly printed materials, can have detectable odours immediately after production, but lose much of the odour during the normal storage before use. Samples of such packaging should only be taken after the normal storage period.

There is also a clear need to ensure that after the packaging material has been sampled, adequate precautions are taken during transportation to the test laboratory. This is to prevent: (i) contamination from external sources, which could adversely influence any subsequent sensory test; and (ii) exposure to environmental conditions, which causes premature release of odorous and/or off-flavour substances. Where a control sample is included with the test samples, it is also important that there is no cross-contamination between these samples. Consequently, samples taken for sensory tests are usually individually wrapped in aluminium foil (preferably washed and cleaned, as some foils can have residues of rolling oil) to provide the necessary protection.

7.3 Printing inks and adhesives

Most food packaging is printed for sales appeal and also for information about the packaged food. The printing can be on: (i) packaging material that is in direct contact with the food, such as with biscuit wrappers and containers for dairy products; (ii) labels attached to the direct packaging; or (iii) secondary packaging such as outer cardboard boxes for confectionery and cereals. Where labels are used, and with some carton board secondary packaging, adhesives are employed. For printed packaging that is in direct contact with the food, the print is almost always on the outside and not itself in contact with the food.

To give the printing added gloss and sometimes to protect it from scuffing and abrasion, an overvarnish coat is applied on top of the printing. The varnishes are usually based on formulations similar to those used for printing inks, but without the pigments and colorants.

Most forms of printing use inks that consist of pigments and resins dissolved or dispersed in solvents. Traditionally the solvents have been of the organic type, but some inks and varnishes now use water as the solvent or part solvent. Many of the organic solvents used are relatively volatile so that drying of the print takes place quickly. A small proportion of a slow-drying solvent is, however, often added to control the drying process and to give the required print quality. The finished printed packaging usually contains small residues of both the more volatile, and the higher boiling solvents.

One form of printing and varnishing now often used for food packaging is solventless and employs ultraviolet radiation to cure and polymerise the resin components. In addition to the monomers or pre-polymers used as the resin components in the formulation, photoinitiators are incorporated to aid the curing process.

Food packaging adhesives vary and include: (i) water-based solvents with dextrins, starch and vinyl acetate, used for example on cartons; (ii) solvent-based adhesives used for laminating plastic layers together; (iii) solventless laminating adhesives; (iv) hot melt adhesives; and (v) cold seal adhesives where the bond is achieved by simple pressure. As with the printing, solvent residues and other volatile substances from the adhesive formulations can remain in the packaging.

Unless there is a functional barrier in the packaging material between the print or the adhesive layers and the food product, then components of the printing or adhesive, such as solvent residues, can permeate through the packaging and migrate into the food. However, normally the levels of any of the print and adhesive substances that transfer to the packaged food are extremely low and do not result in detectable odours or tainting of the food.

7.3.1 Printing ink and varnish types

A variety of printing ink and varnish formulations are used for food packaging depending on the method of printing employed. With each type of ink or varnish, care has to be taken in selecting the components to ensure that the packaged food remains odour and taint-free.

With litho inks the solvents used are principally petroleum distillates. As the name implies these solvents are derived from petroleum products and consist of hydrocarbon fractions. For printing food packaging, 'low odour' petroleum distillates are selected—these contain either low concentrations of aromatic hydrocarbons or are free of these aromatics. The resins in litho inks are also a potential source of odorous and tainting substances. These resins are of the alkyd type, which oxidise in the drying process and can liberate odorous by-products, such as aldehydes and ketones. Careful selection of the resins for litho inks and varnishes for food packaging ensures that any odours are maintained at low and acceptable levels and that the packaged food remains free from tainting.

Flexographic inks use a range of solvents, which are mainly:

- alcohols, such as ethanol, isopropyl alcohol (IPA) and n-propyl alcohol
- esters, such as ethyl acetate, isopropyl acetate and n-propyl acetate

Toluene and xylene have been used in the past but are now very rarely used for food packaging. Modern flexographic ink formulations are largely based on ethyl acetate and ethanol. The ink formulations can include plasticisers and slow-drying solvents, such as propylene glycol monomethyl ether. Resins include polyamides, cellulose nitrate and acrylics. The solvent residues, particularly the acetate esters, are the most likely components to give the flexographic printed packaging an unacceptable odour. It is therefore essential that any solvent residues are reduced at the drying stage to very low levels. It is very unusual for the resins to cause either an odour or taint problem.

Gravure inks and varnishes for food packaging use similar solvents and resins to those used for flexographic inks. Water-based inks and varnishes are often used for food packaging for their low odour and taint properties. A proportion of organic solvent of the alcohol type is usually incorporated.

UV-cured inks and varnishes are basically solventless. They consist of acrylate monomers with a photoinitiator. The acrylate monomers are mainly multifunctional prepolymers, which further polymerise and crosslink under the action of the ultraviolet radiation to give a 'hard' dry print. The photoinitiator—the principal one used being benzophenone—provides the free radicals for the acrylate polymerisation process. There is usually a residue of the photoinitiator left after the curing process is finished.

Some acrylate monomers are highly odorous. However, the acrylate formulations now in use have very low odours and, although there can be a characteristic acrylate-type odour straight off the press, this usually dissipates quickly. Any trace odours associated with UV-cured prints and varnishes are probably due to a combination of minor acrylate residues, residues of the benzophenone or other photoinitiators, and reaction by-products from the polymerisation process.

Benzophenone is a solid at room temperature, but is relatively volatile. Residues remaining in the printed packaging can be released slowly into the packaged food. Benzophenone has a sweet odour, which has been described as geranium-like. Reported minor reaction by-products from UV printing include benzaldehyde and alkyl benzoates. These substances are highly odorous and would be expected to cause food tainting if formed in sufficient quantities (Holman and Oldring, 1988).

Guarino and Ravijst (1988) discuss the development of a low odour UV cure ink system. This incorporated specially purified acrylate monomers/oligomers and a reactive benzophenone derivative with reactive amine co-initiators.

7.3.2 Precautions to prevent odours and tainting from printed packaging

The potential for printed packaging to cause odour and tainting of foods has long been recognised and residues of solvents from the printing ink or varnish have been the main cause of concern. Pilborough (1987) outlined the reasons for high levels of retained solvents. One obvious reason is insufficient drying on the press, arising from faulty heaters or running too fast. Other reasons range from 'case-hardening' or 'skinning' due to applying high temperatures too quickly, to excessive adsorption of solvents into substrates such as paper and board. Changes to the ink formulations on the press to achieve the desired print characteristics can include further additions of slow drying solvents, which can largely remain in the finished printed packaging. It is further pointed out that the base packaging material can itself contain solvent residues, either from coatings or, where the base packaging material is a laminate, from an adhesive. Such solvent residues can be ethyl acetate, toluene and tetrahydrofuran—all of which have strong odours.

Reid (1983) describes the precautions taken by ink makers in the selection of the ink components to ensure that they are suitable for use on food packaging.

Claxton (1981) discusses the problems of ink-related odours from printing of aluminium foil. The printing is carried out either by the gravure or flexographic processes, usually as a continuous web. The unsupported foil used in packaging is often coated with a heatseal layer or with a protective coating on the side intended to contact the food. Laminated products are produced with both paper/board and plastics. The foil is an excellent barrier to odorous solvents from printing on the outer surface of the packaging. Claxton, however, points out that with packaging consisting of foil laminated to paper/board or plastics, components from printing on the outer foil surface can transfer and be adsorbed on to the inner food contact surface when the packaging is reeled up or sheet-stacked prior to use. This problem can occur with most printed packaging and is often overlooked when precautions are taken to prevent odours and taints from printing entering the packaged foods.

Claxton points out that a further potential odour/taint problem from printing inks is the presence of high-boiling impurities in the solvents. Such impurities might not be completely volatilised unless there is extended drying of the printed material. They could then slowly permeate out after the food is packaged. The precaution to be taken to avoid this potential problem is the use of good quality solvents.

There are no generally agreed maximum levels for residual solvents in food packaging. This is because a number of factors determine whether a solvent residue will produce an unacceptable odour or detectable food tainting. These factors include: (i) the nature of the food; (ii) storage time and temperature; (iii) the packaging substrate material; and (iv) whether the packaging is primary or secondary. The packaging substrate material can have a significant influence on the levels of solvents retained, the subsequent rate of release and the related propensity to cause tainting. Adcock and Paine (1974) reported that boards made with a proportion of mechanical pulp had been shown to adsorb and retain residual solvents to a higher degree than boards made only with chemical pulp. This was attributed to the lignin component of the mechanical pulp. In a similar way it was shown that other components of the boards, such as pigments and coatings, also had different abilities to retain any residual solvents.

As a general guide, however, the total level of residual solvents should normally not exceed 30 mg/m^2; for some of the more odorous solvents, such as isopropyl acetate, the individual levels should be less than 5 mg/m^2. The odour threshold levels for the common printing and adhesive solvents range from parts per million to parts per billion (Fazzalari, 1978). The taint threshold levels can also span wide concentrations, with the food itself having a significant influence. It is usually necessary for the packaging printer or converter to establish solvent residue limits for each particular packaging and food product combination and agree them with the customer. Such limits are arrived at by comparing residual solvent levels in the packaging and results from odour/taint tests (described in section 7.2) with the particular food.

7.3.3 Methods for determining levels of residual solvents

The gas chromatography analytical technique is the obvious choice to measure residual solvents in printed packaging and has been described in a number of papers (Kontominas and Voudaris, 1982; Passy, 1983; Pilborough, 1987). The general procedure is to place a portion of the printed packaging in a container, seal the container and then heat it to liberate the residues of solvent into the container headspace. The headspace is sampled, either manually with a gas syringe or by an automatic headspace sampler, and the sample of solvent vapours is injected into a gas chromatograph.

Providing the solvents used in the printing are known, then a gas chromatograph with a standard flame ionisation detector (FID) can be used and the solvent residues quantified by external calibration. The test may not, however, measure the total concentration of the solvent residue in the packaging, as complete volatilisation into the headspace may not occur. The extent of volatilisation of the solvent will depend primarily on the temperature used to heat the sample and the boiling point of the particular solvent but, as already mentioned, some packaging materials have a greater ability to retain some solvents and, in such cases, lower concentrations are released to the headspace.

Adcock and Paine (1974) presented a table showing that after heating, to 115°C, various paper, board and regenerated cellulose film materials containing alcohol print solvents, less than 100% of the alcohols were released to the headspace. The lowest values for ethanol, isopropanol and n-propanol were 41, 48 and 37%, respectively. All these solvents have boiling points well below 115°C. Hydrogen bonding between the alcohols and the cellulose components of the substrates is probably the mechanism which prevents total volatilisation into the headspace.

It is not, however, always necessary for the test method to provide an answer that gives the total quantity of each residual solvent in a packaging material. Where there is a requirement to provide rapid results for production monitoring and process control purposes, a method that primarily gives reliable and reproducible results is acceptable. Such a method was established as a British Standard Method in 1984—BS 6455:1984 *Monitoring the Levels of Residual Solvents in Flexible Packaging Materials* (BSI, 1984).

The method is quoted to be suitable for monitoring the levels of certain residual volatile solvents in unprinted and gravure or flexographic printed paper, films, foils and laminates intended for packaging. There are 13 solvents listed, although it is quoted that with the type of chromatography columns recommended, it is not possible to measure methyl ethyl ketone and tetra-hydrofuran together because of overlapping peaks. It is pointed out however, that these two solvents are rarely used together. The 13 solvents are: acetone, ethanol, isopropanol, ethyl acetate, tetrahydrofuran, methyl ethyl ketone, iso-propyl acetate, *n*-propanol, *n*-propyl acetate, isobutanol, *n*-butanol, toluene and 2-ethoxyethanol.

Either one or two test samples 1 dm^2 in area are used for the analysis. Procedures are described with and without internal standards and the test samples are heated for 20 min at 120°C. The standard recommends packed columns for the gas chromatography, with headspace samples injected manually. The more modern capillary columns could, however, be used equally well, particularly if an automatic headspace sampler is used.

There is an alternative gas chromatography method for determining residual solvents in packaging. This method uses the 'thermal desorption' technique, instead of the headspace technique, to release the residual solvents. A portion of the packaging material is placed inside a thermal desorption tube, which is then heated to release the volatile solvent residues. The released solvents are collected in an intermediate concentration 'cold trap' before being transferred to the gas chromatograph (Perkin Elmer, 1991).

This method is relatively quick with a total analysis time of around 20 min. If an automatic thermal desorption instrument is used, batches of samples can be tested. The method is particularly suitable for production monitoring and process control. The size of the test sample is smaller than that used in the BSI method and is typically 50 cm^2. With some printed films it might be necessary to test a number of samples taken from across the film web to obtain reliable mean values for the solvents determined.

The American ASTM method for the determination of residual solvents in flexible barrier materials is F 151–86 (ASTM, 1986). The test procedure is similar to that described for the BSI method. It is again pointed out that the method does not yield absolute quantitative measurements of solvents retained in flexible barrier materials. Round-robin testing has shown that the test method has a coefficient of variation between laboratory means of ± 15%. It is recommended that optimum heating times and temperatures be determined for each type of sample. A further difference to the BSI method is that the container into which the sample is placed is evacuated to a predetermined vacuum before heating. The solvents released into the headspace are measured by the standard gas chromatography technique.

7.3.4 Some reported odour and taint investigations

Gorrin *et al.* (1987) reported an investigation into a musty off-odour from a printed packaging film. An evaluation of the volatile substances released from

the film after heating, by gas chromatography and simultaneous odour apprai-
sal, detected two substances that were not present with a control film. The
substance with the largest peak had no detectable odour, but the other substance had the musty odour. Using mass spectrometry and infrared spectrometry the largest peak/non-odorous substance was identified as 2-methyl-2,
4-pentanediol. No positive identification was possible for the musty odour
substance, but the data obtained from both the mass spectrometer and the
infrared spectrometer indicated a diether. By synthesising possible compounds
and comparing the spectral data and odour characteristics with the unknown,
the musty odour was identified as the substance 4,4,6-trimethyl-1,3-dioxane.
It was predicted that this substance had been formed by reaction of 2-methyl-2,
4-pentanediol with formaldehyde from an unknown source during storage of
the film. 2-Methyl-2,4-pentanediol had been used in the printing of that batch
of film as an adhesion promoter.

One of the more interesting odours that has been caused by solvent components of printing is that which is described as 'catty'. Such odours have been
attributed to a sulphur-containing compound 4-methyl-4-mercaptopentan-2-
one, which is a hydrogen sulphide adduct of mesityl oxide, and is reported to
have been found in association with meats and other food containing sulphur
compounds (Patterson, 1968). The odour threshold of this hydrogen sul-
phide/mesityl oxide adduct in bland foods is of the order of 0.01 ppb.

Franz et al. (1990) described an extensive investigation to trace the source
of a catty odour in a cook-in-the-bag ham product and showed by analysis and
synthesis experiments that the odour was the previously mentioned hydrogen
sulphide/mesityl oxide adduct. The investigations demonstrated that the mesityl oxide originated from diacetone alcohol, which was present as a residual
solvent in red printing. Formation of the mesityl oxide was thought to have
arisen from dehydration of the diacetone alcohol, promoted by some property
of the ionomer layer in the bag material.

Mesityl oxide can also be formed from acetone, and industrial acetone can
have trace quantities present. Acetone is occassionally used as a solvent in
coatings and adhesives.

Mesityl oxide is not a true oxide. It was assigned this name at an early date
due to an erroneous conception of its chemical nature, and is in fact an $\alpha,
\beta$-unsaturated ketone with the formula $(CH_3)_2C = CHCOCH_3$. It has a pepper-
mint-like odour.

Another reported incident of the catty odour arising from a food packaging
occurred with canned pork, where the mesityl oxide was present as an impurity
in the ketone solvent of a lacquer that had been applied to the soldered seam
to prevent blackening of the meat (Goldenberg and Matheson, 1975).

Passy (1983) reported a number of case studies investigating complaints of
off-flavours. These included the identification of 'high' levels of toluene from
laminated pouches used for fruity soft drinks, originating from the adhesive
used for lamination of the multilayer structure. The levels of toluene in the

pouches that had produced an unpleasant off-flavour were between 1.2 and 2.8 mg/m^2 compared with levels between 0.04 and 0.08 mg/m^2 in 'good' pouches. In a second case with a printed film laminate, residual solvent levels were found to be extremely high. As unprinted samples had similar residual solvent levels, the conclusion on this occasion was that the problem was due to inadequate removal of the solvents from the laminating adhesive.

7.4 Paper and board packaging

Paper used for packaging is made from pulp, which in turn is made from wood. Various additives such as sizing agents and fillers can be present in the paper. There may also be residues of substances used in the manufacturing process, such as defoaming agents and slimicides.

Carton board is also made from pulp and contains similar additives to those used in paper. Some carton board consists of more than one layer and may also have surface coatings for printing and appearance.

The pulp used for both paper and carton board can be: (i) mechanical, where the cellulose fibres are prepared by mechanical processes on the wood; or (ii) chemical, where the cellulose fibres are isolated by means of chemical treatment of the wood. Mechanical pulp still contains most of the lignin and other components of wood, such as resins and inorganics, in addition to the cellulose. Pulps can be bleached with bleaching agents to give improved whiteness to the finished paper or board. The main bleaching agent used in the past has been chlorine. Recently there has been a move away from chlorine due to the tendency to form trace quantities of organochlorines, including dioxins and associated furans, which have well-established toxic properties. Hydrogen peroxide and chlorine dioxide are used as alternatives to chlorine.

Greaseproof paper is manufactured from chemical pulp that has had extended beating to give a product with the necessary grease-resistant properties without need for additional coatings, although, for extra resistance to fats and oils, surface treatments and coatings may be applied. Glassine is paper manufactured by a similar process to greaseproof paper, but is also supercalendered. This process produces a glossy surface to the paper.

The most common surface coatings on boards consist of a filler, such as china clay and calcium carbonate, and a synthetic resin binder, such as a styrene/butadiene copolymer or a styrene/acrylate copolymer.

Although plastics have replaced some of the food packaging that in the past was either paper or board, these two traditional packaging materials are still widely used. Foods such as flour and biscuits are still to be found packaged in kraft paper. Butter, margarine and cooking fats are wrapped in greaseproof paper and chocolates are packaged in glassine. Carton board is used for a variety of carton boxes for packaging foods ranging from confectionery to cereals. With some 'dry' foods, the carton board is in direct contact with the food, but more often it acts as secondary or outer packaging, the food being contained in an inner plastic tray or bag. Any tainting of the packaged food from odorous volatile substances originating from the carton board used as secondary packaging can only occur by transmission through the air space within the carton or, if the inner packaging is sealed, by permeation through the inner packaging material.

With modern packaging, multilayers are often used to give strength, barrier properties, easy sealing and good printability. A typical multilayer structure consists of paper or board laminated to one or more plastics layers with a foil intermediate. Adhesives are used to bond the layers together and the outer surface is printed, sometimes with an overvarnish. With such multicomponent packaging, each layer can contain substances that could be odorous and/or cause tainting in the packaged foods. Odorous and tainting substances that can originate from printing and adhesives have been discussed in section 7.3. Plastics are discussed in section 7.5.

Plastic/board composites are now commonly used for liquid packaging, such as milk. The plastic layer is usually polyethylene. The milk is in contact with the plastic layer but it is possible for substances from the board to permeate through the plastic into the milk. As milk is a 'sensitive' food, extra care is taken in selecting the board, and also the plastic for such packaging.

Where plastic/board composites used for liquid packaging also contain an aluminium foil layer, this provides a barrier to the oxygen permeation of the food or beverage. The aluminium foil layer is between the board and the food or beverage, and consequently also acts as a functional barrier, preventing the migration of any odorous or tainting substances into the food or beverage.

Odours that can be present in paper and board materials have been attributed by Donetzhuber (1982) to substances arising from bacteria/moulds, autoxidation of residual resins, and degradation of processing chemicals. Metallic ions, such as iron and copper, in the inorganic residues left in the pulp can catalyse the oxidation of lipids and other components of the resinous fraction to a range of volatile substances. The reaction mechanisms for these oxidation and similar processes have been postulated by Donetzhuber (1982), who also listed more than 200 volatile substances that have been identified in the gas phases from pulp, paper and board samples. These substances are grouped into 11 main classes:

- *n*-alkanes, up to C15
- branched alkanes, up to C13
- alkenes, up to C13
- aldehydes
- ketones
- alcohols
- esters
- heterocyclics (furans)

- acyclics (benzene and substituted benzenes)
- sulphur compounds (alkyl sulphides)
- terpene hydrocarbons

Although many of these compounds will be odorous, particularly those containing functional groups, such as aldehydes, alcohols, esters and sulphides, and those with unsaturated groups, most are present at very low levels. Only a minor proportion of these compounds can be found in pure fresh wood.

Using the list of odour descriptor terms developed by The Finnish Pulp and Paper Research Institute (see section 7.1), odour descriptors were assigned to a list of volatile compounds found to be present in pulp samples. These are given in Table 7.2 (Soderhjelm and Eskelinen, 1985). Some of the odours listed are not unpleasant, such as the 'green' odour for hexanal, which is sometimes also described as 'grassy'. Of the aldehydes, hexanal is often present at the highest concentration and its predominantly pleasant mild 'natural' odour is accepted.

Table 7.2 Volatile compounds identified in pulp samples

Compound	Description of odour
Pentanal	Sharp
Hexanal	Green
Methyl hexanal	Sweet
Methyl hexanone	Ant-hill
Heptanal	Oil
Benzaldehyde	Bitter almonds
Methyl heptenone	Mushroom
Octenol	Mushroom
Pentylfuran	Moss
Octanal	Rancid fat
Cymene	Medicine
Octenal	Wood bug
Nonanal	Fatty
Undecane	
Nonenal	Rancid
Methyl isoborneol	
Nonadienal	Rancid
Decanal	Rancid
Decenal	Rancid
Cis trans-2,4-decadienal	Rancid
Trans trans-2,4-decadienal	Grill shop (fried oil)

Source: Soderhjelm and Eskelinen (1985)

Unlike many other packaging materials, such as printed plastics, which usually become less odorous when left for a period of time after manufacture, some board materials have been found to become more odorous on storage. This is attributed to ongoing oxidation reactions of the type already mentioned, most probably catalysed by metal ion residues, liberating further quantities of the aldehydes. To reduce the effect of heavy metal ion residues, and to minimise the formation of odorous compounds, complexing agents such as hexametaphosphate and EDTA (ethylene diamine tetraacetic acid) are added to the pulp (Soderhjelm and Eskelinen, 1985).

It was mentioned previously that the surface coatings applied to boards usually consist of a filler such as china clay or calcium carbonate, with a synthetic resin binder. Where the binder is of the styrene/butadiene type, several volatile substances have been identified, including styrene and by-products from the styrene/butadiene polymerisation process. These are:

- vinyl cyclohexene
- ethyl benzene
- styrene
- cumene (isopropyl benzene)
- *n*-propyl benzene
- α-methyl styrene
- 4-phenyl cyclohexene
- other alkyl substituted benzenes

All of these substances are odorous and, if present in sufficient quantities, can make a significant contribution to the general odour of the board. The 4-phenyl cyclohexene, formed by a Diels–Alder condensation reaction between styrene and butadiene, has a particularly strong 'synthetic latex' type odour (Koszinowski *et al.*, 1980). Carton boards for food packaging are, however, manufactured with regular quality control, ensuring that odours from both the board and the coatings are maintained at very low and acceptable levels and do not cause tainting of the packaged foods.

7.5 Plastics

Plastics are the most versatile and widely used materials for packaging of foods. Plastics films are used for wrapping foods such as biscuits, meats, cheeses and vegetables. Plastics bottles are used for beverages, mineral waters, cooking oils and sauces. Plastics containers are used for margarine, fats, dry foods and ready-meals.

To meet the demanding requirements for modern food packaging, the plastics packaging material can be complex and sophisticated, often consisting of a number of different plastics layers, with surface coatings to give improved barrier properties, printability or sealability and, as previously mentioned, sometimes combined with other materials, such as paper, board and aluminium foil.

The basic starting substances for plastics are monomers, which are polymerised to form the polymer component. Usually there are very small quantities of unreacted monomers left in the polymer. With many of the plastics, additives are necessary to act as antioxidants, stabilisers, slip agents and antistatics. For opacity and coloration, white pigments and colorants are added.

The selection of the plastics raw materials and the manufacture into food packaging is closely controlled, and any components that are odorous and have a propensity to produce off-flavours in foods are maintained well below their effective concentrations. Occasions do, however, arise when unacceptable odours and off-flavour substances are present despite all the precautions taken in the manufacture of the plastics and conversion into packaging. Often these substances are reaction by-products or decomposition substances rather than the formulation components. Usually such packaging is prevented from reaching the food packer by the taint and odour quality control tests described in section 7.2.

Many of the monomers used for food packaging plastics are relatively volatile and some also have low odour threshold values. The plastics monomer that is probably most well known for its odour is styrene. This monomer is used to manufacture a range of polystyrene plastics, some of which contain co-monomers, such as butadiene and acrylonitrile. Typical applications of these polystyrene plastics are vending cups, dairy product containers and, in the expanded form, meat trays. Styrene is one of the few monomers that is easily re-formed from the plastics when subjected to excessive heat or when burnt. Consequently, care has to be taken in the processing of the plastics to ensure that the processing temperatures are not excessive.

The odour threshold value for styrene is in the region of 50 ppb (Fazzalari, 1978). The taint threshold values vary according to the food type. Reported values are reproduced in Table 7.3 and illustrate the wide variation with different types of foods. Ramshaw (1984) noted that whereas the higher taint threshold values occur with the higher fat products, migration of the monomer is usually more pronounced when the polystyrene plastics are in contact with the fatty foods. However, although the free styrene monomer levels in polystyrene plastics can be relatively high and typically around 300–400 ppm, the quantities found in food products due to migration are usually in the low ppb range—below the taint threshold values—even with fatty foods (Gilbert *et al.*, 1983).

Passy (1983) reported on a case study where an off-flavour was detected in chocolate and lemon cream cookies. The cookies were packed in polystyrene

Table 7.3 Styrene taint threshold values

Food	Styrene taint threshold value (ppm)	Reference
Sour cream	0.005	Miltz *et al.* (1980)
Water	0.022	Linssen *et al.* (1991)
Tea	0.2	Jenne (1980)
Yogurt	0.5	Jenne (1980)
Whole milk	1.2	Jenne (1980)
Butter	5.0	CSIRO (1969)

trays and wrapped in printed cellophane. The polystyrene trays were found to contain between 0.18 and 0.20% styrene monomer, which was below the specified maximum of 0.5%. Tests carried out with the trays and the printed cellophane placed in glass jars with fresh chocolate and lemon creams, showed that the tainting was concentrated in the cream and was organoleptically recognised as being due to the styrene monomer. A gas chromatography head-space analysis on the fresh cream that had been in contact with the polystyrene trays showed styrene present as a major chromatography peak. No attempt was made to quantify the level of styrene. It was inferred that although the free styrene monomer in the plastic was below the permitted limit, the cream was particularly susceptible to styrene and the small quantity of styrene that migrated was sufficient to cause the off-flavour. Although the levels of free styrene in the plastic were below the specification limit, these levels are much higher than are normally found, as mentioned previously (Gilbert *et al.*, 1983).

A more recent investigation into the sensory effects of styrene monomer migrating from polystyrene plastics into cocoa powder for drinks and choco-late flakes, was carried out in The Netherlands (Linssen *et al.*, 1991). The results obtained with cocoa powder and extrapolated for the cocoa drinks, confirmed the predictions of Ramshaw (1984) that migration of the styrene monomer is higher in products with high fat content. It was also found that milk chocolate flakes were more sensitive than plain chocolate flakes to an off-flavour from styrene monomer. Reference was made in the discussion to results from an earlier publication by the authors on taste threshold values for styrene in water of 22 ppb (0.022 ppm), and in emulsions with 30% fat of 2.3 ppm. These results again illustrate the wide difference in taste sensory effects for the styrene monomer in different food media (Linssen *et al.*, 1990).

A polystyrene plastics packaging supplied to the author's laboratory had a strong 'styrene odour' and was suspected of causing tainting of packaged biscuits. The volatiles from the polystyrene plastic were analysed by gas chromatography/mass spectrometry and the chromatogram obtained is shown in Figure 7.1. The large peak at 6.7 min is the styrene monomer at a level of approximately 1150 ppm (mg/kg) in the plastic. The peak at 6.1 min is ethyl benzene, which is usually present with styrene. In a test to confirm that the free styrene monomer was causing tainting, different biscuits were stored in contact with the polystyrene. Tainting of the biscuits, recognised as due to styrene monomer, was confirmed. The volatiles from the tainted biscuits were analysed by gas chromatography/mass spectrometry and the chromatogram obtained is shown in Figure 7.2. Styrene monomer is easily identified as the peak at 6.3 min. Most of the remaining peaks are due to volatiles from the biscuits, but the peak at 1.3 min was identified as 1-propanol and the peak at 2.8 min as propyl acetate. These substances are printing ink solvents (see section 7.3) and probably originated from the printed plastic packaging in which the test biscuits were originally packaged. The taint threshold values

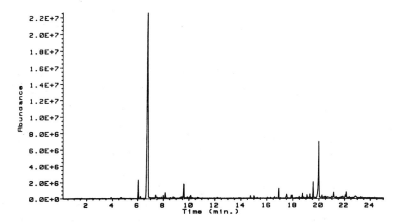

Figure 7.1   Mass spectrometer total ion chromatogram of volatiles from polystyrene plastics with a strong 'styrene odour'.

for these solvents are much higher than styrene and would not have caused detectable tainting.

In the United States an odour and taste sensory evaluation of four different plastics for mineral water bottles showed that PVC (polyvinyl chloride) had the lowest contribution to odour and off-taste of the bottled water. The other three plastics used in the exercise—in increasing order of sensory effect— were PET (polyethylene terephthalate), PC (polycarbonate) and HDPE (high density polyethylene). The tests were carried out with the plastics in the form of bottles and plaques in contact with the water for a period of four weeks at

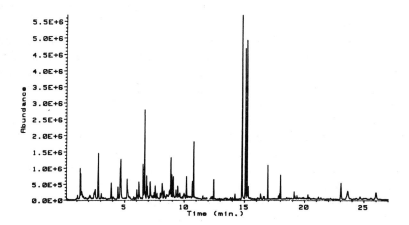

Figure 7.2   Mass spectrometer total ion chromatogram of volatiles from biscuits that had been in contact with polystyrene plastics with a strong 'styrene odour'.

room temperature, and also at 120°F (48.9°C). The odour and off-flavour differences for the water samples exposed to the four plastics were only significant after the four-week period and with the 120°F test temperature (Solin *et al.*, 1988).

An incident of off-flavour in fruit drinks in PVC bottles, reported by Goldenberg and Matheson in their 1975 review paper, was attributed to the use of a metal stabiliser containing mercaptide groups. It was thought that the formation of the mercaptide substance had been due to reaction of hydrogen chloride, liberated from the PVC during processing, with the metal stabiliser. The problem was overcome by changing to a stabiliser that did not contain mercaptide groups.

In addition to bottles, PVC plastics are used for food trays and containers. PVC cling-film, which contains plasticisers, is used for wrapping a wide variety of foods ranging from fresh vegetables to meats and cheeses. The migration of the plasticisers from PVC cling-films has been extensively studied, particularly with fatty foods, but with little emphasis on sensory effects. A sensory investigation with PVC cling-films was carried out in Sweden, but no definite evidence was produced that the migration of plasticisers or other components from the films into cheese produced off-flavours (Sandberg *et al.*, 1982).

Polyethylene terephthalate (PET) is now a widely used plastic for bottles to package beverages, such as mineral water, carbonated drinks, ciders and beers. As PET can withstand relatively high temperatures, another main use is for ovenproof trays for cooking/re-heating foods in either microwave or conventional ovens. The monomers used for the manufacture of PET plastics are either terephthalic acid and ethylene glycol, or dimethylterephthalate and ethylene glycol. These monomers are not particularly odorous and, with the residue levels in PET plastics normally being quite low, they do not cause odours or tainting.

Tainting from PET plastics can, however, occur due to acetaldehyde, which is formed as a degradation product during the manufacture of the PET polymer, and processing into the finished packaging. The odour of acetaldehyde can be detected at a level of 0.2 ppm (Fazzalari, 1978). The off-flavour threshold values for a number of beverages cover quite a wide range. For example, 10–20 ppb for carbonated water, 30–40 ppb for still mineral water, 60–100 ppb for Coca Cola and 1000 ppb for wines and ciders (Lorusso, 1985; Mazzocchi, 1988). The quantity of acetaldehyde present in the plastic has been shown to be directly related to the processing temperature, but the presence of moisture also has an influence (Lorusso, 1985). Levels of acetaldehyde in the PET plastics are usually maintained between 1 and 10 ppm, with some specifications set at the 6–8 ppm level.

Polyolefin plastics, such as polyethylene and polypropylene, are the most common of all the plastics used for food packaging, as wrapping films, containers and bottles. There are various types of polyethylene, including linear low density polyethylene (LLDPE), low density polyethylene (LDPE) and high density polyethylene (HDPE). For some applications ethylene is copolymerised with vinyl acetate to produce ethylene vinyl acetate (EVA), and with acrylic acid to produce ionomers. Polypropylene plastics consist of homopolymers and also copolymers with other alkene monomers.

Odours that can be present with polyethylene and polypropylene and the various copolymer plastics are not usually due to residual monomers. An examination of the volatiles from these polyolefin plastics finds mainly a variety of aliphatic hydrocarbons, as might be expected. Although some of the hydrocarbons will have odorous characteristics, their odour threshold values are often relatively high and, with the concentrations usually present in the plastics, the odours are normally not detectable.

Odours from polyolefin plastics can be due to compounds produced by oxidation. Koszinowski and Piringer (1986) found that trace levels of 1-nonene present in LDPE plastics were oxidised to highly odorous 2-nonenal, probably by a free-radical mechanism. Another odorous oxidation compound identified as 1-hepten-3-one was found to be formed from 2-ethyl-hexanol detected in paper board samples. Both these substances are conjugated unsaturated carbonyl compounds with very low odour threshold values and were sources of odours in LDPE-coated paper board. Experimental work showed that many of the conjugated unsaturated carbonyl compounds have relatively low odour threshold values, C8 and C9 compounds having specially strong odours.

Ionising radiation is used in some countries as a preservation and extended shelf-life treatment for certain foods. Often the most effective way of using the ionising treatment is with the food already packaged. In such applications the packaging is also subjected to the ionising treatment.

Rojas De Gante and Pascat (1990) made an examination of the volatiles present in both polyethylene and polypropylene plastics after treatment with ionising radiation. Various oxidation products were detected, such as ketones, aldehydes and carboxylic acids. It was predicted that these could influence the organoleptic properties of the packaged foods.

In an earlier investigation in Japan, odorous volatiles were found to be formed when polyethylene films were exposed to electron beam irradiation. Of the total quantity of volatiles produced, 26% were aldehydes and ketones, and 18% were carboxylic acids. A synthetic mixture of the substances detected, replicated the odour from the irradiated polyethylene (Azuma, 1983).

7.6 Chlorophenols and chloroanisoles

Tainting and off-flavours in foods due to contamination by chlorophenols and chloroanisoles have been well reported and documented. These substances have very low sensory threshold values with, for example, 2.6-dichlorophenol,

2,4,6-trichlorophenol and pentachlorophenol exhibiting taste threshold values in water of 0.2, 2 and 30 µg/l, respectively (Dietz and Traud, 1978). The odours and taints of chlorophenols are usually described as 'disinfectant' and for chloroanisoles as 'musty'.

Chlorophenols have been used industrially as fungicides, biocides and herbicide intermediates. Pentachlorophenol has been the principal chlorophenol used for antifungal and biocide use, and this has included use in non-food packaging, particularly water-based adhesives. Commercial pentachlorophenol usually contains some 2,4,6-trichlorophenol and 2,3,4, 6-tetrachlorophenol. It is now well recognised that chlorophenols have significant toxicological properties and a recent EC Directive severely restricts usage of pentachlorophenol and its compounds (EC, 1991). In the United States and other countries, pentachlorophenol is also now rarely used as a fungicide or biocide.

Chloroanisoles are not used industrially, but it has been established that they are formed from the chlorophenols due to fungal methylation (Crosby, 1981). The odour and tainting thresholds of chloroanisoles are even lower than for the corresponding chlorophenols. Typical reported odour threshold values in aqueous solutions are: 4×10^{-4} ppm for 2,4-dichloroanisole, 3×10^{-8} ppm for 2,4,6-trichloroanisole and 4×10^{-6} ppm for 2,3,4,6-tetrachloroanisole (Griffiths, 1974).

There have been various reports and investigations of the contamination of foods by chlorophenols and chloroanisoles originating from packaging materials. Many of these have been carried out by Whitfield and coworkers at CSIRO, Australia and a number have been summarised in an article by Whitfield and Last (1985). They include jute sacks used for cereals and flours, where the off-odours were described as 'musty', 'mouldy', 'baggy', etc. Extensive studies showed that the jute sacks contained a number of chloroanisoles with musty odours at concentrations up to $1100 \, \mu g \, kg^{-1}$. Further investigations showed that the corresponding chlorophenols were also present in the jute. It was concluded that the chloroanisoles had been formed from the chlorophenols. The origin of the chlorophenols was unclear, but possible sources were thought to be either chlorophenols used for rot-proofing, or the reaction of chlorine with the jute fibres. Similar chlorophenols are known to be formed from the chlorine bleaching of wood pulp for papermaking.

In another investigation, cocoa powder imported into Australia was found to have a particularly disagreeable odour and taste. The cocoa had been packaged in multi-wall paper sacks. On examination, the sack material and the contents showed the presence of a number of chlorophenols. 2,4-Dichlorophenol was present in the cocoa at a concentration of $520 \, \mu g \, kg^{-1}$, 2,6-dichlorophenol at a concentration of $7 \, \mu g \, kg^{-1}$, 2,4,6-trichlorophenol at $73 \, \mu g \, kg^{-1}$, 2,3,4,6-tetrachlorophenol at $22 \, \mu g \, kg^{-1}$, and pentachlorophenol at $3 \, \mu g \, kg^{-1}$. The corresponding values for the chlorophenols in the sack material were 670, 41, 180, 350 and $520 \, \mu g \, kg^{-1}$, respectively. A further examination of the glued side seams showed that the levels of chlorophenols were generally higher—680, 100, 78, 1200 and 40000 µg kg^{-1}, respectively. The levels of chlorophenols in the cocoa powder were considered sufficient to cause the severe tainting.

Analyses for chloroanisoles in both the cocoa powder and the paper sacks showed these substances to be present in the cocoa and the glued seams of the sacks at significant levels, but the concentrations were negligible in the paper away from the glued seams. The most likely source of chloroanisoles was therefore the adhesive, and their formation was thought to have arisen from the action of moulds isolated in the vicinity of the glued seams on the chlorophenols.

It was concluded that the different concentrations of the individual chlorophenols in the paper and the glued seams of the sacks indicated that they were derived from two sources. 2,4-Dichlorophenol, 2,4,6-trichlorophenol and 2,3,4,6-tetrachlorophenol were known to be formed during the bleaching of wood pulp for paper manufacture, while pentachlorophenol was known to be used as a biocide in adhesives.

A third food product, which had been found to be contaminated by musty taints originating from packaging, and which was discussed by Whitfield and Last (1985), is dried fruit. The tainted fruit had been wrapped in polythene film and packaged in fibreboard cartons for transportation overseas. An analytical investigation on the tainted dried fruit showed the presence of three compounds with musty odours, namely 2,4,6-trichloroanisole, 2,3,4,6-tetrachloroanisole and pentachloroanisole. The concentrations of these chloroanisoles determined in a number of samples of dried fruit showed wide variations, with values of 0.2–12.3 µg kg^{-1} for 2,4,6-trichloroanisole, 0.1–42.0 µg kg^{-1} for 2,3,4,6-tetrachloroanisole and 0.2–6.5 µg kg^{-1} for pentachloroanisole. For all samples, however, the values for 2,4,6-trichloroanisole and/or 2,3,4,6-tetrachloroanisole were in excess of the threshold values.

Further investigation showed the source of these chloroanisoles, formed from the corresponding chlorophenols, to be the fibreboard cartons. It was first thought that the chlorophenols came from the adhesive, but it was ultimately considered that the source was recycled fibre (wastepaper pulp) containing newsprint. Fibreboard cartons made with virgin pulp were shown to have very low concentrations of chlorophenols and did not cause the characteristic musty off-flavour with the dried fruit.

Tindale and Whitfield (1989) recommended that only virgin kraft fibreboard should be used for dried fruit packaging and that the three main chlorophenols should be kept below specified levels, namely 20, 30 and 100 µg kg^{-1} for 2,4,6-trichlorophenol (TCP), 2,3,4,6-tetrachlorophenol (TeCP) and pentachlorophenol (PCP), respectively.

There have been several other incidents where chlorophenols and chloroanisoles responsible for food tainting have been found to be have originated from packaging. In some, the packaging has been the wooden pallets on which carton board has been stacked for delivery to the food manufacturer. The source of the chlorophenols has sometimes been traced to fungicidal preservatives used on the wood (Tindale and Whitfield, 1989).

7.7 Analytical methods for isolating and identifying odorous and tainting substances

The analysis of taints and off-flavours has already been discussed in chapter 3. In this section, therefore, the analysis of taints and off-flavours with particular reference to packaging materials will be described.

Gas chromatography is the most appropriate technique for the analysis of volatile odorous substances from packaging materials. However, where the chemical composition of the individual volatile substance or substances is not known, a mass spectrometer is usually used as the detector instead of the more conventional flame ionisation detector (FID).

The detection limit for a mass spectrometer total ion scan is typically a few nanograms. However, as mentioned in previous sections there is often a requirement to identify volatile substances that are present at very low levels—a concentration procedure is then necessary. The concentration technique most commonly used in reported investigation work for odorous volatiles from packaging materials is the dynamic headspace method (Soderhjelm and Eskelinen, 1985; Wyatt, 1986; Rojas De Gante and Pascat, 1990). With this technique a sample of the packaging material is placed in a vessel, which is closed and then purged for a period of time with an inert gas, such as nitrogen or helium. Volatiles from the packaging material sample are removed with the purge gas, collected and concentrated on a porous polymer. The packaging sample is sometimes heated to a low temperature to accelerate the release of the volatiles. Transfer of the collected volatiles to the gas chromatograph is performed by thermal desorption, or by solvent elution and injection as a solvent solution. Tenax GC (Enka, The Netherlands) is the porous polymer most often used, but for some of the more volatile substances Porapak Q and Porapak R (Waters Associates) are more appropriate.

The combination of the dynamic headspace sampling technique with gas chromatography/mass spectrometry enables very low concentrations of volatiles from packaging materials to be identified. If, however, the identified substances do not have well-established odour characteristics, then it is necessary to determine which are responsible for the odour. To obtain this information a sniffing port is connected to a split at the end of the gas chromatograph column; this enables a 'nose' odour assessment to be made on each of the chromatography peaks.

In recent years a Fourier transform infrared (FTIR) spectrometer has been used as an alternative, or in addition to, the mass spectrometer for identifying the chemical composition of volatile substances. As the FTIR spectrometer is

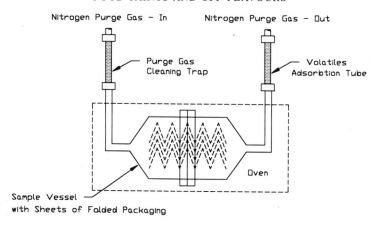

Figure 7.3 Diagram of apparatus for collecting volatile substances released from a sample of packaging material.

a non-destructive detector, in contrast to the mass spectrometer, the two detectors can be connected in series to provide two separate but complementary means of chemical identification. Alternatively, the two detectors can be used in parallel with either: (i) two individual columns for each of the detectors connected to a split at the injector; or (ii) one column, with a split at the end of the column for connection to the two detectors.

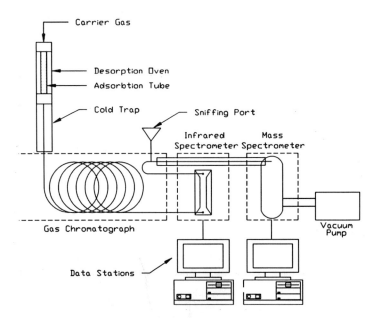

Figure 7.4 Schematic diagram of gas chromatograph with thermal desorption unit and dual Fourier transform infrared spectrometer and mass spectrometer detectors, plus odour sniffing port.

Figure 7.3 is a diagram of the apparatus used for collecting volatile substances released from a sample of packaging by trapping on porous polymer. Figure 7.4 is a diagram of the apparatus used for analysing the collected volatiles. This consists of a thermal desorption oven with cryofocusing cold trap, a gas chromatograph with infrared spectrometer, mass spectrometer and sniffing port detectors, plus data handling stations. The infrared detector is able to provide specific data on chemical functional groups, such as the carbonyl groups in aldehydes, ketones and carboxylic acids, and also on molecular geometry, which helps to distinguish between isomers. Infrared data is, however, poor for the identification of specific members of a homologous series, whereas the mass spectrometric data can often provide the necessary information. On the other hand, the mass spectrometer is not usually able to distinguish between isomers. Detection limits with the infrared spectrometer are usually not as good as those obtained with the mass spectrometer, but can still be in the low nanogram range.

Gas chromatograms of volatiles collected from a UV printed carton board and analysed by means of a combined Fourier transform infrared spectrometer (FTIR) and mass spectrometer (MS) in series are shown in Figures 7.5 and 7.6. Figure 7.5 contains the MS chromatogram, together with the mass spectrum of the peak at 16.4 min, which is benzophenone. Figure 7.6 contains the corresponding FTIR chromatogram, with the FTIR spectrum of the 16.35 min peak, which is again benzophenone. The slight difference in retention times is due to the time taken for the substance to transfer to the second detector. The patterns of peaks in the two chromatograms are similar, but the relative heights of some peaks differ. This is due to the different spectral characteristics of the substances and is most noticeably seen with the peak at 8.2 min in the

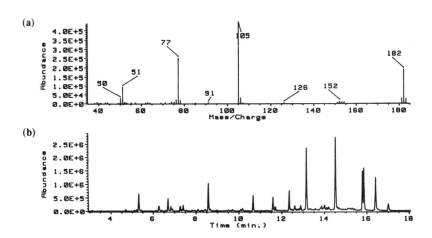

Figure 7.5 (a)   Mass spectrometer total ion chromatogram of volatiles from UV printed carton board; (b) mass spectrum of benzophenone.

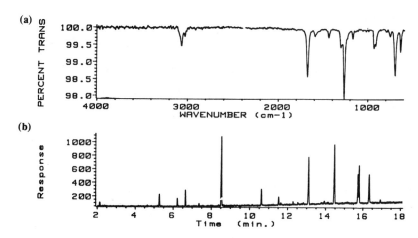

Figure 7.6 **(a)** Infrared spectrometer total response chromatogram of volatiles from UV printed carton board; **(b)** infrared spectrum of benzophenone.

FTIR chromatogram and the peak at 8.25 min in the MS chromatogram, which is benzaldehyde.

A variety of other concentration techniques can be used, including steam distillation, solvent extraction, solid phase isolation and the Likens–Nickerson combined distillation extraction procedure. Some of these techniques are more applicable to the less volatile substances. High-performance liquid chromatography has been used in some analytical investigations of tainting substances, particularly the less volatile and where the identity of the substance is known, such as the chlorophenols (Shanz-zhi and Stanley, 1983). High-performance liquid chromatography can be coupled to mass spectrometry for identification of the tainting and odorous substances, but this combination is less well-established than the corresponding gas chromatography/mass spectroscopy technique.

Jennings and Filsoof (1977) carried out a comparative study on various sampling techniques used to isolate trace volatiles for subsequent gas chromatography analysis. Their conclusion was that no single sampling procedure gave satisfactory results for all types of samples.

Koszinowski and Piringer (1986) described a procedure for the isolation of odorous substances originating from packaging where a large proportion of the total volatiles consisted of relatively non-odorous hydrocarbons. Water was used to preferentially absorb the odorous substances, such as carbonyl compounds. The water-absorbed volatiles were removed with gas stripping in a countercurrent column and re-absorbed into a solvent, such as hexane. After concentration of the hexane solution, analysis and identification was carried out by gas chromatography/mass spectrometry.

References

Adcock, L. H. and Paine, F. A. (1974). Some aspects of odour retention by laminated and coated packaging materials. *Report IR309*, Pira International, UK.

Anon (1989). Intelligent odour discriminating nose. *Electronics and Wireless World* February, 178–180.

ASTM (1986). *Designation: F 151–86. Standard Test Method for Residual Solvents in Flexible Barrier Materials.*

ASTM (1988). *Standard: E 460–88. Standard Practice for Determining Effect of Packaging on Food and Beverage Products During Storage.*

ASTM (1989a). *Standard: E 462–84 (Reapproved 1989). Standard Test Method for Odor and Taste Transfer from Packaging Film.*

ASTM (1989b). *Standard: E 619–84 (Reapproved 1989). Standard Practice for Evaluating Foreign Odors in Paper Packaging.*

Azuma, K. (1983). Identification of the volatiles from low density polyethylene film irradiated with an electron beam. *Agri. Biol. Chem. J.* **47** (4), 855–860.

British Carton Association (1956). *Odour from Paper and Board—Notes for Guidance.*

BSI (1971). *Methods of Test for Assessment of Odour from Packaging Materials used for Foodstuffs.* British Standard 3755: 1964 (amended March 1971).

BSI (1984). *Monitoring the Levels of Residual Solvents in Flexible Packaging Materials.* British Standard 6455: 1984.

Cadbury (1991), private communication.

Claxton, R. (1981). The problem of ink-caused odors in the printing of foil. *Package Printing* **28** (11), 14 and 40.

Crosby, D. G. (1981). Environmental chemistry of pentachlorophenol. *Pure and Applied Chemistry* **53**, 1052–1080.

CSIRO Division of Dairy Research (1969). Examination of taints caused by shipping containers. *CSIRO Australia Annual Reports*, p. 35.

Dietz, F. and Traud, J. (1978). Geruchs- und Geschmacks-Schewellen-Konzentrationen von Phenolkorpen, Gas-Wasserfach. *Wasser-Abwasser* **119**, 318–325.

DIN (1983). *Sensory Analysis, Testing of Container Materials and Containers for Food Products.* DIN 10955, April 1983.

Donetzhuber, A. (1982). Characterization of pulp and paper with respect to odour. *Conference Proceedings of the International Symposium on Wood and Pulping Chemistry.* Vol. 4. The Ekman Dags, Stockholm, pp. 136–138.

Dravniecks, A. and Bock, F. C. (1978). Comparison of odors directly and through profiling. *Chemical Senses and Flavour* **2**, 191–225.

European Communities (1976). Council Directive 76/893/EEC relating to materials and articles intended to come into contact with foodstuffs. *Official Journal of the European Communities* No. L340, 9 December.

European Communities (1989). Council Directive 89/109/EEC relating to materials and articles intended to come into contact with foodstuffs. *Official Journal of the European Communities* No. L40, 11 February.

European Communities (1991). Council Directive 91/73/EEC amending for the ninth time Directive 76/769/EEC on the approximation of the laws, regulations and administrative provisions of the Member States relating to restrictions on the marketing and use of certain dangerous substances and preparations. *Official Journal of the European Communities* No. L85 5 April.

Fazzalari, F. A. (ed.) (1978). *Compilation of Odor and Taste Threshold Values Data.* American Society for Testing and Materials DS 48A, Philadelphia.

Finnish Pulp and Paper Research Institute (1987). *Method used at the Finnish Pulp and Paper Research Institute and at the Technical Research Centre of Finland,* VTT-4317-87.

Franz, R., Kluge, S., Lindner, A. and Piringer, O. (1990). Cause of catty odour formation in packaged food. *Packaging Technology and Science* **3** 89–95.

van Gemert, L. J. (1984). *Compilation of Odour Threshold Values in Air.* Report No. A 84.220/090070. CIVO-TNO, Zeist, The Netherlands.

Gilbert, J. and Startin, J. R. (1983). A survey of styrene monomer levels in foods and plastic packaging by coupled mass spectrometry–automatic headspace gas chromatography. *J. Food Sci. Agric.* **34** 647–652.

Goldenberg, N. and Matheson, H. R. (1975). 'Off-flavours' in foods, a summary of experience: 1948–74. *Chemistry and Industry*, 551–557.

Gorrin, R. J., Pofahl, T. R. and Croasmun, W. R. (1987). Identification of the musty component from an off-odor packaging film. *Anal. Chem.* **59** (18), 1109A–1112A.

Griffiths, N. M. (1974). Sensory properties of the chloroanisoles. *Chemical Senses and Flavor* **1**, 187–195.

Guarino, J. P. and Ravijst J. P. (1988). Low odor UV cure coatings. *J. Radiation Curing* July, 2–10.

Harvey, H. G. (1963). *Survey of Odour in Packaging of Foods*. The Institute of Packaging, UK.

Holman, R. and Oldring, P. (1988). *UV and EB Curing Formulations for Printing Inks, Coatings and Paints*. Publ. SITA–Technology, London, pp. 72 and 164.

Jenne, H. (1980). Polystyrol und polypropylen als Spritzguss und Teifziehmaterial für Molkerei-produkte. *Deutsche Molkerei Zeitung* **101**, 1906–1910.

Jennings, W. G. and Filsoof, M. (1977). Comparison of sample preparation techniques for gas chromatographic analysis. *J. Agric. Food Chem.* **25** (3), 440–445.

Kontominas, M. G. and Voudouris, E. (1982). The determination of residual solvents in plastics packaging materials in relation to off odors developed in packaged bakery products. *Chimika Chronika New Series* **11**, 215–223.

Koszinowski, J. and Piringer, O. (1986). Evaluation of off-odors in food packaging—The role of conjugated unsaturated carbonyl compounds. *J. Plastic Film and Sheeting* **2** (January), 40–50.

Koszinowski, J., Muller, H. and Piringer, O. (1980). Identifizierung und quantitative Analyse von Geruchsstoffen in Latex-gestrichenen Papieren. *Coating* **13**, 310–314.

The Institute of Packaging, UK (1960). *Odour in Packaging—Conference Report*.

ISO (1983). ISO Standard 4120–1983: *Sensory Analysis—Methodology—Triangular Test*.

Linssen, J. P. H., Legger-Huysman, A. and Roozen, J. P. (1990). Threshold concentrations of migrants from food packaging: styrene and ethyl benzene. *Flavour Science and Technology* (proceedings of the 6th Weurman Symposium, Geneva). Eds Y. Bessiere and A. F. Thomas, John Wiley and Sons, Chichester, pp. 359–362.

Linssen, J. P. H., Janssens, J. L. G. M., Reitsma, J. C. E. and Roozen, J. P. (1991). Sensory analysis of polystyrene packaging material taint in cocoa powder for drinks and chocolate flakes. *Food Additives and Contaminants* **8** (1), 1–7.

Lorusso, S. (1985). Formalising an approach to acetaldehyde. *Food Processing* **54** (11), 43–44.

Mazzocchi, B. (1988). Faut-il se mefier du PET. *Emballage Digest* No. 322, May 1988, pp. 166–168.

Miltz, J., Elisha, C. and Mannheim, C. H. (1980). Sensory threshold of styrene and the monomer migration from polystyrene food packages. *J. Food Processing and Preservation* **4**, 281–289.

Passy, N. (1983). Off-flavours from packaging materials in food products, some case studies. In *Instrumental Analysis of Foods*. Vol. 1. Academic Press, Inc., New York, pp. 413–421.

Patterson, R. L. S. (1968). Catty odours in food: confirmation of the identity of 4-mercapto-4-methyl-pentan-2-one by gas chromatography and mass spectrometry. *Chemistry and Industry*, 48.

Perkin Elmer Ltd., UK (1991). *Thermal Desorption Applications, No. 26—Thermal Desorption of Volatiles from Food Packaging Film*.

Pilborough, D. J. (1987). The printers' curse: odour and taint from packaging print. *Packaging News (Aust)*, November, 28–29.

Ramshaw, E. H. (1984). Off-flavour in packaged foods. *CSIRO Food Research Quarterly* **44**, (4), 83–88.

Reid, T. L. (1983). Inks for food packaging. *Folding Carton Industry* April, 4–5.

Robinson, L. (1964). *Transfer of Packaging Odours to Cocoa and Chocolate Products*. Analytical Methods of the Office International du Cacao et du Chocolat, 12-E.

Rojas De Gante, C. and Pascat, B. (1990). Effects of β-ionizing radiation on the properties of flexible packaging materials. *Packaging Technology and Science* **3**, 97–115.

Sandberg, E., Vaz, R., Albanus, L., Mattsson, P. and Nilsson, K. (1982). Migration of plasticisers from PVC films to food. *Var Foda* **34** 470–482.

Shang-zhi, S. and Stanley, G. (1983). High performance liquid chromatographic analysis of chloro-phenols in cardboard food containers and related materials. *J. Chromatogr.* **267**, 183–189.

Shurmer, H. V. (1987). Development of an electronic nose. *Physics in Technology* **18**, 170–176.

Soderhjelm, L. and Eskelinen, S. (1985). Characterization of packaging materials with respect to taint and odour. *Appita* **38** (3), 205–209.

Soderhjelm, L. and Parssinen, M. (1985). The use of descriptors for the characterisation of odour in packaging materials. *Paperi ja Puu—Papper och Tra* No.8, pp. 412–416.

Solin, C. F., Fazey, A. C. and McAllister, P. R. (1988). Which plastic package has the lowest contribution to taste and odor? *J. Vinyl Technol.* **10** (1), 30–32.

Tindale, C. R. and Whitfield, F. B. (1989). Production of chlorophenols by reaction of fibreboard and timber components with chlorine based cleaning agents. *Chemistry and Industry*, 835–836.

UK Statutory Instrument (1987). *The Materials and Articles in Contact with Foodstuffs Regulations 1987*. No. 1523. HMSO, London.

Whitfield, F. R. and Last, J. H. (1985). Off-flavours encountered in packaged foods. In *The Shelf Life of Foods and Beverages*. Ed. G. Charalambous. Elsevier, Amsterdam, pp. 483–499.

Wyatt, D. M. (1986). Analytical analysis of tastes and odors imparted to foods by packaging. *J. Plastic Film and Sheeting* **2**, 144–152.

8 A retailer's perspective

D. A. CUMMING, K. SWOFFER and R. M. PASCAL

8.1 Introduction

Previous sections of this book have dealt with the technical distinction between a taint and an off-flavour. To consumers, however, it does not matter how the failure is described. The product is equally unacceptable.

The retailer is in the front line; consumers will invariably refer complaints about taint or off-flavour to the retailer, whether the product is a retailer's own brand or a supplier's branded product. Despite the many procedures operated to ensure that 'own brand' products are produced only from suppliers working to good manufacturing practice and that products branded are sourced only from reputable suppliers, problems of taint and off-flavour do occur; fortunately, however, these problems are comparatively rare. The experience within Safeway is that approximately four taint/off-flavour problems are dealt with each year. Some of these problems are transient in nature and are never satisfactorily resolved. Others are brought to a satisfactory conclusion with full identification of the origins of the taint/off-flavour.

Retailers have a legal obligation to provide foods of the nature, substance and quality demanded by the consumer; clearly, tainted or off-flavoured food does not meet that obligation. Prevention of taint or off-flavour becomes an integral part of the 'all reasonable precautions' and 'due diligence' provisions in food law. Inevitably some complaints are received through the Enforcement Agencies and this generally increases the profile of the complaint. Often there is the implied threat of legal proceedings even before the investigation has begun, and this can lead to the risk of over-reaction. Current trends in consumer reaction to food complaints frequently result in the involvement of police or media, which can escalate the reaction and promote 'panic' withdrawals.

The commercial implications are, therefore, highly significant and should not be under-estimated. Any complaint on an 'own brand' product will have the potential to depress sales on other such products. Any product problem that proves to be serious and widespread, necessitating both shelf withdrawal and recall from consumers, carries with it the stigma of failure on the part of the retailer, regardless of the cause or origins of the taint/off-flavour. Inevitably this has an adverse effect on sales volume, at least temporarily, both of the brand and the particular product. The effect on individual consumers will be varied. Many consumers quite genuinely develop an aversion to the food in question or heightened sensitivity to certain flavours; indeed many solicitors acting on behalf of aggrieved consumers place much emphasis on the trauma of the event.

All this points the way to ensuring that adequate procedures are in place to prevent failures, or to pick up potential problems at an early stage. The importance of organoleptic testing at critical stages must not be underestimated and must form part of the overall hazard analysis critical control point (HACCP) procedures for the product. Testing must extend from raw materials, through processing stages, to acceptance testing on receipt at the retail distribution point. Unfortunately, many problems that could have been prevented have been missed because organoleptic tests have received far less priority than compositional values or other physical properties. It is all too easy to assume that if it looks acceptable and the laboratory results are satisfactory, then it must taste right. This has often proven to be flawed thinking. Yet, the majority of taint incidents could have been prevented by tasting at the critical control point.

8.2 Product recall

Product failure, for whatever reason, is an inevitable risk, and any food business must be prepared for it, plan for it and have robust procedures to effectively deal with the problem. The stages in dealing with problems are dealt with in Safeway in the following manner:

(i) Establish if a problem exists.
(ii) Identify what the problem is.
(iii) Estimate the extent of the problem.
(iv) As soon as a potential problem is identified, stop further stock movement from supplier and from distribution centres.
(v) If the problem is serious enough, remove from sale in stores/branches.
(vi) Decide if recall from consumers is necessary—if so set in motion the major incident routine.

Current communications technology makes it relatively easy to deal with product stockholds and withdrawal from shelf in stores/branches.

In practice the procedures for dealing with taint/off-flavour would be as outlined in the following sections.

8.2.1 Does a problem exist?

For Safeway there are six principal sources of information that may identify taint or off-flavour incidents (and any other cause of customer complaint). These are:

  (i) Customer complaints to stores
 (ii) Customer complaints to head office
(iii) Information from supplier
(iv) Store notifications to head office
 (v) Routine intake product checks at distribution centres.
(vi) Enforcement Agencies.

It is essential that all customer complaints about products are recorded, no matter how minor they might appear, and that this information is collated and analysed frequently by product, by supplier, and by nature of complaint. Complaints via stores and central office departments must be cross-referenced and interpreted for trends that may reveal the existence of a flavour problem.

In addition to the complaint record system, Safeway operates a quality assurance hotline for stores, which encourages early communication of potential problems including 24 h coverage via radio pager. All incidents and reports are recorded and have to include all details of coding, product, origin, etc. Single complaints of flavour problem will not be immediately acted upon (unless of a very serious nature), but two or more similar reports on the same product will result in follow-up action.

8.2.2 What is the problem?

Follow-up action will require products to be returned to the technical centre by the quickest method (e.g. datapost, courier, collection by member of staff, etc.). Where sufficient sample is available, full evaluation will take place, covering both organoleptic and analytical parameters. Samples of product from the same codes, same store, or randomly from the same supplier will be drawn in for comparative evaluation.

At this stage, the supplier will be notified and the nature and seriousness of the problem will be identified if possible. In this respect mention must be made of the excellent 'emergency services' available from the food research associations and other specialist test houses, which provide rapid response facilities for microbiological or chemical analysis where in-house facilities are not available or appropriate. They can also provide specialist sensory panels for specific categories of taint, which generally the retailer cannot provide. The first priority is to establish whether or not a taint or off-flavour exists. It is then necessary to identify wherever possible what has given rise to the problem so that remedial action can be taken.

8.2.3 How widespread?

The benefits of clear batch or lot marking are never more evident than when faced with potential recall. If all complaint samples relate to one code or one batch number, and the supplier can give assurance that the nature of the problem is likely to be related to a batch, or short production period, the product withdrawal can be confined to those specific codes. If any doubt exists, however, then all of that product, i.e. all codes and sizes, will be withdrawn.

8.2.4 Stockholding

As soon as a product failure is identified suppliers will be told to prevent deliveries and, if the problem appears serious, distribution centres will prevent further product issues to stores.

8.2.5 Store/branch product removal

Computer messages will be sent to all stores, requiring immediate withdrawal of code, batch or all stocks as appropriate. Removed product will be isolated in the warehouse and identified, by use of hazard warning tape, as being stock not for sale.

8.2.6 Major incidents

For the very serious failures a full major incident plan would be put into operation, involving notification to media, local authorities, etc. In these instances it is critical that proper arrangements are set up to collate information about returned product. Knowledge of quantities produced and in circulation need to be matched against returns to gauge the success of the recall coverage, with further media messages where necessary. Adequate briefing of store management, customer relations departments and technical departments are essential to reassure consumers and co-ordinate information.

8.3 Case histories

Some of the taint/off-flavour incidents that Safeway have dealt with over the past five years are discussed in the following sections.

UHT milk phenolic/disinfectant flavour. Spasmodic complaints in clusters over a 2–3-year period. Vague description by consumers of TCP or disinfectant-like flavour. There were no more than 12 complaints in any one incident and only very sensitive palates could detect the off-flavour. Analysis did not show the presence of phenolic substances (at the level of detection), no reasons were identified at suppliers, and there was no resolution of the problem. There were specific code withdrawals but none was classed as a major incident.

Sterilised milk chocolate drink. Complaints of off-flavour, and burning sensation when consumed on two separate occasions. These resulted in specific code withdrawals and the problem was traced to 'burnt' taint developing from the product recycle tank (feedback tank from steriliser).

Smoked salmon.   Off-flavours resulting from microbiological problems. All codes from one supplier were withdrawn over the peak Christmas selling period. Poor raw material, inadequate curing and poor salt distribution were factors contributing to this problem.

Mince pies—frozen.   Large quantities exhibiting a 'cardboard' taint withdrawn from stores. The product was frozen down in its final packaging. When 'recovered' prior to distribution, cardboard taints were found. This was not satisfactorily resolved and the taints could only be picked up by certain panellists.

Mince pies—ambient.   Large quantities withdrawn due to a 'solvent' taint. The problem was traced to inadequate 'airing' of the carton before packing the product (residual solvent from the ink present).

Canned beans in tomato sauce.   Two incidents of taint involving significant code withdrawals. One was traced to contamination of raw beans in transit, the other to can coating problems. The transit problems originated from a dirty shipping container. The can coating problem arose from reversal of the tin plate during forming, i.e. a higher tin coating weight was on the outside of the can.

Frozen peas.   Phenolic taint resulting in multi-code withdrawal, traced to unsuitable process equipment. Rubber hosing that was not food grade was used between tanks, and reacted with chlorinated water.

Canned spaghetti in tomato sauce.   Metallic taint due to lower than normal tin coating on the internal can body—easily stripped by the product.

Soft drinks in PVC.   Many taint problems in the early days of PVC bottles. Metallic impact modifiers that were incorporated in the plastic pre-mix reacted with the product.

Potatoes—pre-packed in polythene bags.   Solvent taint from ink used on the bags—not adequately 'aired' before use.

The taint that wasn't.   As an example of how incidents can escalate, a call on a Saturday from a store was received by radio pager. The problem reported was that a child had consumed some canned blackcurrant drink that tasted 'funny' and caused serious swelling to lips, tongue and throat. Doctors and the local hospital were informed. The store was asked to check other cans of the same code of product, and reported a strange flavour. Suppliers were contacted, and all product was taken off sale.

The supplier collected the part-consumed product and, after microbiological and chemical analysis and organoleptic testing, could find no problem. Investigation at the consumer's home revealed that the child occasionally reacted adversely to blackcurrant products. Further checks at the store revealed that the member of staff who tasted the product did not normally consume blackcurrant products—this was the reason for the comment on the flavour.

In summary, therefore, there was nothing actually wrong with any of this product. Although a good test of the incident routine, this example highlights how easily a taint complaint can generate over-reaction.

8.4 Avoidance of taint

Procedures for avoidance of taint and off-flavour at the manufacturing stages are dealt with elsewhere in this book. However, it is appropriate to reiterate the need for identification of the critical control points relating to product flavour. Organoleptic tasting must be given high profile in any HACCP approach to food manufacturing.

Having said this, the retailer has a significant role to play. In the first instance they must deal with reputable suppliers and those known to be operating to GMP. Establishing close relationships with suppliers is essential, as is the establishment of detailed and realistic product specifications for 'own brand' products. This includes detailed raw material and packaging specifications (sourcing and grade), since many taint/off-flavour problems have originated from these sources. The working relationships must extend to a good understanding of supplier capability, such that demands on delivery times or volume of product do not encourage compromise on production parameters.

Most major retailers operate supplier quality assurance procedures designed to ensure that proper selection of suppliers is made and that adequate technical liaison exists between the producer and retailer. The Safeway procedure is to source 'own brand' product only from suppliers that pass a stringent audit prior to initial approval. Product specifications are then agreed, and full sensory evaluation and shelf-life testing completed before product launch approval is given. Technical personnel responsible for the initial audit revisit the supplier at intervals (on a risk assessed frequency basis dependent on the knowledge of supplier and risk presented by the product). The same technical personnel deal with interpretation of complaint statistics and feedback of relevant information to their suppliers.

The development stages of an 'own brand' product include full shelf-life evaluation—an area that requires very careful consideration to avoid problems with oxidative or microbial action during the life of the product. For retailers

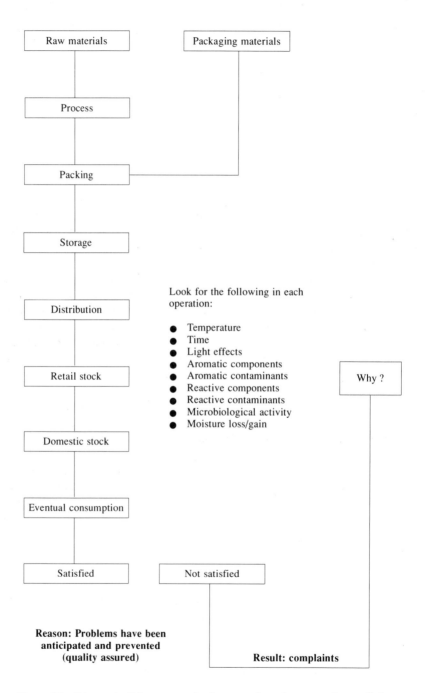

Figure 8.1 Taints and off-flavours—a simple process hazard assessment or prediction.

the additional responsibilities come into play once the product leaves the supplier, with the additional hazards of transportation and distribution. Cross-contamination between incompatible products in vehicles is not unknown, and both food (e.g. onions and pineapple) and non-food products (detergents etc.) have been known to produce off-flavours. Transportation of foodstuffs in vehicles not normally used for foods is an obvious hazard and poorly cleaned vehicles can be a source of contamination.

Temperature abuse during transportation can contribute to microbiological problems. It is essential, therefore, for retailers to carry out stringent checks on receipt of goods at distribution or at store. These checks must extend to vehicle condition and suitability and, for temperature-controlled goods, product temperature history. During storage in retail warehouses, and during order picking and collation, consideration must be given to adequate segregation of foods from non-foods likely to give rise to odour transfer, e.g. detergents. Regular cleaning of distribution centres and storage areas, spillage control, careful pallet selection, and vehicle interior cleaning are all part of the good distribution practice necessary to minimise taint/off-flavour.

In stores/branches the same attention to detail concerning product segregation, cleaning and spillage control is essential. Within stores, the selection of janitorial chemicals is as important as it is in manufacturing, particularly where any non-prepackaged foods are handled. Temperature control on perishable goods is essential to minimise risks of microbial action. Adequate training of checkout personnel can also contribute to minimising off-flavour risks if a sensible approach is taken when 'bagging up' consumer purchases. Retailers can also undertake consumer education to complete the chain from manufacture to consumption, although it would be true to say that very few problems of taint/off-flavour have been traced to domestic origins.

The retailer is one link in the food chain. Prevention of product problems requires continual vigilance at all stages, from raw material production through to display of product in the retail store, and requires close working relationships between all sections of the industry. Figure 8.1 shows a simple process hazard assessment in which none of the stages can be ignored and to which the retailer must contribute at every stage.

Acknowledgement

The authors wish to acknowledge Safeway Stores plc for help during the preparation of this chapter.

9 Formation of off-flavours due to microbiological and enzymic action

M.B. SPRINGETT

9.1 Introduction

Off-flavours which arise in foods due to microbiological or enzymic activity fall into three main categories. These are described in the following sections.

9.1.1 Off-flavours preformed in the food

Off-flavours may be preformed in the food due to normal biochemical metabolism or as stress metabolites. These types of off-flavour are dependent to a large extent on agronomic factors such as varietal differences, feeding or fertilizer regimes, level of water used, spacing, etc.

9.1.2 Off-flavours formed as a result of cellular disruption

After harvest or slaughter the structural integrity of foods begins to break down due to damage from handling and normal decay processes. This results in the mixing of normally compartmentalized enzymes and substrates, and the generation of flavour compounds. In many cases the formation of compounds in this way is essential for the typical flavour characteristics of the product, e.g. onions, garlic, peppers, etc. However, in many instances this leads to the formation of off-flavours and spoilage of the product.

9.1.3 Off-flavours arising as a consequence of microbial deterioration

Microbial spoilage is a major cause of quality loss in foods after harvest or slaughter. As a by-product of growth and metabolism microorganisms produce a range of chemicals which alter the quality attributes of food, including flavour, ultimately rendering the product inedible.

This chapter reviews certain off-flavours which arise as a consequence of microbial or enzymic action in milk and dairy products, fruit and vegetables, wine, beer, meat and fish.

9.2 Milk and dairy products

Extended refrigerated storage and transport has resulted in new quality problems for milk, related to the growth and metabolic activities of microorganisms at low temperatures. Psychrotrophic microorganisms are ubiquitous in nature and common contaminants in milk. Although they have an optimum growth temperature of 20–30°C, psychrotrophic bacteria will grow at temperatures as low as − 10°C, with moulds growing at temperatures as low as − 18°C (Kraft and Rey, 1979). During microbial growth heat-stable enzymes are formed. These biochemically alter the milk, eventually causing spoilage. Two types of enzymes are particularly important in the formation of off-flavours in milk and milk products. These are (i) lipases; and (ii) proteinases.

9.2.1 Lipases

Lipase activity has been reported for most psychrotrophs isolated from milk and milk products. *Pseudomonas, Flavobacteria* and *Alcaligenes* species are the most active lipolytic bacteria (Muir *et al.*, 1979). Microbial lipases are heat stable and it has been reported that 40% of *Pseudomonas, Alcaligenes* and *Aerobacter* species retained 75% of activity after 2 min at 90°C (Kishonti, 1975). Lipase activity in milk leads to the preferential release of medium- and short-chain fatty acids from triglycerides (Olivecrona and Bengtsson-Olivecrona, 1991), hydrolysis of as little as 1–2% triglycerides leading to rancid off-flavours. Milk naturally contains high levels of indigenous lipase, with 1 l of bovine milk containing 1 mg of pure lipase (Olivecrona and Bengtsson-Olivecrona, 1991). It is therefore extremely likely that indigenous as well as microbial lipases are important in the development of lipolytic rancidity in milk (Allen, 1989).

9.2.2 Proteinases

The major cause of bitterness in milk and milk products is the formation of bitter peptides due to the action of proteinases. Although bitterness is usually indicative of microbial spoilage, indigenous milk enzymes and the activity of starter microorganisms used in cultured milk products may also lead to bitter off-flavours. Proteinase activity has been detected in many bacterial species, in particular *Pseudomonas, Aeromonas, Serratia* and *Bacillus* species (Sorhaug and Stepaniak, 1991). Heat stability of proteinases from several bacterial species was investigated by Griffiths *et al.* (1981). They found that 20–60% of original proteinase activity remained after heating cell-free culture supernatants for 5 s at 140°C. Strict quality control is therefore critical in UHT milk products to ensure that heat-stable proteinases do not cause bitter off-flavours.

The most investigated source of bitter peptides is the caseins. Ney (1979) showed bitterness to be related to the hydrophobicity of casein-derived showed bitterness to be related to the hydrophobicity of casein-derived peptides rather than to specific amino acids or chain length. As a measure of hydrophobicity Ney further suggested the Q value, which is the average free energy for the transfer of the amino acid side chains from ethanol to water. Peptides with a Q value over $+1400$ are bitter, those with a Q value below $+1400$ kcal/mole are non-bitter. This rule has been assessed and verified for small peptides, although there are some exceptions; in particular, peptides containing high levels of glycine tend not to obey the rule (Adler-Nissen, 1986).

9.3 Fruit and vegetables

9.3.1 Citrus fruits

A flavour defect which constitutes a major problem worldwide to the citrus industry is bitterness. Maier and Beverley (1968), and Maier and Margdeth (1969) determined the mechanism of formation of bitter limonin in citrus fruits. They showed that intact fruits were not bitter and did not contain limonin itself but a non-bitter precursor, limonoate A-ring lactone. When juice is extracted this non-bitter precursor is slowly converted to limonin under acidic conditions and is accelerated by the presence of limonin D-ring lactonase (EC 3.1.1.36) (Maier *et al.*, 1969). Maier *et al.* (1980) suggest that the precursor limonoate A-ring lactone is synthesized in the leaves, from where it is transported to the seeds and converted into limonin. It is likely that other limonoids, and flavanone glycosides, e.g. naringin, may also contribute to bitter off-flavours in citrus fruits, although in grapefruit the bitter note due to naringin is essential to its expected flavour. The structures of limonin and naringin are shown in Table 9.1.

Medicinal-like off-odours have also been detected in certain lemon juices, whenever small quantities of Meyer lemon juice oil are present (Wenzel *et al.*, 1958). This has been attributed to high levels of thymol, which occurs as a normal volatile aroma component in the Meyer lemon (Moshonas *et al.*, 1972).

9.3.2 Legumes

The enzyme lipoxygenase (EC 1.13.11.12) is believed to be ubiquitous amongst eukaryotic organisms (Whitaker, 1991). This enzyme poses a particular problem in legumes such as soy beans, winged beans, lentils, green beans, etc., giving rise to a range of off-flavours described variously as beany, grassy and rancid. Flavour problems associated with lipoxygenase have been summarized recently by O'Conner and O'Brien (1991).

Table 9.1 Summary of bitter compound structures

Product	Bitter compounds
Milk and milk products	Peptides
Citrus	Limonin
	Naringin
Cucumber, squash, etc.	Cucurbitacin E
Lettuce	Lactucin
	R=H
	Lactucopicrin
	R = -COCH₂ — ⬡ — OH
Brussels sprouts	Goitrin

R = -COCH$_2$ —〈 〉— OH

There has been considerable interest in lipoxygenase in soy beans, related to the formation of volatile aroma compounds. Soy beans contain high levels of lipoxygenase, constituting 1–2% of the protein (Axelrod et al., 1981). The oil fraction contains 55% linoleic acid and 8% linolenic acid, which are substrates for lipoxygenase. Volatiles produced lead to a range of off-flavours described variously as grassy, beany, rancid, etc. The initial products of lipoxygenase activity are hydroperoxides, which undergo rapid degradation to give a range of compounds including aldehydes, ketones and alcohols. In soy beans, formation of n-hexanal, n-hexanol, n-pentanol and n-heptanol are the major causes of quality loss (O'Conner and O'Brien, 1991).

The 'viney' off-flavour which develops in green peas (Pisum sativum) is thought to be due to volatile products of lipoxygenase activity (Lee and Wagenknecht, 1958), while the strong beany aromas in winged beans are due to lipoxygenase-catalysed formation of hexanal and 2-pentylfuran (Mtebe and Gordon, 1987).

Off-flavours produced due to lipoxygenase activity are potentially a major problem for the commercial development of many legume products. Heat treatments to inactivate the enzyme are the most widely used control measures, both canned and freeze-stored legumes being water or steam blanched prior to storage.

9.3.3 Brassicas

Perhaps the most important group of compounds to the flavour of Brassicas are the volatile degradation products of glucosinolates. To date, all members of the genus Brassica are known to contain glucosinolates (Kjaer, 1960). Brassicas also contain thioglucosidase (myrosinase) (EC 3.2.3.1) enzymes, which are released when the cells of the plant are disrupted and which then come into contact with the glucosinolates. This leads to a range of potent flavour compounds dependent upon: (i) the specific glucosinolate involved; and (ii) the conditions of reaction. In many instances this gives rise to desirable flavour notes; for example 2-propenylisothiocyanate, derived from the thioglucosidase-initiated breakdown of 2-propenyl glucosinolate (sinigrin), is an important flavour component of black pepper. However, the undesirable bitter note in Brassica oleraceae cultivars in particular Brussels sprouts has been attributed to the formation of goitrin (5-vinyloxazolidine-2-thione) from the glucosinolate progoitrin (2-hydroxy-3-butenyl glucosinolate) (Fenwick and Griffiths, 1981; Fenwick et al., 1983). This reaction is summarized in Figure 9.1.

9.3.4 Potato

Off-odours described as faecal, cheese-like, pigsty, cowpat and drain have been detected in samples of freeze-processed chips (Whitfield et al., 1982).

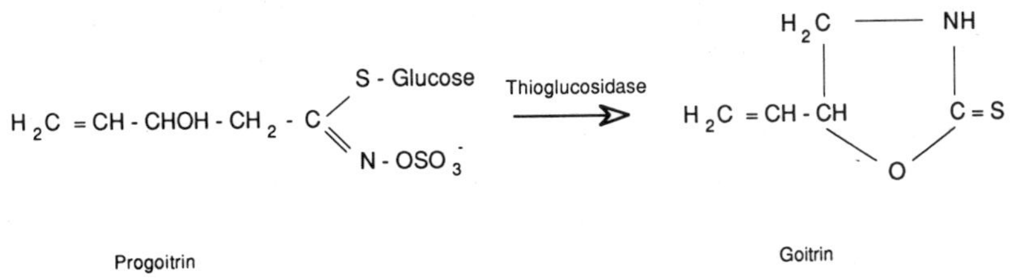

Figure 9.1 Formation of the bitter compound goitrin in Brussels sprouts.

Three compounds, *p*-cresol, indole and skatole, were identified as being responsible for this off-flavour. The source of these compounds was traced back to tubers infected with *Erwinia carotovora*, *Erwinia chrysanthemi* and *Clostridia* species, which cause soft rot. Metabolism of the amino acids tyrosine and tryptophan was the cause of the off-flavour.

9.3.5 Cucumber and lettuce

Bitterness is also a major off-flavour problem in members of the *Cucurbitaceae* and, to a lesser degree, the *Lactuceae*. In the case of the *Cucurbitaceae* there are around 750 species worldwide, many of which are rendered inedible due to the formation of the bitter tetracyclic terpenes – the cucurbitacins (Guha and Sen, 1975; Hutt and Herrington, 1985). These compounds are also toxic and there have been several cases of poisoning due to eating squashes (Stoewsand *et al.*, 1985; Rymal *et al.*, 1984). With lettuce a similar bitter off-flavour may arise, in this instance due to the sesquiterpene lactones, lactucin and lactucopicrin (Bachelor and Ito, 1973). The structures of these compounds are shown in Table 9.1.

9.4 Wine and beer

Enzymes are essential for the conversion of grape juice into wine, and malt extract into beer. This includes the formation of many flavour compounds, together with the typical ethanol component of these drinks. Off-flavours which arise are invariably associated with fermentation problems or microbial contamination.

9.4.1 Wine

Grass-like tastes which may arise in wines are caused by the formation of hexanal, *cis*-hexen-3-al and *trans*-hexen-2-al. These compounds are formed during the juice extraction phase due to the breakdown of membrane lipids. During the fermentation stage these aldehydes are converted to their equivalent alcohols. Although sensorically the alcohols are not as potent as the aldehydes, it is believed they can also contribute to the grass-like aroma (Villettaz and Dubourdieu, 1991). It has been shown that four enzymes are involved in the formation of C6 volatiles (Crouzet et al., 1985). These are:

1. Acyl hydrolases—act on membrane lipids to form linoleic and linolenic acids.
2. Lipoxygenase—forms hydroperoxides from the fatty acids. In particular C13 hydroperoxide from linoleic acid.
3. Hydroperoxide-cleaving enzyme—acts only on C13 peroxide to give the C6 aldehydes.
4. Alcohol dehydrogenase—catalyses the conversion of the C6 aldehydes to C6 alcohols.

A range of off-flavours, collectively known as 'cork taint', may arise in wines and spirits. The incidence of 'cork taint' is extremely high with estimates ranging from 2–6% of wines being rejected (Heimann et al., 1983; Carey, 1988). It is generally accepted that this type of taint arises due to microbial contamination of the cork, followed by leaching of products into the wine. The major cause of cork taint is believed to be 2,4,6-trichloroanisole (TCA) (Tanner et al., 1981). TCA can be formed from the microbial degradation and methylation of pentachlorophenol—a component of some insecticides or wood preservatives (Maarse et al., 1985). Alternatively, 2,4,6-trichlorophenol may be produced during the hypochlorite treatment of corks; subsequent microbial methylation would lead to TCA formation (see Figure 9.2). A range of microflora has been implicated in carrying out these changes, including *Streptomyces* species, *Aspergillus versicolor, Aspergillus repens, Pencillium* species and *Trichodermas* species (Lefebvre et al., 1983; Daly et al., 1984).

Bacterial contaminants present in brewers yeast can lead to off-flavours in the final beer product. Free-living *Klebsiella, Enterobacter* and *Hafnia* species have been shown to have decarboxylase activity, acting on compounds such as ferulic acid and *p*-coumaric acid to give volatile phenolics which cause a phenolic, medicinal-like off-flavour (Lindsay and Priest, 1975).

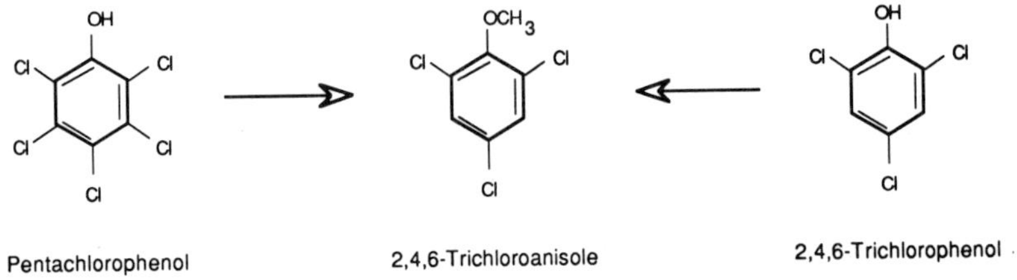

Pentachlorophenol   2,4,6-Trichloroanisole   2,4,6-Trichlorophenol

Figure 9.2 Formation of the musty compound 2,4,6-trichloroanisole.

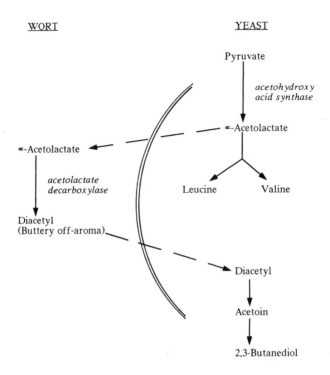

Figure 9.3 Diacetyl formation during beer fermentation. From Slaughter and Priest (1991).

Perhaps one of the most studied off-flavour problems in beer is the formation of the butter-like compound diacetyl. α-Acetolactate is an intermediate in the biosynthetic pathway from pyruvate to leucine and valine, within a normal metabolizing yeast. However, it is possible for α-acetolactate to pass into the bulk of the wort where chemical oxidative decarboxylation converts it into diacetyl. Given sufficient time the yeast will absorb the diacetyl and convert it via acetoin to 2,3-butanediol. However, if the yeast is removed, diacetyl will accumulate leading to the off-flavour. A summary of these reactions based on that of Slaughter and Priest (1991) is shown in Figure 9.3. Several approaches to the removal or prevention of diacetyl off-flavour have been suggested. These include the use of mutant yeast strains which do not possess acetohydroxy acid synthase activity, so preventing α-acetolactate formation (Slaughter and Priest, 1991). An alternative approach is to accelerate the decomposition of α-acetolactate using bacterial enzymes, such as α-acetolactate decarboxylase from *Enterobacter aerogenes*. In this case the appropriate gene for this enzyme has been incorporated into a lager yeast. Although the resulting organism showed no change in fermentation rate, the level of diacetyl produced was considerably lower (Sone *et al.*, 1988).

9.5 Meat and fish

Off-flavours which develop due to surface microbial contamination are major causes of spoilage in meat. Soil and faeces which accompany livestock into the slaughter house are undoubtedly the main source of contamination. During the subsequent slaughter, skinning, eviscerating, etc. there is ample opportunity for the spread and growth of these organisms (Dainty, 1991). Dainty *et al.* (1985) showed the first signs of spoilage to be caused by the formation of fruity, sweet-smelling esters, followed by the formation of putrid sulphur compounds. These workers identified *Pseudomonas* species as the main bacterial contamination. Sour flavours may also arise due to the formation of lactic acid from carbohydrate metabolism by several bacteria, including *Lactobacilli* and *Leuconostoc* species (Kramlich *et al.*, 1975). Many putrid odours arise due to decomposition of proteins and amino acids by anaerobic bacteria. Volatiles produced include indole, methanethiol, dimethyl disulphide and ammonia (Lawrie, 1974; Bailey and Murdoch, 1991). Rancidity problems which occur are usually due to oxidation of unsaturated lipids and are not associated with microbial growth (Kramlich *et al.*, 1975).

In the case of fish, Huss (1988) describes a four-phase pattern for the changes in flavour quality after harvest. In phase 1 the fish is fresh and has the delicate flavour of the particular fish type. In phase 2 there is a general loss of odour and taste. In phase 3 off-flavours such as sour, sweet, fruity, cabbage-like, ammoniacal, sulphury and strong-fish, begin to develop. Finally, in phase 4 these off-flavours reach an unacceptable level and the fish is spoiled. Initial microbial contamination and growth is by aerobes, which act on carbohydrates giving carbon dioxide and water. As the surface becomes covered and slime builds up, conditions become more favourable to the growth of anaerobes. Reduction of trimethylamine oxide to the unpleasant, fish-smelling trimethylamine, catalysed by trimethylamine-*N*-oxide reductase (EC 1.6.6.10), is carried out by many bacteria including *Escherichia coli, Proteus* species and *Alteromonas* species. Many off-flavours are also associated with the breakdown of sulphur-containing amino acids. Typical volatile products are hydrogen sulphide, methyl mercaptan and dimethyl sulphide. Bacteria often associated with fish spoilage include *Alteromonas putrefaciens, Pseudomonas* species, *Vibrio* species and *Aeromonas* species (Ringo *et al.*, 1984; Huss, 1988).

Metabolites of *Actinomycetes* and algae, in particular geosmin (1,10-dimethyl-9-decalol), 2-methylisoborneol and 2-isopropyl-3-methoxypyrazine, may impart musty, earthy odours to their aquatic environment. Fish exposed to this become tainted and unacceptable to consumers (Gerber, 1979; Persson, 1979; Yurkowski and Tabachek, 1980). It is important to note that, since the contaminants are waterborne, these types of taint may arise in any food grown on contaminated water or using contaminated water during processing.

Garlic-like off-flavours have been detected in the deep sea royal red prawn *Hymenopenaeus sibogae*. Two compounds, bis-(methylthio)methane and trimethylarsine were identified as the causative agents (Whitfield *et al.*, 1981; Whitfield *et al.*, 1983). Proposed schemes for the formation of both of these compounds, involving microorganisms, are summarized by Whitfield and Tindale (1984). Microbial metabolism of cysteine and trimethylamine oxide leads to the formation of bis-(methylthio)methane, while trimethylarsine arises as a result of reductive microbial methylation of arsenic acid, which naturally accumulates in prawns.

9.6 Concluding remarks

The interrelationship between microbial enzymes, endogenous enzymes and off-flavours can be complex. Although some aspects, such as bitterness in milk and diacetyl in beer, have been well studied, formation of off-flavours due to enzymes, both microbial and endogenous, is often poorly understood. Progress in the area of genetic manipulation and plant breeding offers real opportunities for food scientists to eliminate undesirable characteristics and enhance desirable ones. However, only by a fuller understanding of the biochemistry of flavour and off-flavour formation can these advances be fully exploited.

Note

Information emanating from Campden Food and Drink Research Association is given after the exercise of all reasonable care and skill in its compilation, preparation and issue, but is provided without liability in its application and use.

References

Adler-Nissen, J. (1986). A review of food protein hydrolysis—general issues. In *Enzymic Hydrolysis of Food Proteins*. Elsevier Applied Science, Essex, pp. 25–26.
Allen, J.C. (1989). Rancidity in dairy products. In *Rancidity in Foods*. Eds J.C. Allen and R.J. Hamilton. Elsevier Applied Science, Essex, pp. 199–209.
Axelrod, B., Cheesbrough, T.M. and Laasko, S. (1981). In *Methods in Enzymology 71*. Ed. J.M. Lowenstein. Academic Press, New York, pp. 441–451.
Bachelor, F.W. and Ito, S. (1973). A revision of the stereochemistry of lactucin, *Can. J. Chem.* **51**, 3626–3630.
Bailey, M.E. and Murdoch jr, F.A. (1991). Indigenous and exogenous enzymes of meat. In *Food Enzymology*. Ed. P.F. Fox. Elsevier Applied Science, Essex, pp. 237–264.
Carey, R. (1988). Natural cork: the 'new' closure for wine bottles. *Vineyard and Winery Management* **14**, 5–38.
Crouzet, J., Nicolas, M., Molina, I. and Valentin, G. (1985). Enzymes occurring in the formation of six-carbon aldehydes and alcohols in grapes. In *Progress in Flavour Research 1984*. Ed. J. Adda. Elsevier Applied Science, Amsterdam, pp. 401–408.

Dainty, R.H. (1991). Spoilage microbes on meat and poultry. *Food Sci. and Technol. Today* **3**(4), 250–251.

Dainty, R.H., Edwards, R.A. and Hibbard, C.M. (1985). Time course of volatile compound formation during refrigerated storage of naturally contaminated beef in air. *J. Appl. Bact.* **59**(4), 303–309.

Daly, N.M., Lee, T.H. and Fleet, G.H. (1984). Growth of fungi on wine corks and its contribution to corky taints in wine. *Food Technology in Australia* **36**(1), 22–24.

Fenwick, G.R. and Griffiths, N.M. (1981). The identification of the goitrogen (-) 5-vinyloxazo-lidine-2-thione (goitrin) as a bitter principle of cooked Brussels sprouts (*Brassica oleraceae* L. var. *gemmifera*). *Z. Lebensm Unters Forsch* **172**, 90–92.

Fenwick, G.R., Griffiths, N.M. and Heaney, R.K. (1983). Bitterness in Brussels sprouts (*Brassica oleraceae* L. var. *gemmifera*): The role of glucosinolates and their breakdown products. *J. Sci. Food Agric.* **34**, 73–80.

Gerber, N.N. (1979). Odorous substances from *Actinomycetes*. *Developments in Industrial Microbiology* **20**, 225–238.

Griffiths, M.W., Phillips, J.D. and Muir, D.D. (1981). Thermostability of proteases and lipases from a number of species of psychrotrophic bacteria of dairy origin. *J. Appl. Bacteriol.* **50**(2), 289–303.

Guha, J. and Sen, S.P. (1975). The cucurbitacins – a review. *Plant Biochem. J.* **2**, 12–28.

Heimann, W., Rapp, A., Volter, I. and Knisper, W. (1983). Beitrag zur Enstehung des Korktones in Wein. *Dtsch. Lebensm. Rundschau* **79**, 103–107.

Huss, H.H. (1988). Fresh Fish – Quality and Quality Changes. A Training Manual for the FAO/DANIDA training programme on Fish Technology and Quality Control. FAO Fisheries Series, No. 29.

Hutt, T.F. and Herrington, M.E. (1985). The determination of bitter principles in zucchinis. *J. Sci. Food Agric.* **36**, 1107–1112.

Kishonti, E. (1975). Influence of heat resistant lipases and proteases in psychrotrophic bacteria on product quality. *Int. Dairy Federation Annual Bulletin* **36**, 121–124.

Kjaer, A. (1960). Naturally derived isothiocyanates (mustard oils) and their parent glucosides. *Progress in the Chemistry of Organic Natural Products* **18**, pp. 122–176.

Kraft, A.A. and Rey, C.R. (1979). Psychrotrophic bacteria in foods: An update. *Food Technol.* **33**, 66–71.

Kramlich, W.E., Pearson, A.M. and Tauber, F.W. (1975). Processed meat deterioration. In *Processed meats*. AVI, Westport, Connecticut, pp. 331–341.

Lawrie, R.A. (1974). The spoilage of meat by infecting organisms. In *Meat Science*. Pergamon Press, Oxford, pp. 153–189.

Lee, F.A. and Wagenknecht, A.C. (1958). Enzyme action and off-flavour in frozen peas. II The use of enzymes prepared from garden peas. *Food Research* **23**, 584–590.

Lefebvre, A., Riboulet, J.M., Boidran, J.N. and Ribéreau-Gayon, P. (1983). Incidence des microorganisms du liège sur les alterations olfactives du vin. *Sci. Aliment.* **3**, 265–278.

Lindsay, R.F. and Priest, F.G. (1975). Decarboxylation of substituted cinnamic acids by *Enterobacteria*: the influence on beer flavour. *J. Appl. Bact.* **39**(2), 181–187.

Maarse, H., Nijssen, L.M. and Jetten, J. (1985). Chloroanisoles: a continuing story. In *Topics in Flavour Research. Proceedings of the International Conference, 1–2 April 1985*. Eds R.G. Berger, S. Nitz, and P. Schreier. Freising Weihenstephan, Marzling-Hangeham, Germany, pp. 241–250.

Maier, U.P. and Beverley, G.D. (1968). Limonin monolactone, the non bitter precursor responsible for delayed bitterness in certain citrus juices *J. Food Sci.* **33**, 488–492.

Maier, U.P., and Margdeth, D.A. (1969). Limonoic acid A-ring lactone a new limonin derivative in citrus. *Phytochem.* **8**, 243–248.

Maier, U.P., Hasegawa, S. and Hera, E. (1969). Limonin D-ring lactone hydrolase. A new enzyme from citrus seeds. *Phytochem.* **8**, 405–407.

Maier, V.P., Hasegawa, S., Bennett, R.D. and Echols, L.C. (1980). Limonin and limonoids: chemistry, biochemistry and juice bitterness. In *Citrus Nutrition and Quality*. Eds S. Nagy and J.A. Attaway. ACS Symposium Series 143. American Chemical Society, Washington DC, pp. 63–82.

Moshonas, M.G., Shaw, P.E. and Veldhuis, M.K. (1972). Analysis of volatile constituents from Meyer lemon oil. *J. Agric. Food Chem.* **20**(4), 751–752.

Mtebe, K. and Gordon, M.H. (1987). Volatiles derived from lipoxygenase-catalysed reactions in winged beans (*Psophocarpus tetragonolobus*). *Food Chem.* **23**(3), 175–182.

Muir, D.D., Phillips, J.D. and Dalgleish, D.G. (1979). The lipolytic and proteolytic activity of bacteria isolated from blended raw milk. *J. Soc. Dairy Technol.* **31**, 137–144.

Ney, K.H. (1979). Amino acid composition and chain length. In *Food Taste Chemistry*. Ed. J.C. Boudreau. ACS Symposium Series 115. American Chemical Society, Washington DC, pp. 149–173.

O'Connor, T.P. and O'Brien, N.M. (1991). Significance of lipoxygenase in fruits and vegetables. In *Food Enzymology*. Ed. P.F. Fox. Elsevier Applied Science, Essex, pp. 337–372.

Olivecrona, T. and Bengtsson-Olivercrona, G. (1991). Indigenous enzymes in milk. Il Lipase. In *Food Enzymology*. Ed. P.F. Fox. Elsevier Applied Science, Essex, pp. 62–78.

Persson, P.E. (1979). The source of muddy odour in bream (*Abramis brama*) from the Porvoo Sea area (Gulf of Finland). *J. Fish Res. Board Can.* **36**, 883–890.

Ringo, E., Stenberg, E. and Strom, A.R. (1984). Amino acid and lactate metabolism in trimethylamine oxide respiration of *Alteromonas putrefaciens* NCMB 1735. *Appl. Environ. Micro.* **47**(5), 1084–1089.

Rymal, K.S., Chambliss, O.L., Bond, M.D. and Smith, D.A. (1984). Squash containing toxic cucurbitacin compounds occurring in California and Alabamo. *J. Food Prot.* **47**, 270–271.

Slaughter, J.C. and Priest, F.G. (1991). Significance and use of enzymes in brewing. In *Food Enzymology*. Ed. P.F. Fox. Elsevier Applied Science, Essex, pp. 47–68.

Sone, H., Fujii, T., Kondo, K., Shimizu, F., Tanaka, J. and Inoue, T. (1988). Nucleotide sequence and expression of the *Enterobacter aerogenes* α-acetolactate decarboxylase gene in brewers yeast. *Appl. Environ. Micro.* **54**(1), 38–42.

Sorhaug, T. and Stepaniak, L. (1991). Microbial enzymes in the spoilage of milk and dairy products. In *Food Enzymology*. Ed. P.F. Fox. Elsevier Applied Science, Essex, pp. 169–218.

Stoewsand, G.S., Jaworski, A., Shannon, S. and Robinson, R.W. (1985). Toxicological response in mice fed cucurbita fruit. *J. Food Prot.* **48**(1), 50–51.

Tanner, H., Zannier, C. and Buser, H.R. (1981). 2,4,6-Trichloroanisol: Eine dominierende Komponente des korkgeschmockes separatdruk aus der Schweiz. *Zeitschrift fur Obst und Weinbrau* **117**(90), 97–103.

Villetaz, J.C. and Dubourdieu, D. (1991). Enzymes in wine making. In *Food Enzymology*. Ed. P.F. Fox. Elsevier Applied Science, New York, pp. 427–453.

Whitaker, J.R. (1991). Lipoxygenases. In *Oxidative Enzymes in Foods*. Ed. D.S. Robinson and N.A.M. Eskin. Elsevier Applied Science, Essex, pp. 175–216.

Whitfield, F.B. and Tindale, C.R. (1984). The role of microbial metabolites in food off-flavours. *Food Technology in Australia* **36**(5), 204–209, 213.

Whitfield, F.B., Freeman, D.J., Last, J.H. and Bannister, P.A. (1981). Bis-(methylthio)methane: An important off-flavour component in prawns and sand lobsters. *Chem. Ind. (London)*, 158–159.

Whitfield, F.B., Last, J.H. and Tindale, C.R. (1982). Skatole, indole and *p*-cresol: Components in off-flavoured frozen French fries. *Chem. Ind. (London)*, 662–663.

Whitfield, F.B., Freeman, D.J. and Shaw, K.J. (1983). Trimethylarsine: An important off-flavour component in some prawn species. *Chem. Ind. (London)*, 786–787.

Wenzel, F.W., Olsen, R.W., Barron, R.W., Huggart, R.L., Patrick, R. and Hill, E.C. (1958). Use of Florida lemons in frozen concentrate for lemonade. *Proc. Fla. State Hort. Soc.* **71**, pp. 129–132.

Yurkowski, M. and Tabachek, J.A.L. (1980). Geosmin and 2-methylisobomeol implicated as a cause of muddy odor and flavor in commercial fish from Cedar Lake, Manitoba. *Can. J. Fish Aquat. Sci.* **37**, 1449–1450.

Index